T0296158

ANALYTICAL GEOMETRY
OF
THREE DIMENSIONS

ANALYTICAL GEOMETRY
OF
THREE DIMENSIONS

BY

D. M. Y. SOMMERVILLE

CAMBRIDGE

AT THE UNIVERSITY PRESS

1959

CAMBRIDGE
UNIVERSITY PRESS

University Printing House, Cambridge CB2 8BS, United Kingdom

Cambridge University Press is part of the University of Cambridge.

It furthers the University's mission by disseminating knowledge in the pursuit of
education, learning and research at the highest international levels of excellence.

www.cambridge.org
Information on this title: www.cambridge.org/9781316601907

© Cambridge University Press 1959

First edition 1934
Reprinted 1939, 1943, 1947, 1951, 1959
First paperback edition 2015

A catalogue record for this publication is available from the British Library

ISBN 978-1-316-60190-7 Paperback

Cambridge University Press has no responsibility for the persistence or accuracy
of URLs for external or third-party internet websites referred to in this publication,
and does not guarantee that any content on such websites is, or will remain,
accurate or appropriate.

CONTENTS

Chap. III. General homogeneous or projective
coordinates

Chap. IV. The sphere

Chap. V. The cone and cylinder

Chap. VI. Types of surfaces of the second order

Chap. IX. Generating lines and parametric representation

Chap. X. Plane sections of a quadric

Chap. XI. Tangential equations

Chap. XII. Foci and focal properties

Chap. XIII. Linear systems of quadrics

Chap. XIV.   Curves and developables

Chap. XVII. Algebraic surfaces

PREFACE

Until recent years there has been a tendency, in England at least, to regard Geometry as if it were a mine which had been worked out and exhausted. Mathematical interest was largely transferred to analysis. The great stimulus given by Cayley, Salmon and Clifford in the 'sixties and 'seventies of last century had dissipated, and no great successor to these pioneers had appeared. But if their influence in Britain had become weakened, it grew upon the Continent, especially in Italy, and it is from Italy, largely through the medium of Professor H. F. Baker, that once more a renewed interest in geometry has arisen and is flourishing in England.

It is seventy-one years since Salmon's *Treatise on the Analytic Geometry of Three Dimensions* was first published. It has been translated into German, French and Italian, and has been expanded into two volumes in later English editions.* In its first form it embodied the results of many very recent researches, and, brought up to date and including several new topics, it is still recognised as the standard work in the English language. There seems, however, to be room for a text-book written more in accordance with the tendencies of the present "cosmic epoch", to apply a suggestive term of Whitehead's. Fashions in mathematics, as in other things, alter. The facts remain but their values change. Rather, perhaps, new principles, wider and more unifying, are discovered, leading to different treatment and a different emphasis being put on the various developments.

In some ways the present text-book should be regarded as an introduction to Professor Baker's inspiring volumes on the *Principles of Geometry*. This work, especially in the two recent volumes, shows strongly the Italian influence, and the same must be acknowledged in the case of the present text-book. It is natural that the Italian school, which has been responsible for a great part of modern geometrical research, should have produced also some of the finest text-books, such as Bianchi's *Lezioni di geometria analitica* (Pisa, 1920), Castelnuovo's book with the same title (6th ed. Milan, 1924), Berzolari's two Hoepli manuals entitled *Geometria analitica* (Milan: 1, 3rd ed. 1925; II, 2nd ed. 1922), and Comessati's *Lezioni di geometria analitica e proiettiva* (Milan, 1930). To these, as well as to Salmon, Baker, the Collected Mathematical Papers of Cayley and of Klein,

* Vol. 1, 7th ed. revised by R. A. P. Rogers and edited by C. H. Rowe, 1928 Vol. 2, 5th ed. revised by Rogers, 1915.

Hudson's classic on *Kummer's Quartic Surface*, Pascal's *Repertorium*, the *Enzyklopädie der mathematischen Wissenschaften*, and other sources difficult to particularise, I have to express my indebtedness.

Being an elementary text-book it may be used by beginners. For such it may be useful to indicate a first course of reading: Chap. I (omitting 1·8*), Chap. II (omitting 2·34, 2·35, 2·36, 2·41, 2·5, 2·731, 2·8, 2·91), Chap. IV (omitting 4·2, 4·31, 4·51, 4·81), Chap. V (omitting 5·122, 5·123, 5·23, 5·6), Chap. VI, Chap. VII (omitting 7·38, 7·73), Chap. VIII (omitting 8·11–8·13, 8·24, 8·41–8·432, 8·53–8·54, 8·64, 8·72, 8·74, 8·9), Chap. IX (omitting 9·3, 9·4, 9·6, 9·7), Chap. X (omitting 10·71).

Except for a few insignificant references the subject-matter of differential geometry has been excluded from this book. On the other hand free use has been made of homogeneous co-ordinates, tangential coordinates and line-coordinates. In the case of metrical geometry the circle at infinity is used wherever it is applicable; this is especially the case in the treatment of foci, which follows somewhat closely on the lines of Berzolari. There are several illustrative references to Non-Euclidean Geometry, and much use has been made, as in Baker's volumes, of geometry of higher dimensions, especially in the exposition of line-geometry. In the enumeration of types of linear systems of quadrics opportunity has been taken to explain the notation of invariant-factors. No exhaustive treatment of the theory of algebraic curves and surfaces has been attempted; the two chapters which have been devoted to these are intended rather to be suggestive, and are confined practically to curves of the third and fourth orders, and to ruled and rational cubic and quartic surfaces.

I have to express my grateful thanks to Mr F. P. White for much encouragement in the preparation of the book; to Professor W. Saddler, D.Sc., Christchurch, who read the entire manuscript, for many helpful and valuable suggestions; and to Mr F. F. Miles, M.A., Lecturer in Mathematics at Victoria University College, for great assistance in reading the proof-sheets and in checking the examples.

I have also to acknowledge with thanks the courtesy and close attention of the Staff of the Cambridge University Press.

---

* This is to be understood as including all further subdivisions, as 1·81, 1·82, etc.

D. M. Y. SOMMERVILLE

VICTORIA UNIV. COLL.
WELLINGTON, N.Z.

October 1933

ANALYTICAL GEOMETRY
OF THREE DIMENSIONS

CHAPTER I

CARTESIAN COORDINATE-SYSTEM

1·1. Cartesian coordinates.

In a plane the position of a point P is determined by two co-ordinates, x and y, referred to two straight lines OX, OY, the coordinate-axes; viz. if $NP \parallel OX$ and $MP \parallel OY$, so that we have a parallelogram $OMPN$, then $x = OM = NP$, $y = ON = MP$.

To fix the position of a point in space we take three planes. These have a point O in common and intersect in pairs in three lines $X'OX$, $Y'OY$, $Z'OZ$. O is called the origin, the three lines the coordinate-axes, and the three planes the coordinate-planes.

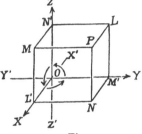

Fig. 1

Let P be any point. Through P draw PL parallel to XOX' cutting the plane YOZ in L, and similarly PM and PN parallel to the other axes. Let the plane MPN cut OX in L', and similarly obtain M' and N'. We obtain then a parallelepiped whose faces are parallel to the coordinate-planes, and edges parallel to the coordinate-axes, and OP is a diagonal. The figure is determined by the lengths of OL', OM', ON'.

In the usual way we attach signs to lengths measured along the axes, defining by a convention that distances measured in one direction are positive, distances measured in the opposite direction being negative. With these conventions we then define the *coordinates* of the point P as the three lengths

$$OL' = x, \quad OM' = y, \quad ON' = z.$$

To every point P there corresponds uniquely a set of three numbers $[x, y, z]$, and conversely to every set of three numbers, positive or negative, there corresponds a unique point.

1·11. Convention of signs.

Let the positive directions along the axes of x and y be defined arbitrarily, say OX and OY; then in the plane XOY we may pass from OX to OY by a rotation through an angle XOY less than two right angles. Viewed from one side of the plane this rotation is clockwise, and from the other side it appears counter-clockwise. We define that side of the plane from which the rotation appears to be counter-clockwise as the *positive side of the plane*. Then the positive direction of the axis of z is defined to be that which lies on the positive side of the plane XOY. This relation then holds for each of the axes, viz. the positive direction of the axis of x is on the positive side of the plane YOZ, and the positive direction of the axis of y is on the positive side of the plane ZOX. This is called a *right-handed* system of cartesian coordinates.

1·2. When the planes are mutually at right angles we call it a *rectangular* system, otherwise it is *oblique*. We shall confine our attention for the present to rectangular coordinates.

OP is called the *radius-vector* of P, denoted by r. Since PN is perpendicular to the plane XOY, ON is the orthogonal projection, or simply the projection, of OP on the plane XOY. Again, since the plane $PML'N \perp OX$, $PL' \perp OX$ and OL' is the projection of OP on the line OX.

Let the angles which OP makes with the positive directions of the axes be α, β, γ, then

$$x = r\cos\alpha, \quad y = r\cos\beta, \quad z = r\cos\gamma.$$

The position of P is determined by the angles α, β, γ and the radius-vector r, for these then determine x, y, z. The angles by themselves determine only the direction of the line OP. We call them the *direction-angles* of the line OP. As the cosines of these angles occur repeatedly it is convenient to call them the *direction-cosines*, and frequently we denote them by single letters l, m, n. There is a redundance in fixing the position of a point by the radius-vector and its direction-angles, for these are four numbers, and three, x, y, z, are sufficient to fix the position. We shall find that the three direction-cosines are connected by an identical relation.

1·21. The direction-angles are not uniquely defined since each is indeterminate to an added multiple of 2π, but the direction-cosines are unique. In fact, r *being always positive*, the direction-cosines l, m, n are uniquely defined as

$$l = x/r, \quad m = y/r, \quad n = z/r.$$

1·22. *To determine the radius-vector in terms of the rectangular coordinates.*

By the theorem of Pythagoras

$$OP^2 = ON^2 + NP^2 = OL'^2 + L'N^2 + NP^2.$$

Hence $\qquad\qquad r^2 = x^2 + y^2 + z^2.$

1·23. *Identity connecting the direction-cosines.*

Putting $x = r\cos\alpha$, $y = r\cos\beta$, $z = r\cos\gamma$, we find, on dividing by r^2, $\qquad \cos^2\alpha + \cos^2\beta + \cos^2\gamma = 1.$

It is often convenient to speak of a line whose direction-cosines are proportional to three given numbers l, m, n. The actual values of the direction-cosines are obtained by dividing each by $\sqrt{(l^2 + m^2 + n^2)}$. For suppose the actual values to be kl, km, kn; then $\qquad k^2l^2 + k^2m^2 + k^2n^2 = 1$,

hence $\qquad\qquad k = (l^2 + m^2 + n^2)^{-\frac{1}{2}}.$

Ex. Find the direction-cosines of the line joining the origin to the point $(-1, 2, 2)$.

Here $\qquad\qquad r^2 = 1 + 4 + 4 = 9$, hence $r = 3$.

Then $\qquad\qquad l = -\tfrac{1}{3}, m = \tfrac{2}{3}, n = \tfrac{2}{3}.$

1·3. Change of origin.

Let a new coordinate-system be constructed with origin $O' \equiv [X, Y, Z]$ and coordinate-planes parallel to the old ones. Let the coordinates of a point P referred to the two systems be $[x, y, z]$ and $[x', y', z']$. Draw through P a line parallel to the axis of x cutting the planes yOz and $y'O'z'$ in L and L', and let the plane $y'O'z'$ cut Ox in K. Then since the parallel planes yOz, $y'O'z'$ intercept

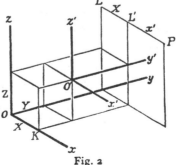

Fig. 2

equal segments on parallel lines, $LL' = OK$. But $LP = x$, $L'P = x'$, $OK = X$, hence

$$\text{similarly} \quad \begin{rcases} x = x' + X \\ y = y' + Y \\ z = z' + Z \end{rcases} \quad \text{and} \quad \begin{rcases} x' = x - X \\ y' = y - Y \\ z' = z - Z \end{rcases}.$$

x', y', z' are the coordinates of P relative to $O' \equiv [X, Y, Z]$.

This may be described in vector language thus. The step from O to P can be broken up into the two steps O to O' and O' to P. Further the step from O to O' can be broken up into three steps of lengths X, Y, Z parallel to the axes, and similarly for the steps from O to P and O' to P. Parallel steps are then simply added.

1·4. Distance between two points.

Let $P_1 \equiv [x_1, y_1, z_1]$, $P_2 \equiv [x_2, y_2, z_2]$. Then the coordinates of P_2 referred to parallel axes through P_1 are

$$x_2 - x_1, \quad y_2 - y_1, \quad z_2 - z_1,$$

and $P_1 P_2$ is the relative radius-vector. Hence

$$(P_1 P_2)^2 = (x_2 - x_1)^2 + (y_2 - y_1)^2 + (z_2 - z_1)^2.$$

1·5. Angle between two lines.

Let the two lines pass through O and have direction-angles $[\alpha_1, \beta_1, \gamma_1]$ and $[\alpha_2, \beta_2, \gamma_2]$. Take two points $P_1 \equiv [x_1, y_1, z_1]$ and $P_2 \equiv [x_2, y_2, z_2]$, one on each; let $OP_1 = r_1$, $OP_2 = r_2$ and let the angle $P_1 O P_2 = \theta$. Then

$$(P_1 P_2)^2 = OP_1^2 + OP_2^2 - 2 OP_1 . OP_2 \cos \theta,$$

therefore $\qquad \Sigma (x_2 - x_1)^2 = r_1^2 + r_2^2 - 2 r_1 r_2 \cos \theta,$

hence $\quad \Sigma x_2^2 + \Sigma x_1^2 - 2\Sigma x_1 x_2 = \Sigma x_1^2 + \Sigma x_2^2 - 2 r_1 r_2 \cos \theta.$

But $\qquad\qquad x_1 = r_1 \cos \alpha_1, \quad x_2 = r_2 \cos \alpha_2, \text{ etc.}$

Therefore $\qquad \Sigma r_1 r_2 \cos \alpha_1 \cos \alpha_2 = r_1 r_2 \cos \theta,$

i.c. $\quad \cos \theta = \cos \alpha_1 \cos \alpha_2 + \cos \beta_1 \cos \beta_2 + \cos \gamma_1 \cos \gamma_2$

$$= \Sigma \cos \alpha_1 \cos \alpha_2 = \Sigma l_1 l_2.$$

1·51. If the direction-cosines are only proportional to the numbers $[l_1, m_1, n_1]$ and $[l_2, m_2, n_2]$, then

$$\cos \theta = \frac{\Sigma l_1 l_2}{\sqrt{(\Sigma l_1^2 \Sigma l_2^2)}}.$$

1·52. A useful expression may be found also for $\sin^2\theta$.

We have
$$\sin^2\theta = 1 - (\Sigma l_1 l_2)^2 = \Sigma l_1^2 \Sigma l_2^2 - (\Sigma l_1 l_2)^2$$
$$= \Sigma(m_1^2 n_2^2 + m_2^2 n_1^2) - 2\Sigma m_1 m_2 n_1 n_2$$
$$= \Sigma(m_1 n_2 - m_2 n_1)^2.$$

These results may be applied to any two lines. When two lines do not intersect we define the angle determined by them as the angle between two lines through O parallel to the given lines. All parallel lines are then considered as having the same direction-angles.

1·61. Condition for perpendicularity.

If two lines $[l_1, m_1, n_1]$ and $[l_2, m_2, n_2]$ are perpendicular or orthogonal, their angle $\theta = \frac{1}{2}\pi$, and $\cos\theta = 0$, hence

$$l_1 l_2 + m_1 m_2 + n_1 n_2 = 0$$

or
$$\Sigma \cos\alpha_1 \cos\alpha_2 = 0.$$

Conversely, if this condition is satisfied, $\cos\theta = 0$ and $\theta = \frac{1}{2}\pi$.

The expression $\Sigma l_1 l_2$ is linear in each of the two sets of direction-cosines l_1, m_1, n_1 and l_2, m_2, n_2, and also symmetrical as regards these two sets. It is the *bilinear symmetrical* expression associated with the quadratic expression Σl^2.

1·62. Conditions for parallelism.

By definition, two lines are parallel when they have the same direction-angles. It follows then that $\sin\theta = 0$ and $\theta = 0$.

Conversely, if $\theta = 0$, $\sin\theta = 0$ and $\Sigma(m_1 n_2 - m_2 n_1)^2 = 0$. For real values of the direction-cosines this can be true only if

$$m_1 n_2 - m_2 n_1 = 0, \quad n_1 l_2 - n_2 l_1 = 0, \quad l_1 m_2 - l_2 m_1 = 0,$$

hence
$$\frac{l_1}{l_2} = \frac{m_1}{m_2} = \frac{n_1}{n_2}.$$

If l_1, m_1, n_1 and l_2, m_2, n_2 are the actual direction-cosines we have also
$$l_1^2 + m_1^2 + n_1^2 = 1 = l_2^2 + m_2^2 + n_2^2.$$

Putting each of the equal ratios equal to t and substituting

$$l_1 = tl_2, \quad m_1 = tm_2, \quad n_1 = tn_2$$

we get
$$t^2(l_2^2 + m_2^2 + n_2^2) = l_2^2 + m_2^2 + n_2^2.$$

Hence
$$t = \pm 1.$$

If $t = +1$, the direction-cosines are identical; if $t = -1$, they are equal but of opposite sign. The latter case is interpreted to mean that the lines are in opposite senses. In both cases they are parallel.

If l_1, m_1, n_1 and l_2, m_2, n_2 are only numbers proportional to the direction-cosines, the necessary and sufficient conditions for parallelism are
$$l_1 : m_1 : n_1 = l_2 : m_2 : n_2,$$
equivalent to *two* conditions only.

1·7. Position-ratio of a point with regard to two base-points.

The formulae for the coordinates (rectangular or oblique) of a point $P \equiv [x, y, z]$ dividing the join of two points $P_1 \equiv [x_1, y_1, z_1]$ and $P_2 \equiv [x_2, y_2, z_2]$ in a given ratio $l : m$ or $k : 1$ are exactly the same as in plane geometry. For if the planes through P_1, P_2, P parallel to the plane of yz cut Ox in L_1, L_2, L, then L cuts $L_1 L_2$ in this same ratio, and $OL_1 = x_1$, $OL_2 = x_2$, $OL = x$. Hence

$$x = \frac{lx_2 + mx_1}{l + m} = \frac{kx_2 + x_1}{k + 1},$$

with similar formulae for y and z. The ratio $l/m = k$ is called the *position-ratio* of P with regard to P_1 and P_2. The formulae are sometimes referred to as Joachimsthal's formulae.

1·71. Another set of section-formulae is useful. If

$$P_1 P = t . P_1 P_2,$$

we have
$$\left.\begin{array}{l} x = x_1 + t(x_2 - x_1) \\ y = y_1 + t(y_2 - y_1) \\ z = z_1 + t(z_2 - z_1) \end{array}\right\}.$$

1·72. If two particles of masses m_1, m_2 are placed at P_1 and P_2 their centre of mass divides $P_1 P_2$ in the ratio $m_2 : m_1$, hence the coordinates of the centre of mass are

$$x = \frac{m_1 x_1 + m_2 x_2}{m_1 + m_2}, \text{ etc.}$$

This point is also called the *mean point* for the multiples m_1, m_2.

1·73. Similarly, if three masses m_1, m_2, m_3 are placed at three points P_1, P_2, P_3 the coordinates of the centre of mass are

$$x = \frac{m_1 x_1 + m_2 x_2 + m_3 x_3}{m_1 + m_2 + m_3}, \text{ etc.}$$

By admitting negative masses any point in the plane can be represented in this way.

1·8. General cartesian coordinates.

In the general cartesian system the planes are not necessarily at right angles. The system will be determined by the angles between the coordinate-axes, viz. $\angle YOZ = \lambda$, $\angle ZOX = \mu$, $\angle XOY = \nu$.

The direction-angles are then no longer convenient, but in place of the direction-cosines we define the *direction-ratios* as follows $l = x/r, \quad m = y/r, \quad n = z/r$.

1·81. To find the radius-vector r we have (Fig. 1)

$$r^2 = OP^2 = ON^2 + NP^2 + 2ON \cdot NP \cos NOZ.$$

Now the projection of ON on OZ is equal to the sum of the projections of OM' and $M'N$, hence

$$ON \cos NOZ = OM' \cos YOZ + M'N \cos ZOX$$
$$= y \cos \lambda + x \cos \mu.$$

Also $\qquad ON^2 = OL'^2 + L'N^2 + 2OL' \cdot L'N \cos XOY$
$$= x^2 + y^2 + 2xy \cos \nu.$$

Hence

$$r^2 = x^2 + y^2 + z^2 + 2yz \cos \lambda + 2zx \cos \mu + 2xy \cos \nu.$$

1·82. Then substituting from (1·8) we have

$$l^2 + m^2 + n^2 + 2mn \cos \lambda + 2nl \cos \mu + 2lm \cos \nu = 1,$$

as the identity connecting the direction-ratios.

1·83. The square of the distance between two points (x_1, y_1, z_1) and (x_2, y_2, z_2) is found by substituting in (1·81) $x_1 - x_2$ for x, etc.

1·84. The angle between two straight lines (l_1, m_1, n_1) and (l_2, m_2, n_2) is found as in 1·5,

$$\cos \theta = \frac{\Sigma l_1 l_2 + \Sigma (m_1 n_2 + m_2 n_1) \cos \lambda}{\sqrt{(\Sigma l_1^2 + 2\Sigma m_1 n_1 \cos \lambda)(\Sigma l_2^2 + 2\Sigma m_2 n_2 \cos \lambda)}}.$$

The numerator is the bilinear symmetrical expression associated with the quadratic expressions which occur in the denominator.

1·9. EXAMPLES.

1. Find the direction-angles of the lines joining the origin to the following points: (i) $[\sqrt{2}, 1, 1]$, (ii) $[-1, 2, 2]$, (iii) $[2, -3, 6]$.

Ans. (i) $[45°, 60°, 60°]$, (ii) $[\pi - \cos^{-1}\frac{1}{3}, \cos^{-1}\frac{2}{3}, \cos^{-1}\frac{2}{3}]$, (iii) $[\cos^{-1}\frac{2}{7}, \pi - \cos^{-1}\frac{3}{7}, \cos^{-1}\frac{6}{7}]$.

2. A line makes angles 60° and 45° with the positive axes of x and y respectively; what angle does it make with the positive axis of z?

Ans. 60° or 120°.

3. Show that the point $[3, -1, 2]$ is the centre of the sphere which passes through the four points $[2, 1, 4], [5, 1, 1], [4, -3, 0], [1, -3, 3]$, and find its radius.

Ans. $r = 3$.

4. Find the centre and the radius of the sphere which passes through the four points $[-2, -2, 3], [1, -5, 3], [1, -2, 0], [0, -6, -1]$.

Ans. $[-1, -4, 1], r = 3$.

5. A regular tetrahedron is placed with a vertex at the origin O, the altitude through O making equal angles with each of the three rectangular axes, and each of the edges through O lying in the same plane with the altitude and the corresponding coordinate-axis. Find the direction-cosines of the edges through O.

Ans. $[4, 1, 1], [1, 4, 1], [1, 1, 4]$ or $[0, 1, 1], [1, 0, 1], [1, 1, 0]$.

6. Find the actual direction-cosines of the line joining the origin to the point $[2u, 2v, u^2 + v^2 - 1]$.

Ans. Each divided by $u^2 + v^2 + 1$.

7. Show that the four points $[1, -1, -1], [-1, 1, -1], [-1, -1, 1], [1, 1, 1]$ form the vertices of a regular tetrahedron and find the length of the edge.

Ans. $2\sqrt{2}$.

8. Show that $[-3, -3, -3], [5, -1, -1], [-1, 5, -1], [-1, -1, 5]$ are the vertices of a regular tetrahedron whose centre is at the origin.

9. Prove that $[a, b, c]$, $[c, a, b]$, $[b, c, a]$, $[d, d, d]$ are the vertices of a regular tetrahedron with its centre at the origin when

$$a = t^2 + 3t + 1, \quad b = t^2 - t - 1, \quad c = -t^2 - t + 1, \quad d = -t^2 - t - 1,$$

t being any parameter.

10. Show that the four points $[2, 9, 12]$, $[1, 8, 8]$, $[-2, 11, 8]$, $[-1, 12, 12]$ are the vertices of a square.

11. Show that the six points $[0, 1, -1]$, $[0, -1, 1]$, $[1, 0, -1]$, $[1, -1, 0]$, $[-1, 0, 1]$, $[-1, 1, 0]$ form the vertices of a regular hexagon; and so also the points whose coordinates are $[a, b, c]$ and the permutations of these, where a, b, c are in arithmetical progression.

12. Show that the six points $[-1, 2, 2]$, $[2, -1, 2]$, $[2, 2, -1]$, $[1, -2, -2]$, $[-2, 1, -2]$, $[-2, -2, 1]$ form the vertices of a regular octahedron.

13. Show that the six points $[1, 5, 6]$, $[4, 2, 6]$, $[4, 5, 3]$, $[3, 1, 2]$, $[0, 4, 2]$, $[0, 1, 5]$ form the vertices of a regular octahedron.

14. OP, OQ are lines in the planes of zx, xy, bisecting the angles between the positive directions of the axes in these planes. Prove that the angle $POQ = 60°$. Hence show that six regular octahedra and eight regular tetrahedra will exactly fill up the space about a point.

15. Show that the 12 points $[0, \pm 1, \pm 1]$, taking all permutations, form the vertices of a polyhedron bounded by 6 squares and 8 equilateral triangles.

16. Show that the 24 points $[0, \pm a, \pm b]$, taking all permutations, form the vertices of a polyhedron bounded by 6 squares and 8 hexagons, and that the hexagons are regular if $a = 2b$.

17. Show that the 24 points $[\pm a, \pm b, \pm b]$, taking all permutations, form the vertices of a polyhedron bounded by 6 squares, 12 rectangles, and 8 equilateral triangles; and find the relation between a and b if the rectangles are squares. $(a > b.)$

Ans. $a^2 - 2ab - b^2 = 0$.

18. Show that the 48 points $[\pm 1, \pm(1+\sqrt{2}), \pm(1+2\sqrt{2})]$, where all permutations are taken, form the vertices of a polyhedron bounded by 6 regular octagons, 8 regular hexagons, and 12 squares.

19. Show that if $a^2 - ab - b^2 = 0$ the 12 points $[0, \pm a, \pm b]$, $[\pm b, 0, \pm a]$, $[\pm a, \pm b, 0]$ are vertices of a regular icosahedron, and the 20 points $[0, \pm b, \pm(a+b)]$, $[\pm(a+b), 0, \pm b]$, $[\pm b, \pm(a+b), 0]$, $[\pm a, \pm a, \pm a]$ are vertices of a regular dodecahedron.

20. If the position of a point P is determined with reference to a rectangular coordinate-system by its radius-vector ρ, the angle $ZOP = \phi$, and the angle XOL, or θ, which the plane ZOP makes with ZOX, show that

$$x = \rho \sin\phi \cos\theta, \quad y = \rho \sin\phi \sin\theta, \quad z = \rho \cos\phi,$$

and
$$ds^2 = d\rho^2 + \rho^2 d\phi^2 + \rho^2 \sin^2\phi \, d\theta^2.$$

21. If A, B, C are three consecutive vertices of a parallelogram show that the coordinates of the fourth vertex are

$$x = x_A + x_C - x_B, \text{ etc.}$$

CHAPTER II

THE STRAIGHT LINE AND PLANE

2·1. Since a point in space requires three coordinates to fix its position we say that it has three degrees of freedom. Similarly in a plane a point has two degrees of freedom. More generally, if it is confined to any surface it has two degrees of freedom, and if it moves only on a line or curve it has one degree of freedom. A point is deprived of one degree of freedom when its coordinates are connected by any relation. Hence an equation in x, y, z represents a surface. Two such equations deprive the point of two degrees of freedom and limit it to a curve. If three equations are given no freedom is left; the values of x, y, z can be found by solving the equations and only a finite number of positions are possible for the point.

2·11. The equations of a straight line.

Let the straight line pass through the point $A \equiv [x_1, y_1, z_1]$ and have direction-cosines (or ratios) $[l, m, n]$. Then if $P \equiv [x, y, z]$ is any point on the line, and $AP = r$, we have

2·111.
$$\left.\begin{aligned} x &= x_1 + lr \\ y &= y_1 + mr \\ z &= z_1 + nr \end{aligned}\right\}.$$

2·112. Eliminating r, we get the (two) equations
$$\frac{x - x_1}{l} = \frac{y - y_1}{m} = \frac{z - z_1}{n}.$$

(2·112) is adopted as the standard form for the equations of a straight line. (2·111) are called *freedom-equations* of the line in terms of the *parameter r*. $[x_1, y_1, z_1]$ is an arbitrary point of reference on the line.

2·113. The coordinates of any point on the join of $[x_1, y_1, z_1]$ and $[x_2, y_2, z_2]$ can be written
$$\left.\begin{aligned} x &= x_1 + t(x_2 - x_1) \\ y &= y_1 + t(y_2 - y_1) \\ z &= z_1 + t(z_2 - z_1) \end{aligned}\right\},$$

(see 1·71). These are freedom-equations in terms of the parameter t.

2·12. Equation of a plane.

If P_1, P_2, P_3 are three points, the coordinates of any point on the plane determined by them are (1·73)

$$x = \frac{m_1 x_1 + m_2 x_2 + m_3 x_3}{m_1 + m_2 + m_3}, \text{ etc.}$$

Put $m_2 = t\Sigma m$ and $m_3 = u\Sigma m$, so that

$$m_1 = \Sigma m - m_2 - m_3 = (1 - t - u)\Sigma m.$$

We have then

2·121.
$$\left. \begin{aligned} x &= x_1 + (x_2 - x_1)t + (x_3 - x_1)u \\ y &= y_1 + (y_2 - y_1)t + (y_3 - y_1)u \\ z &= z_1 + (z_2 - z_1)t + (z_3 - z_1)u \end{aligned} \right\}.$$

These are freedom-equations involving two parameters t and u, corresponding to the two degrees of freedom in the plane.

2·122. Eliminating t and u we get an equation in x, y, z, which is equivalent to

$$\begin{vmatrix} x & x_1 & x_2 & x_3 \\ y & y_1 & y_2 & y_3 \\ z & z_1 & z_2 & z_3 \\ 1 & 1 & 1 & 1 \end{vmatrix} = 0.$$

The equation is thus of the first degree in x, y, z. (2·122) is also the condition that the four points $[x, y, z]$, $[x_1, y_1, z_1]$, $[x_2, y_2, z_2]$, $[x_3, y_3, z_3]$ should be coplanar.

2·123. *An equation of the first degree always represents a plane.*

The general equation of the first degree is

$$lx + my + nz + p = 0. \qquad \ldots\ldots(1)$$

The characteristic property of a plane is that if it contains two points P_1, P_2 of a straight line it contains all points of the straight line. Let then $P_1 \equiv [x_1, y_1, z_1]$ and $P_2 \equiv [x_2, y_2, z_2]$ be two points on the plane. Then

$$lx_1 + my_1 + nz_1 + p = 0 \atop lx_2 + my_2 + nz_2 + p = 0 \Bigg\} \quad \dots\dots(2)$$

Substituting in (1) the coordinates (1·71) of any point on the line P_1P_2,

$$lx + my + nz + p = (lx_1 + my_1 + nz_1 + p)(1 - t)$$
$$+ (lx_2 + my_2 + nz_2 + p)t,$$

which vanishes by (2). The theorem is therefore proved.

2·124. Orientation of a plane.

The orientation or lie of a plane is determined by the direction of any straight line perpendicular or normal to the plane. A plane may be determined by the direction $[\alpha, \beta, \gamma]$ of its normal and its distance p from a fixed point, say the origin. We shall call α, β, γ the direction-angles, and their cosines the direction-cosines, of the plane. Let N be the foot of the normal from O to the plane, and $P \equiv [x, y, z]$ any point on the plane. Then the projection of OP is equal to the sum of the projections of $OL', L'N, NP$, hence projecting on ON we have

$$x \cos\alpha + y \cos\beta + z \cos\gamma = p.$$

This will be adopted as the normal or canonical form for the equation of a plane. The special property of this equation is that the sum of the squares of the coefficients of x, y, z is equal to unity.

2·125. Equation of a plane in terms of intercepts on the axes.

Let the plane cut the axes in A, B, C; let $OA = a$, $OB = b$, $OC = c$; these are the intercepts. Let $P \equiv [x, y, z]$ be any point on the plane. Then the tetrahedron $OABC$ is divided into three tetrahedra $OABP, OBCP, OCAP$. The volume of $OABC = \frac{1}{6}abc$, that of $OBCP = \frac{1}{6}bcx$, etc. Hence

$$\tfrac{1}{6}bcx + \tfrac{1}{6}cay + \tfrac{1}{6}abz = \tfrac{1}{6}abc,$$

or

$$\frac{x}{a} + \frac{y}{b} + \frac{z}{c} = 1.$$

This equation has the same meaning when the axes are oblique.

2·13. Intersection of two planes.

Two planes intersect in a straight line. Hence the equations of a straight line may be given in the form

$$\left.\begin{array}{l} lx + my + nz + p = 0 \\ l'x + m'y + n'z + p' = 0 \end{array}\right\} \cdot \qquad \ldots\ldots(1)$$

If $[x_1, y_1, z_1]$ is any point on the line, it lies on both planes, therefore

$$\left.\begin{array}{l} lx_1 + my_1 + nz_1 + p = 0 \\ l'x_1 + m'y_1 + n'z_1 + p' = 0 \end{array}\right\}, \qquad \ldots\ldots(2)$$

hence, subtracting the corresponding equations of (1) and (2)

$$l(x - x_1) + m(y - y_1) + n(z - z_1) = 0,$$
$$l'(x - x_1) + m'(y - y_1) + n'(z - z_1) = 0.$$

Solving for the ratios of $x - x_1$, $y - y_1$, $z - z_1$ we have

$$\frac{x - x_1}{mn' - m'n} = \frac{y - y_1}{nl' - n'l} = \frac{z - z_1}{lm' - l'm}.$$

The denominators are therefore proportional to the direction-cosines (or ratios) of the line (1).

2·14. Note on vectors.

A line of definite length and direction is a vector. Following the usual convention we represent vectors in Clarendon (heavy) type; the same symbol in ordinary italic type is taken to represent the length of the vector. Thus **v** represents a vector whose length is v. All parallel vectors of the same length are equivalent. If $P \equiv [x, y, z]$ is any point, the vector OP has x, y, z for its rectangular components. The direction-cosines $[l, m, n]$ are rectangular components of a unit-vector. The sum of two vectors is a vector determined by the "triangle of vectors", viz. if the two vectors are placed consecutively as **AB** and **BC** their sum **AB** + **BC** = **AC**. Subtraction is the inverse of addition, and a negative vector is equivalent to a positive vector taken in the opposite sense. Addition and subtraction are associative and commutative.

The multiplication of vectors may be defined in various ways, and two distinct kinds of products of vectors are found to be of special importance. If $[x_1, y_1, z_1]$ and $[x_2, y_2, z_2]$ are the rectangular components of two vectors $\mathbf{v}_1$, $\mathbf{v}_2$, the sum

$$x_1 x_2 + y_1 y_2 + z_1 z_2$$

is called their *scalar product* and is written

$$\mathbf{v}_1 \cdot \mathbf{v}_2 = x_1 x_2 + y_1 y_2 + z_1 z_2.$$

A common example of this is the work done by a force F. If X, Y, Z are the rectangular components of the force, and x, y, z those of the displacement s of its point of application, the work done by the force is represented by $W = Xx + Yy + Zz$. If θ is the angle between the line of action of the force and the direction of the displacement, $W = Fs \cos\theta = \mathbf{F} \cdot \mathbf{s}$. The scalar product of two unit vectors $[l, m, n]$ and $[l', m', n']$ is

$$ll' + mm' + nn',$$

and vanishes when the vectors are at right angles.

The *vector-product* of two vectors $\mathbf{v}_1 \equiv [x_1, y_1, z_1]$ and $\mathbf{v}_2 \equiv [x_2, y_2, z_2]$ is a vector whose components are $y_1 z_2 - y_2 z_1$, $z_1 x_2 - z_2 x_1, x_1 y_2 - x_2 y_1$, and which is therefore perpendicular to the two vectors. It is denoted by $\mathbf{v}_1 \times \mathbf{v}_2$. Evidently $\mathbf{v}_1 \times \mathbf{v}_2 = -\mathbf{v}_2 \times \mathbf{v}_1$, so that vector multiplication is not commutative. The direction of the vector-product is defined as in the figure. For example, in a right-handed coordinate-system if X and Y are vectors along the positive axes of x and y, $X \times Y$ is along the positive axis of z.

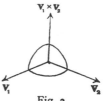

Fig. 3

2·2. Angles.

The angle between two planes is equal to the angle between their normals. The angle between the two planes

$$lx + my + nz + p = 0,$$
$$l'x + m'y + n'z + p' = 0,$$

(the axes being rectangular) is therefore given by

$$\cos\theta = \Sigma ll' / \sqrt{(\Sigma l^2 \Sigma l'^2)}.$$

The two planes are orthogonal if $\Sigma ll' = 0$, and parallel if $l : m : n = l' : m' : n'$.

The angle between a straight line and a plane is equal to the complement of the angle between the straight line and the normal. Hence if θ is the angle between the plane

$$lx + my + nz + p = 0,$$

and the straight line

$$\frac{x - x'}{l'} = \frac{y - y'}{m'} = \frac{z - z'}{n'},$$

$$\sin\theta = \Sigma ll' / \sqrt{(\Sigma l^2 \Sigma l'^2)}.$$

The straight line is parallel to the plane if $\Sigma ll' = 0$ and perpendicular if $l : m : n = l' : m' : n'$.

2·31. Intersection of a straight line and a plane.

Let the equation of the plane be

$$lx + my + nz + p = 0, \qquad \ldots\ldots(1)$$

and the freedom-equations of the line

$$\left.\begin{aligned} x &= x' + l't \\ y &= y' + m't \\ z &= z' + n't \end{aligned}\right\} . \qquad \ldots\ldots(2)$$

Then substituting for x, y, z in the equation of the plane, we have

$$t(ll' + mm' + nn') + lx' + my' + nz' + p = 0. \qquad \ldots\ldots(3)$$

This gives in general one value for t, and this value substituted in (2) gives the coordinates of a single point.

If, however, $\Sigma ll' = 0$, (3) cannot be satisfied unless

$$lx' + my' + nz' + p = 0. \qquad \ldots\ldots(4)$$

In this case the line is parallel to the plane and has the point $[x', y', z']$ common with the plane, hence it lies entirely in the plane.

If $\Sigma ll' = 0$, while (4) is not satisfied, the line is parallel to the plane and has no point in common with it. (This is true also if the axes are oblique.)

2·32. Points, lines, and plane at infinity.

Two straight lines in the same plane either intersect or are parallel. When they intersect they have a common point, and this point belongs to each of a single infinity of lines forming a pencil; when they are parallel they have a common *direction*, and this direction belongs to a single infinity of lines all mutually parallel. Two distinct points determine a unique line, the line which possesses the two points. A point and a direction also determine a unique line, the line which possesses the given point and the given direction.

In three dimensions a point is common to a doubly infinite system of lines forming a bundle; a direction is common to a doubly infinite system of lines all mutually parallel. Two planes either intersect or are parallel. When they intersect they have a common line, their line of intersection, and this line belongs to each of a single infinity of planes forming a pencil; when they are parallel they have a common *orientation*, and this orientation belongs to a single infinity of planes all mutually parallel. A straight line possesses a single infinity of points and one direction; a plane possesses a double infinity of points and a single infinity of directions, the directions of all the lines in the plane. If a line and a plane intersect they have a common point; if they are parallel they have a common direction.

Three distinct points determine a unique plane, the plane which possesses the three points. A plane is also uniquely determined by two given points and a given direction, or by a point and a straight line, or by a point and an orientation, or by a given point and two given directions. Two directions alone determine an orientation, since a system of parallel planes can be drawn all parallel to two given lines.

Thus in determining lines and planes, we may in certain cases replace points and lines by directions and orientations respectively. This connection is emphasised by using the suggestive terms "point at infinity" for "direction" and "line at infinity" for "orientation".

A line possesses one point at infinity; a plane possesses one line at infinity but a single infinity of points at infinity. Two points at infinity determine uniquely a line at infinity, and we say that the points at infinity on a plane lie on the line at infinity of this plane. Two lines at infinity determine uniquely a point at infinity (the direction of the line of intersection of two planes with the given orientations). Since two lines which determine a point also determine a plane, we say that two lines at infinity, a, b, determine a "plane at infinity" α. A third line at infinity c determines a point at infinity with each of the lines at infinity a, b, and we conclude that c also belongs to the plane at infinity α. Hence there is just one plane at infinity, the assemblage of all points at infinity and all lines at infinity.

2·33. Homogeneous cartesian coordinates. A point at infinity, or direction, is represented by the ratios of the three direction-cosines, but it is convenient to modify the coordinate-system to admit of the representation of ordinary points and points at infinity equally. This is done by the introduction of *homogeneous cartesian coordinates*. If $[X, Y, Z]$ are the ordinary non-homogeneous coordinates (rectangular or oblique), let

$$X = x/w, \quad Y = y/w, \quad Z = z/w,$$

then $[x, y, z, w]$ are called the homogeneous cartesian coordinates. If $w \neq 0$ every set of values of x, y, z, w uniquely represents a point, and for every value of k, not zero, the values kx, ky, kz, kw represent the same point.

2·34. The equations of the straight line through $[x', y', z', w']$, with direction-ratios $[l, m, n]$, become

$$\frac{w'x - x'w}{l} = \frac{w'y - y'w}{m} = \frac{w'z - z'w}{n},$$

and freedom-equations are obtained by equating each of these ratios to t:

$$w'x = x'w + lt,$$
$$w'y = y'w + mt,$$
$$w'z = z'w + nt,$$

where w may have any value. Introducing another parameter u, writing $\rho w = w'u$, and replacing t by $w't/\rho$, the equations become

$$\rho x = x'u + lt,$$
$$\rho y = y'u + mt,$$
$$\rho z = z'u + nt,$$
$$\rho w = w'u,$$

where ρ is a factor of proportionality. These are homogeneous freedom-equations in terms of two homogeneous parameters. Any point on the line is determined by the ratio t/u.

When $w = 0$ the cartesian coordinates x/w, etc., in general all become infinite, and we get the point at infinity on the line. The homogeneous coordinates are $[l, m, n, 0]$. These are therefore the coordinates of the point at infinity in the direction $[l, m, n]$.

The equations of the straight line through the points $[x_1, y_1, z_1, w_1]$ and $[x_2, y_2, z_2, w_2]$ are

$$\frac{w_1 x - w x_1}{w_1 x_2 - w_2 x_1} = \frac{w_1 y - w y_1}{w_1 y_2 - w_2 y_1} = \frac{w_1 z - w z_1}{w_1 z_2 - w_2 z_1}.$$

Equate each of these to u/ρ, and write $\rho w = w_1 t + w_2 u$, and we get as the homogeneous freedom-equations

$$\rho x = x_1 t + x_2 u,$$
$$\rho y = y_1 t + y_2 u,$$
$$\rho z = z_1 t + z_2 u,$$
$$\rho w = w_1 t + w_2 u.$$

Eliminating ρ, t, and u between these equations, taken three at a time, we obtain four equations, of which only two are independent. These may be represented by the notation

$$\left\| \begin{array}{cccc} x & y & z & w \\ x_1 & y_1 & z_1 & w_1 \\ x_2 & y_2 & z_2 & w_2 \end{array} \right\|_3 = 0,$$

which means that each of the determinants of the third order formed from this matrix vanishes.

Ex. Show that the ratio in which $[x, y, z, w]$ divides the join of $[x_1, y_1, z_1, w_1]$ and $[x_2, y_2, z_2, w_2]$ is $w_2 u / w_1 t$.

2·341. Matrices.

We shall frequently have occasion to use the matrix notation. A matrix is a set of numbers arranged in m rows and n columns, and is denoted by

$$\begin{bmatrix} a_{11} & a_{12} \ldots a_{1n} \\ \cdots\cdots\cdots\cdots\cdots \\ a_{m1} & a_{m2} \ldots a_{mn} \end{bmatrix} \text{ or } [a_{mn}].$$

Any determinant which is formed by striking out $m - r$ rows and $n - r$ columns is called a determinant of the matrix. If all the determinants of order $r + 1$ vanish, but at least one of the determinants of order r does not vanish, the matrix is said to be of *rank r*.

The condition that three points $[x_1, y_1, z_1, w_1]$, $[x_2, y_2, z_2, w_2]$, $[x_3, y_3, z_3, w_3]$ should be collinear can be expressed by saying that the matrix

$$\begin{vmatrix} x_1 & y_1 & z_1 & w_1 \\ x_2 & y_2 & z_2 & w_2 \\ x_3 & y_3 & z_3 & w_3 \end{vmatrix}$$

is of rank 2. If this matrix is of rank 1 the three points all coincide.

2·35. The equation of a plane in homogeneous coordinates becomes

$$lx + my + nz + pw = 0.$$

The condition that four points $[x_1, y_1, z_1, w_1]$, $[x_2, y_2, z_2, w_2]$, $[x_3, y_3, z_3, w_3]$, $[x_4, y_4, z_4, w_4]$ should be coplanar is

$$\begin{vmatrix} x_1 & y_1 & z_1 & w_1 \\ x_2 & y_2 & z_2 & w_2 \\ x_3 & y_3 & z_3 & w_3 \\ x_4 & y_4 & z_4 & w_4 \end{vmatrix} = 0,$$

or that this matrix is of rank 3. If the matrix is of rank 2 the four points are collinear; if of rank 1, the four points coincide.

2·351. The freedom-equations of the plane through three points $[x_1, \ldots]$, $[x_2, \ldots]$, $[x_3, \ldots]$ are

$$\rho x = x_1 t + x_2 u + x_3 v,$$
$$\rho y = y_1 t + y_2 u + y_3 v,$$
$$\rho z = z_1 t + z_2 u + z_3 v,$$
$$\rho w = w_1 t + w_2 u + w_3 v.$$

2·36. Consider the intersection of the line

$$\rho x = x' u + l t,$$
$$\rho y = y' u + m t,$$
$$\rho z = z' u + n t,$$
$$\rho w = w' u,$$

which passes through the point $[x', y', z', w']$ and has direction-ratios $[l, m, n]$, with the plane

$$l'x + m'y + n'z + p'w = 0.$$

Substituting for x, y, z, w in terms of t, u we have

$$(ll' + mm' + nn')t + (l'x' + m'y' + n'z' + p'w')u = 0. \quad \dots(1)$$

When the line is parallel to the plane, $\Sigma ll' = 0$. Then either $l'x' + m'y' + n'z' + p'w' = 0$, which expresses that the point $[x', y', z', w']$ lies in the plane and then the whole line lies in the plane, since (1) is then true for all values of t and u; or else $u = 0$, and we get as the coordinates of the point of intersection

$$x = lt, \ y = mt, \ z = nt, \ w = 0,$$

or, since only the ratios are significant, $[l, m, n, 0]$. These are thus the homogeneous coordinates of the point at infinity on the line whose direction-ratios are $[l, m, n]$.

The coordinates of any point at infinity satisfy the equation $w = 0$. Since this is an equation of the first degree we consider that it represents a plane which we call the *plane at infinity*.

The equations of the parallel planes

$$lx + my + nz + pw = 0,$$

$$lx + my + nz + p'w = 0$$

are satisfied simultaneously only if $w = 0$, i.e. by the coordinates of points at infinity. They have then in common a *straight line at infinity*. On every plane (except the plane at infinity) there is one straight line at infinity; the equations of the straight line at infinity on the plane $lx + my + nz + pw = 0$ being

$$\left.\begin{array}{c} lx + my + nz = 0 \\ w = 0 \end{array}\right\}.$$

2·41. Intersection of three planes.

Three planes

$$l_1 x + m_1 y + n_1 z + p_1 w = 0,$$

$$l_2 x + m_2 y + n_2 z + p_2 w = 0,$$

$$l_3 x + m_3 y + n_3 z + p_3 w = 0$$

intersect in a point whose coordinates satisfy the three equations simultaneously. These coordinates are therefore given by

$$\frac{x}{\begin{vmatrix} m_1 & n_1 & p_1 \\ m_2 & n_2 & p_2 \\ m_3 & n_3 & p_3 \end{vmatrix}} = \frac{-y}{\begin{vmatrix} l_1 & n_1 & p_1 \\ l_2 & n_2 & p_2 \\ l_3 & n_3 & p_3 \end{vmatrix}} = \frac{z}{\begin{vmatrix} l_1 & m_1 & p_1 \\ l_2 & m_2 & p_2 \\ l_3 & m_3 & p_3 \end{vmatrix}} = \frac{-w}{\begin{vmatrix} l_1 & m_1 & n_1 \\ l_2 & m_2 & n_2 \\ l_3 & m_3 & n_3 \end{vmatrix}},$$

which may also be expressed as

$$[x \quad y \quad z \quad w] = \begin{bmatrix} l_1 & m_1 & n_1 & p_1 \\ l_2 & m_2 & n_2 & p_2 \\ l_3 & m_3 & n_3 & p_3 \end{bmatrix},$$

i.e. x, y, z, w are proportional to the determinants of the matrix taken with the proper signs. If the matrix on the right is of rank 2 the three planes have a common line; if of rank 1 they coincide. If the matrix

$$\begin{bmatrix} l_1 & m_1 & n_1 \\ l_2 & m_2 & n_2 \\ l_3 & m_3 & n_3 \end{bmatrix}$$

is of rank 2 the point of intersection is at infinity, and the three planes are parallel to one line; if of rank 1 they are mutually parallel.

The condition that four planes should have a point in common is that the matrix

$$\begin{bmatrix} l_1 & m_1 & n_1 & p_1 \\ l_2 & m_2 & n_2 & p_2 \\ l_3 & m_3 & n_3 & p_3 \\ l_4 & m_4 & n_4 & p_4 \end{bmatrix}$$

should be of rank 3, i.e. the determinant vanishes. If the matrix is of rank 2 the planes have a line in common; if of rank 1 they all coincide.

2·42. Pencil of planes.

If
$$u_1 \equiv l_1 x + m_1 y + n_1 z + p_1 = 0,$$
$$u_2 \equiv l_2 x + m_2 y + n_2 z + p_2 = 0$$
represent two planes, the equation
$$u_1 + \lambda u_2 = 0$$

is of the first degree, therefore it represents a plane. Also it is satisfied if $u_1 = 0$ and $u_2 = 0$, i.e. by the coordinates of any point on the line common to u_1 and u_2. Hence it represents a plane through the line of intersection of u_1 and u_2.

All the planes through a given line form a *sheaf* or *axial pencil* of planes; the line is called the *axis*.

Similarly, if u_1, u_2, u_3 are three planes,

$$u_1 + \lambda u_2 + \mu u_3 = 0$$

represents a plane through their common point. By giving all values to λ and μ we obtain a doubly infinite system of planes through a point, forming a *bundle*.

2·43. *Condition that three lines through the origin should lie in one plane.*

Let the direction-ratios of the three lines be $[l_1, m_1, n_1]$, etc., and let
$$lx + my + nz = 0$$
be the equation of the plane containing the lines. Then since the point at infinity on each line lies in this plane,

$$ll_1 + mm_1 + nn_1 = 0,$$
$$ll_2 + mm_2 + nn_2 = 0,$$
$$ll_3 + mm_3 + nn_3 = 0.$$

Hence eliminating l, m, n,

$$\begin{vmatrix} l_1 & m_1 & n_1 \\ l_2 & m_2 & n_2 \\ l_3 & m_3 & n_3 \end{vmatrix} = 0.$$

This is also the condition that three lines should be parallel to one plane, or that the three points at infinity $[l_1, m_1, n_1, 0]$, etc., should be collinear.

2·44. Intersection of two straight lines.

Since a straight line is the intersection of two planes, and four planes do not in general have a point in common, two straight lines do not in general intersect. In this general case they are said to be *skew*.

Condition that two straight lines should intersect.

Let $[x_1, y_1, z_1]$ and $[x_2, y_2, z_2]$ be points of reference on the two lines, $[l_1, m_1, n_1]$ and $[l_2, m_2, n_2]$ their direction-ratios, then the freedom-equations are

$$x = x_1 + l_1 t_1, \qquad x = x_2 + l_2 t_2,$$
$$y = y_1 + m_1 t_1, \qquad y = y_2 + m_2 t_2,$$
$$z = z_1 + n_1 t_1, \qquad z = z_2 + n_2 t_2.$$

If t_1, t_2 are the respective parameters of the point of intersection, we have, equating the coordinates,

$$x_1 + l_1 t_1 = x_2 + l_2 t_2,$$
$$y_1 + m_1 t_1 = y_2 + m_2 t_2,$$
$$z_1 + n_1 t_1 = z_2 + n_2 t_2.$$

Between these three equations we can eliminate t_1, t_2 and we get

$$\begin{vmatrix} x_1 - x_2 & l_1 & l_2 \\ y_1 - y_2 & m_1 & m_2 \\ z_1 - z_2 & n_1 & n_2 \end{vmatrix} = 0.$$

This expresses that the two lines and a line which cuts both lines are all parallel to one plane. Writing the condition in the form

$$\begin{vmatrix} x_1 & x_2 & l_1 & l_2 \\ y_1 & y_2 & m_1 & m_2 \\ z_1 & z_2 & n_1 & n_2 \\ 1 & 1 & 0 & 0 \end{vmatrix} = 0,$$

it expresses that the two points $[x_1, y_1, z_1, 1]$, $[x_2, y_2, z_2, 1]$ and the two points at infinity on the lines, $[l_1, m_1, n_1, 0]$, $[l_2, m_2, n_2, 0]$, are coplanar.

2·5. Number of data which determine a point, a plane, and a straight line.

A point is determined by *three* data, its three coordinates.

A plane is determined by a single equation of the first degree, which contains four constants. Only the ratios of these constants, however, are significant, hence a plane is determined by *three* data.

A straight line is determined by two points, and each point requires three data to fix it; hence we have six data. But each point has one degree of freedom on the line; hence the number of necessary data reduces to *four*. Otherwise, the line is determined by the two points in which it cuts two of the co-ordinate-planes, and each of these points requires two data to fix it, hence again the line requires four data.

The number of constants required to fix a figure is called its *constant-number*. Thus the constant-number of a point or a plane is 3, while the constant-number of a straight line is 4. In plane geometry the constant-number of a circle is 3, and of a conic 5.

2·51. Coordinates of a plane.

The coordinates of a figure are any set of independent numbers serving to fix the figure. In the simplest case the number of coordinates is equal to the constant-number of the figure, but sometimes it is convenient to employ a greater number of co-ordinates connected by certain relations. Such coordinates are said to be superabundant. In choosing coordinates for a figure two conditions are desirable:

(1) to each figure there should correspond a unique set of values of the coordinates, and

(2) to each set of values of the coordinates there should correspond a unique figure,

i.e. between the figures and the sets of values of the coordinates there should exist a (1, 1) or biunivocal correspondence.

In the case of the plane these conditions are satisfied if we take as coordinates the ratios of the coefficients in the equation of the plane. The four coefficients $[l, m, n, p]$ may then be called the homogeneous coordinates of the plane.

As a set of non-homogeneous coordinates we may take the three ratios $l/p, m/p, n/p$, provided that $p \neq 0$, i.e. provided that the plane does not pass through the origin. The geometrical meaning of these coordinates is readily obtained. They are in fact equal to the reciprocals of the intercepts which the plane makes on the coordinate-axes, each with reversed sign.

An equation in point-coordinates represents a two-dimensional locus of points or a surface; in particular an equation of the first degree represents a plane. An equation in plane-coordinates represents a two-dimensional assemblage of planes. (We shall see later that such an assemblage may envelop either a surface or a curve.)

The equation $\qquad lx + my + nz + pw = 0$

represents the condition that the point $[x, y, z, w]$ lies on the plane $[l, m, n, p]$. If l, m, n, p are fixed, while x, y, z, w are variable, this equation represents a locus of points, in particular a plane. If x, y, z, w are fixed, while l, m, n, p vary, it represents the assemblage of planes which pass through the fixed point $[x, y, z, w]$, i.e. it represents a bundle of planes.

2·511. We may here explain a notation which it will sometimes be convenient to use. Instead of using the letters x, y, z, w for the coordinates of a point, and l, m, n, p for those of a plane, we may economise letters by representing a point by the coordinates x_0, x_1, x_2, x_3, and a plane by $\xi_0, \xi_1, \xi_2, \xi_3$. Here x_0 replaces w, so that $x_0 = 0$ represents the plane at infinity, and $\xi_0 = 0$ represents the origin (bundle of planes through O). To distinguish two points we may use different letters, e.g. the point $[y_0, y_1, y_2, y_3]$. The point whose coordinates are $[x_0, x_1, x_2, x_3]$ may be referred to simply as the point (x), and similarly the plane (ξ) for the plane whose coordinates are $[\xi_0, \xi_1, \xi_2, \xi_3]$ or whose equation is

$$\xi_0 x_0 + \xi_1 x_1 + \xi_2 x_2 + \xi_3 x_3 = 0.$$

2·52. Coordinates of a line.

A set of four coordinates for a line may be found as follows. Draw a plane through the line perpendicular to the plane of xz. This cuts the plane of xz in a line

$$y = 0, \quad x = \lambda z + p,$$

and $x = \lambda z + p$ is the equation of this plane. Similarly the plane which passes through the line and is perpendicular to the plane of yz is of the form

$$y = \mu z + q.$$

The straight line is then represented by the two equations

$$\left. \begin{aligned} x &= \lambda z + p \\ y &= \mu z + q \end{aligned} \right\},$$

and the four numbers λ, μ, p, q may be taken as the coordinates of the line. They satisfy the conditions of 2·51, but they are lacking in symmetry.

A more symmetrical set of coordinates for a line can be obtained as follows. Using the notation of 2·511 the line is determined either by two points (x), (y), or by two planes (ξ), (η). In each case we have 8 constants which reduce to 6, since only the ratios of each set of 4 are significant. The straight line may be represented by either of the two matrices

$$\begin{bmatrix} x_0 & x_1 & x_2 & x_3 \\ y_0 & y_1 & y_2 & y_3 \end{bmatrix} \quad \text{or} \quad \begin{bmatrix} \xi_0 & \xi_1 & \xi_2 & \xi_3 \\ \eta_0 & \eta_1 & \eta_2 & \eta_3 \end{bmatrix},$$

and is, as we shall prove, completely determined by the ratios of the second-order determinants of either matrix.

Let $\qquad x_i y_j - x_j y_i = p_{ij}$ and $\xi_i \eta_j - \xi_j \eta_i = \varpi_{ij}$,

so that $\qquad p_{ij} = -p_{ji}$ and $\varpi_{ij} = -\varpi_{ji}$.

The straight line is completely determined by the two points (x) and (y), and their coordinates then determine p_{ij}. The line is also determined by any two points on the same range, say $(x + \lambda y)$ and $(x + \mu y)$. Now

$$(x_i + \lambda y_i)(x_j + \mu y_j) - (x_j + \lambda y_j)(x_i + \mu y_i) = (\mu - \lambda)(x_i y_j - x_j y_i).$$

Hence the ratios of p_{ij} are the same whatever pair of points are chosen on the line. Similarly we may prove that the ratios of ϖ_{ij} are the same whatever pair of planes are chosen from the pencil whose axis is the given line.

2·521. The six p_{ij} are connected by an identical homogeneous relation. We have in fact

$$p_{01}p_{23} + p_{02}p_{31} + p_{03}p_{12} = \sum_{1,2,3} (x_0 y_i - x_i y_0)(x_j y_k - x_k y_j)$$

$$= x_0 \begin{vmatrix} y_1 & y_2 & y_3 \\ x_1 & x_2 & x_3 \\ y_1 & y_2 & y_3 \end{vmatrix} - y_0 \begin{vmatrix} x_1 & x_2 & x_3 \\ x_1 & x_2 & x_3 \\ y_1 & y_2 & y_3 \end{vmatrix} = 0,$$

and similarly $\qquad \varpi_{01}\varpi_{23} + \varpi_{02}\varpi_{31} + \varpi_{03}\varpi_{12} = 0.$

2·522. We can now prove that the ratios of any six numbers p_{ij} which satisfy the identical relation uniquely determine a line, such that if (x) and (y) are two points on the line

$$\rho p_{ij} = x_i y_j - x_j y_i,$$

where ρ is a factor of proportionality. Choose (x) as the point at infinity on the line, so that $x_0 = 0$, then the ratios of x_1, x_2, x_3 are determined by the equations

$$\rho p_{i0} = x_i y_0,$$

i.e. the direction-cosines of the line are $[p_{10}, p_{20}, p_{30}]$. Similarly the coordinates of the point of intersection with the plane $x_1 = 0$ are $[p_{01}, 0, p_{21}, p_{31}]$, etc.

2·523. The two sets of line-coordinates p_{ij} and ϖ_{ij} are connected by simple equations. Since (x) and (y) are two points on the line, and (ξ) and (η) are two planes through the line, each of these points lies on each of the planes. Hence

$$\Sigma\xi x = 0, \quad \Sigma\xi y = 0, \quad \Sigma\eta x = 0, \quad \Sigma\eta y = 0.$$

Eliminate ξ_0 between the first two equations and we get

$$\xi_1 p_{10} + \xi_2 p_{20} + \xi_3 p_{30} = 0.$$

Similarly, eliminating η_0 between the last two, we get

$$\eta_1 p_{10} + \eta_2 p_{20} + \eta_3 p_{30} = 0.$$

Hence $\qquad p_{10} : p_{20} : p_{30} = \varpi_{23} : \varpi_{31} : \varpi_{12};$

and in a similar way we prove the complete set of relations

$$p_{10} : p_{20} : p_{30} : p_{23} : p_{31} : p_{12} = \varpi_{23} : \varpi_{31} : \varpi_{12} : \varpi_{10} : \varpi_{20} : \varpi_{30}.$$

These superabundant coordinates are called the *Plücker coordinates* of a line.

If $[l, m, n]$ are the direction-cosines of the line, and $[x', y', z']$ any point on it, the matrix which defines the coordinates of the line is

$$\begin{bmatrix} x' & y' & z' & 1 \\ l & m & n & 0 \end{bmatrix},$$

and $p_{01} = l$, $p_{02} = m$, $p_{03} = n$, $p_{23} = ny' - mz'$, etc. We shall sometimes use the notation $[l, m, n; l', m', n']$ for the coordinates of a line whether l, m, n are simply the direction-cosines or the more general coordinates. The identical relation is then

$$ll' + mm' + nn' = 0.$$

In representing a line by its coordinates we shall adhere to the conventional order $[p_{01}, p_{02}, p_{03}; p_{23}, p_{31}, p_{12}]$. Thus

$$[2, 3, 6; 3, -6, 2]$$

represents a line with direction-cosines [2, 3, 6] and passing through the point [0, −1, −3]; while the line [3, −6, 2; 2, 3, 6] has direction-cosines [3, −6, 2] and passes through the point [0, −2, 1].

Ex. 1. Prove that the coordinates of the point of intersection of the line (p) with the plane (ξ) are

$$\xi_1 p_{10} + \xi_2 p_{20} + \xi_3 p_{30},$$
$$\xi_0 p_{01} \qquad + \xi_2 p_{21} + \xi_3 p_{31},$$
$$\xi_0 p_{02} + \xi_1 p_{12} \qquad + \xi_3 p_{32},$$
$$\xi_0 p_{03} + \xi_1 p_{13} + \xi_2 p_{23},$$

and that if the line lies in the plane these four expressions all vanish.

Ex. 2. Prove that the line-coordinates of the plane containing the line (ϖ) and the point (x) are

$$x_1 \varpi_{01} + x_2 \varpi_{02} + x_3 \varpi_{03},$$
$$x_0 \varpi_{10} \qquad + x_2 \varpi_{12} + x_3 \varpi_{13},$$
$$x_0 \varpi_{20} + x_1 \varpi_{21} \qquad + x_3 \varpi_{23},$$
$$x_0 \varpi_{30} + x_1 \varpi_{31} + x_2 \varpi_{32},$$

and that if these four expressions vanish the point (x) lies on the line (ϖ).

Ex. 3. Prove that the conditions that the line (p) should pass through the point (x) are any two of the equations

$$x_i p_{jk} + x_j p_{ki} + x_k p_{ij} = 0,$$

and that the line (p) should lie in the plane (ξ) are any two of the equations

$$\xi_i \varpi_{jk} + \xi_j \varpi_{ki} + \xi_k \varpi_{ij} = 0,$$

where i, j, k are given any three of the values 0, 1, 2, 3.

2·524. *Condition that two lines (p), (q) should intersect.*

Let the lines be determined by the pairs of points (x), (x') and (y), (y'), so that $p_{ij} = x_i x_j' - x_j x_i'$, $q_{ij} = y_i y_j' - y_j y_i'$. The condition required is that the four points should be coplanar, i.e.

$$\begin{vmatrix} x_0 & x_1 & x_2 & x_3 \\ x_0' & x_1' & x_2' & x_3' \\ y_0 & y_1 & y_2 & y_3 \\ y_0' & y_1' & y_2' & y_3' \end{vmatrix} = 0.$$

Expanding this determinant we find the condition in the form

$$p_{01}q_{23} + p_{23}q_{01} + p_{02}q_{31} + p_{31}q_{02} + p_{03}q_{12} + p_{12}q_{03} = 0.$$

The condition is linear and homogeneous in the coordinates of each of the lines. The left-hand side of the equation is the bilinear symmetrical expression which reduces to the identity

$$p_{01}p_{23} + p_{02}p_{31} + p_{03}p_{12} = 0$$

when the two lines coincide.

2·6. Imaginary elements.

In applying algebra to geometry we must frequently deal with equations having imaginary roots. The number-system in algebra having been extended to include complex numbers, an interpretation of the corresponding elements in geometry is required. A purely geometrical treatment of imaginary elements was given by VON STAUDT on the basis of elliptic involutions, but we shall simply define an imaginary point as the thing which corresponds to a set of values of the coordinates when some of them at least are complex numbers. The two sets of values

$$[x+ix', \ y+iy', \ z+iz']$$

and
$$[x-ix', \ y-iy', \ z-iz']$$

represent "conjugate imaginary points", and the two equations

$$(l+il')x + \ldots = 0, \ (l-il')x + \ldots = 0$$

represent "conjugate imaginary planes". The line joining conjugate imaginary points, and the line of intersection of conjugate imaginary planes, are real. Through an imaginary point there is just one real line, the line joining the point to its conjugate; and on an imaginary plane there is just one real line, the line of intersection with the conjugate plane.

A line is determined either by two points or by two planes, and we obtain the corresponding Plücker coordinates. In order that the line should be real the ratios of its Plücker coordinates must be all real. In general the line (l) determined by two imaginary points is imaginary, and then the line (l') determined by the two corresponding conjugate points is the conjugate imaginary. Two sorts of imaginary lines arise, according as the line does or does not meet its conjugate.

If l is of the *first species* it meets its conjugate l'. Then we have one real point on l and one real plane through l.

If l is of the *second species* it does not meet its conjugate. It contains no real points, and lies in no real plane.

As an example let A, A' (Fig. 4) be two conjugate imaginary points, and B, C two real points, such that the real lines BC and AA' do not intersect. Then AC, $A'C$ and AB, $A'B$ are pairs of conjugate imaginary lines of the first species. The planes $AA'B$ and $AA'C$ are real, while ABC and $A'BC$ are conjugate imaginaries.

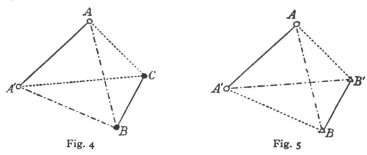

Fig. 4   Fig. 5

If A, A' and B, B' (Fig. 5) are two pairs of conjugate imaginary points such that the real lines AA' and BB' do not intersect, then AB', $A'B$ and AB, $A'B'$ are pairs of conjugate imaginary lines of the second species. The planes $AA'B$ and $AA'B'$, $BB'A$ and $BB'A'$ are pairs of conjugate imaginaries.

2·71. Distance from a point to a plane.

Let $P \equiv [x', y', z']$ be the given point and

$$x \cos \alpha + y \cos \beta + z \cos \gamma = p$$

the given plane, so that the direction-angles of the normal are α, β, γ and p is its distance from the origin. Let d be the distance of P from the plane. The projection of OP upon the normal is $p \pm d$ according as P is on the opposite side of the plane from O or on the same side. But this is equal to the sum of the projections of the coordinates of P. Hence

$$p \pm d = x' \cos\alpha + y' \cos\beta + z' \cos\gamma,$$

i.e.

$$\pm d = x' \cos\alpha + y' \cos\beta + z' \cos\gamma - p.$$

If the equation of the plane is

$$lx + my + nz + p = 0,$$

then

$$d = \pm \frac{lx' + my' + nz' + p}{\sqrt{(l^2 + m^2 + n^2)}}.$$

2·72. Distance of a point from a straight line.

Let the equations of the line be

$$\frac{x - X}{l} = \frac{y - Y}{m} = \frac{z - Z}{n}$$

and the given point $P \equiv [x', y', z']$.

Let $A \equiv [X, Y, Z]$.

Then if p is the length of the perpendicular PN, and $AP = r$, and $\angle PAN = \theta$,

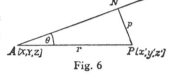

Fig. 6

$$p = r \sin\theta,$$
and $$r^2 = \Sigma(x' - X)^2.$$

Also, projecting AP on AN, we have

$$r \cos\theta = \Sigma l (x' - X)(\Sigma l^2)^{-\frac{1}{2}}.$$

From these three equations, by eliminating r and θ, we can obtain p.

Ex. Prove that

$$p^2 = \Sigma \{m (z' - Z) - n (y' - Y)\}^2/(\Sigma l^2).$$

2·73. Shortest distance between two straight lines.

Let the freedom-equations of the two lines be

$$x = X + lt, \text{ etc.,} \quad \text{and} \quad x = X' + l't', \text{ etc.}$$

Take a point $P \equiv [x, y, z]$ on the former, and $P' \equiv [x', y', z']$ on the latter. We have then to find P and P' so that PP' may be a minimum.

We have $$PP'^2 = \Sigma\{(X + lt) - (X' + l't')\}^2.$$

The conditions for a minimum are found by differentiating separately with regard to t and t' and equating to zero. Hence

$$\Sigma l \{(X + lt) - (X' + l't')\} = 0,$$
$$\Sigma l' \{(X + lt) - (X' + l't')\} = 0.$$

But if λ, μ, ν are the direction-cosines of PP' these equations are equivalent to

$$\Sigma l\lambda = 0, \quad \Sigma l'\lambda = 0.$$

Hence PP' is perpendicular to both lines.

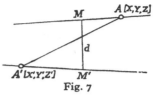

Fig. 7

Let MM', $=d$, be the common perpendicular, and let θ be its inclination to AA'. The direction-cosines $[\lambda, \mu, \nu]$ of MM' are found from the equations

$$\Sigma l\lambda = 0, \quad \Sigma l'\lambda = 0,$$

hence $\qquad \lambda : \mu : \nu = mn' - m'n : nl' - n'l : lm' - l'm.$

Then, d being the projection of AA' on MM',

$$d = \Sigma\lambda(X - X')/(\Sigma\lambda^2)^{\frac{1}{2}}.$$

The equations of the common perpendicular MM' are most symmetrically expressed by forming the equations of the two planes AMM' and $A'MM'$. The plane AMM' is expressed as the plane through A which is parallel to the two directions $[l, m, n]$ and $[\lambda, \mu, \nu]$. Its direction-cosines are therefore $m\nu - n\mu$, etc. Similarly for the plane $A'MM'$. Lastly, the coordinates of M' and M are found by the intersection of the plane AMM' with the line $A'M'$, and the plane $A'MM'$ with the line AM.

2·731. Let $[l, m, n]$, $[l', m', n']$ be the actual values of the direction-cosines and i the angle between the two lines, then since $\sin^2 i = \Sigma(mn' - m'n)^2$, we have, defining the sign conventionally,

$$-d\sin i = \Sigma(X - X')(mn' - m'n)$$

$$= \begin{vmatrix} X - X' & l & l' \\ Y - Y' & m & m' \\ Z - Z' & n & n' \end{vmatrix}.$$

This can be expressed in terms of the coordinates of the two lines. We have

$$-d\sin i = \Sigma l'(nY - mZ) + \Sigma l(n'Y' - m'Z')$$

$$= \Sigma(p'_{01}p_{23} + p_{01}p'_{23}).$$

The expression $d\sin i$ is called the *moment* of the two lines. Its sign depends on the directions of the two lines. To bring the one line into coincidence with the other, with the proper sense, we may combine a translation through d with a rotation through i; if these displacements are effected simultaneously we have a screwing motion, and the moment is positive if, when $i < 180°$, the screw is right-handed. Thus if $\mathbf{a}$ is along the positive x-axis, $\mathbf{a}'$ parallel to the positive y-axis, and $d = +1$ along the z-axis, the coordinates of the two lines are $[1, 0, 0; 0, 0, 0]$ and $[0, 1, 0; -1, 0, 0]$ and their moment $= +1$. The vanishing of the moment is the condition that the two lines should intersect or be parallel.

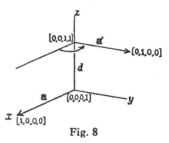

Fig. 8

2·8. Volume of tetrahedron.

Consider the tetrahedron with one vertex at the origin, and the other vertices numbered 1, 2, 3, so that the coordinates of the vertices are $[x_1, y_1, z_1]$, etc.

The volume $= \frac{1}{3}$ base × altitude. Take the base as the plane (123). Its equation is

$$\begin{vmatrix} x & y & z & 1 \\ x_1 & y_1 & z_1 & 1 \\ x_2 & y_2 & z_2 & 1 \\ x_3 & y_3 & z_3 & 1 \end{vmatrix} = 0.$$

Now the coefficient of x is

$$\begin{vmatrix} y_1 & z_1 & 1 \\ y_2 & z_2 & 1 \\ y_3 & z_3 & 1 \end{vmatrix}.$$

This is equal to twice the area of the triangle in the plane of yz with coordinates in that plane $(y_1, z_1), (y_2, z_2), (y_3, z_3)$. But this triangle is the projection of the triangle (123), $= (123) \cos \alpha$. Similarly for the coefficients of y and z. Hence dividing the equation by (123) it reduces to the normal form

$$x \cos \alpha + y \cos \beta + z \cos \gamma + p = 0.$$

Therefore $p \times (123) = 6V = \begin{vmatrix} x_1 & y_1 & z_1 \\ x_2 & y_2 & z_2 \\ x_3 & y_3 & z_3 \end{vmatrix}$.

For a general tetrahedron with vertices 1, 2, 3, 4, if we take new axes parallel to the old through the vertex 1, the relative coordinates of the vertices are $x_2 - x_1$, etc., and

$$6V = \begin{vmatrix} x_2 - x_1 & y_2 - y_1 & z_2 - z_1 \\ x_3 - x_1 & y_3 - y_1 & z_3 - z_1 \\ x_4 - x_1 & y_4 - y_1 & z_4 - z_1 \end{vmatrix} = \begin{vmatrix} x_1 & y_1 & z_1 & 1 \\ x_2 & y_2 & z_2 & 1 \\ x_3 & y_3 & z_3 & 1 \\ x_4 & y_4 & z_4 & 1 \end{vmatrix} .$$

2·81. Corresponding to the expression

$$\tfrac{1}{2}r_1 r_2 \sin(\theta_2 - \theta_1) = \tfrac{1}{2}r_1 r_2 \sin \phi$$

for the area of a triangle we have an expression for the volume of a tetrahedron in terms of the lengths of three coterminous edges and the angles between them.

Thus $6 \text{ Vol. } (0123) = r_1 r_2 r_3 \begin{vmatrix} \cos \alpha_1 & \cos \beta_1 & \cos \gamma_1 \\ \cos \alpha_2 & \cos \beta_2 & \cos \gamma_2 \\ \cos \alpha_3 & \cos \beta_3 & \cos \gamma_3 \end{vmatrix}$.

The square of this determinant is

$$\begin{vmatrix} 1 & \cos(12) & \cos(13) \\ \cos(21) & 1 & \cos(23) \\ \cos(31) & \cos(32) & 1 \end{vmatrix}$$

$$= 1 - \Sigma \cos^2(23) + 2 \cos(23) \cos(31) \cos(12).$$

The square root of this expression is sometimes called the sine of the solid angle of the tetrahedron.

2·82. An expression for the volume of a tetrahedron can also be obtained in terms of the edges r_{ij}. Multiply together the two determinants

$$6V = \begin{vmatrix} x_1 & y_1 & z_1 & 1 & 0 \\ x_2 & y_2 & z_2 & 1 & 0 \\ x_3 & y_3 & z_3 & 1 & 0 \\ x_4 & y_4 & z_4 & 1 & 0 \\ 0 & 0 & 0 & 0 & 1 \end{vmatrix}, \quad -6V = \begin{vmatrix} x_1 & y_1 & z_1 & 0 & 1 \\ x_2 & y_2 & z_2 & 0 & 1 \\ x_3 & y_3 & z_3 & 0 & 1 \\ x_4 & y_4 & z_4 & 0 & 1 \\ 0 & 0 & 0 & 1 & 0 \end{vmatrix}$$

by rows. In the diagonal we get the terms $r_1^2, ..., r_4^2$, o, and in the place (i, j) we have

$$\Sigma x_i x_j = \tfrac{1}{2}(r_i^2 + r_j^2 - r_{ij}^2) \quad (i, j = 1, 2, 3, 4).$$

The last row and the last column are each 1, 1, 1, 1, o. Now multiply the last row by $\tfrac{1}{2} r_i^2$ and subtract from the ith, and the last column by $\tfrac{1}{2} r_i^2$ and subtract from the ith. Finally, taking out the factors $-\tfrac{1}{2}$ we get

$$8 \times 36 V^2 = \begin{vmatrix} 0 & r_{12}^2 & r_{13}^2 & r_{14}^2 & 1 \\ r_{21}^2 & 0 & r_{23}^2 & r_{24}^2 & 1 \\ r_{31}^2 & r_{32}^2 & 0 & r_{34}^2 & 1 \\ r_{41}^2 & r_{42}^2 & r_{43}^2 & 0 & 1 \\ 1 & 1 & 1 & 1 & 0 \end{vmatrix}.$$

This is the three-dimensional analogue of Heron's formula for the area of a triangle

$$\Delta^2 = s(s-a)(s-b)(s-c) = -\tfrac{1}{16} \begin{vmatrix} 0 & c^2 & b^2 & 1 \\ c^2 & 0 & a^2 & 1 \\ b^2 & a^2 & 0 & 1 \\ 1 & 1 & 1 & 0 \end{vmatrix}.$$

Ex. 1. If the line-coordinates (p) are the actual values of the determinants of the matrix

$$\begin{bmatrix} x_1 & y_1 & z_1 & 1 \\ x_2 & y_2 & z_2 & 1 \end{bmatrix},$$

where $[x_1, y_1, z_1]$, $[x_2, y_2, z_2]$ are the rectangular coordinates of two points on the line, and r is the distance between the two points, show that the expression (2·731)

$$\Sigma(p_{01}' p_{23} + p_{01} p_{23}') = r r' d \sin i.$$

Ex. 2. Show by expanding the determinant

$$\begin{vmatrix} x_1 & y_1 & z_1 & 1 \\ x_2 & y_2 & z_2 & 1 \\ x_3 & y_3 & z_3 & 1 \\ x_4 & y_4 & z_4 & 1 \end{vmatrix}$$

in terms of the minors formed from the first two and the last two rows that if r, r' are the lengths of two opposite edges of a tetrahedron, d the distance and i the angle between these edges, the volume

$$V = \tfrac{1}{6} r r' d \sin i.$$

2·9. Transformation of coordinates.

In 1·3 we have already considered the effect of changing the origin alone. We shall now consider the transformation to new axes with the same origin. Let the direction-cosines of the new axes of x', y', z' with respect to the old be respectively $[l_1, m_1, n_1]$, $[l_2, m_2, n_2]$, $[l_3, m_3, n_3]$, and assume that both sets of axes are orthogonal. Let P be any point whose old coordinates are $[x, y, z]$ and new coordinates $[x', y', z']$. Then, pro-

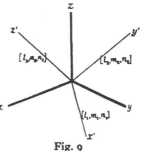

Fig. 9

jecting OP on each of the original axes in succession, we have, since the projection of OP is equal to the sum of the projections of x', y' and z',

$$\left. \begin{aligned} x &= l_1 x' + l_2 y' + l_3 z' \\ y &= m_1 x' + m_2 y' + m_3 z' \\ z &= n_1 x' + n_2 y' + n_3 z' \end{aligned} \right\}.$$

Also, since $[l_1, l_2, l_3]$ are the direction-cosines of Ox with respect to the new axes,

$$\left. \begin{aligned} x' &= l_1 x + m_1 y + n_1 z \\ y' &= l_2 x + m_2 y + n_2 z \\ z' &= l_3 x + m_3 y + n_3 z \end{aligned} \right\}.$$

These two sets of equations, which represent *inverse* transformations, can be represented by the scheme

	x'	y'	z'
x	l_1	l_2	l_3
y	m_1	m_2	m_3
z	n_1	n_2	n_3

The nine coefficients are connected by a number of equations. Since l_1, m_1, n_1 are a set of three direction-cosines, we have

Similarly

$$\left. \begin{aligned} l_1{}^2 + m_1{}^2 + n_1{}^2 &= 1. \\ l_2{}^2 + m_2{}^2 + n_2{}^2 &= 1, \\ l_3{}^2 + m_3{}^2 + n_3{}^2 &= 1. \end{aligned} \right\}$$

Further, since the new axes are mutually at right angles,

$$\left.\begin{array}{l} l_2 l_3 + m_2 m_3 + n_2 n_3 = 0 \\ l_3 l_1 + m_3 m_1 + n_3 n_1 = 0 \\ l_1 l_2 + m_1 m_2 + n_1 n_2 = 0 \end{array}\right\}.$$

These six equations reduce the nine constants to three.

Another set of six equations can be written down by considering the old axes in terms of the new, and we have

$$l_1{}^2 + l_2{}^2 + l_3{}^2 = 1, \qquad m_1 n_1 + m_2 n_2 + m_3 n_3 = 0,$$
$$m_1{}^2 + m_2{}^2 + m_3{}^2 = 1, \qquad n_1 l_1 + n_2 l_2 + n_3 l_3 = 0,$$
$$n_1{}^2 + n_2{}^2 + n_3{}^2 = 1, \qquad l_1 m_1 + l_2 m_2 + l_3 m_3 = 0.$$

2·91. This transformation is equivalent to a rotation through a definite angle about some definite line. The angle θ and the ratios of the direction-cosines $[\lambda, \mu, \nu]$ of the axis of rotation form the three independent constants which are required, and the equations of transformation could be expressed in terms of these. Thus, let OC be the axis of rotation, and consider a vector

$$OP = r = [x, y, z]$$

which is transformed into

$$OQ = r' = [x', y', z']$$

by a rotation through the angle $-\theta$

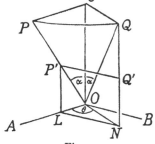

Fig. 10

about OC. This is equivalent to a rotation of the coordinate-system through the angle $+\theta$. In the plane POC take $OA \perp OC$, and take $OB \perp$ the plane AOC. Draw $QN \perp AOB$. Then $AON = \theta$. Let $POC = \alpha = QOC$. Draw $NL \| BO$, $LP' \| OC$ and $P'Q' \| LN$. Then the vector
Now
$$OQ = OP' + P'Q' + Q'Q.$$

$$OP' = OL \operatorname{cosec}\alpha = ON\cos\theta \operatorname{cosec}\alpha = OQ\cos\theta = OP\cos\theta.$$

Therefore $\qquad OP' = r\cos\theta.$

$$P'Q' = LN = ON\sin\theta = OQ\sin\alpha \sin\theta,$$

therefore $\qquad P'Q' = a \times r\sin\theta,$

where a is the unit vector in the direction OC.

$$Q'Q = NQ - LP' = (OQ - OP')\cos\alpha = OP(1 - \cos\theta)\cos\alpha,$$

therefore $\qquad Q'Q = a\,(a.r)\,(1 - \cos\theta).$

Hence we have the vector equation

$$r' = r\cos\theta + (a \times r)\sin\theta + (a.r)a(\mathrm{1} - \cos\theta).$$

Writing this in terms of the rectangular components referred to the original axes we have

$$x' = x\cos\theta + (vy - \mu z)\sin\theta + \lambda(\mathrm{1} - \cos\theta)(\lambda x + \mu y + vz),$$

$$y' = y\cos\theta + (\lambda z - vx)\sin\theta + \mu(\mathrm{1} - \cos\theta)(\lambda x + \mu y + vz),$$

$$z' = z\cos\theta + (\mu x - \lambda y)\sin\theta + v(\mathrm{1} - \cos\theta)(\lambda x + \mu y + vz).$$

These are known as Euler's equations of transformation.

The inverse equations are obtained by interchanging accented and unaccented letters and changing the sign of θ.

2·92. In the general orthogonal transformation the determinant

$$D \equiv \begin{vmatrix} l_1 & m_1 & n_1 \\ l_2 & m_2 & n_2 \\ l_3 & m_3 & n_3 \end{vmatrix}$$

is called the *modulus* of the transformation. Squaring it and using the equations of 2·9 we find $D^2 = \mathrm{1}$, and therefore $D = \pm\,\mathrm{1}$. For the identical transformation $l_1 = \mathrm{1} = m_2 = n_3$ and the other elements all vanish, so that $D = +\,\mathrm{1}$. The sign is negative when the two coordinate-systems are one right-handed and the other left-handed.

2·95. EXAMPLES.

1. Find the equation of the plane through the line $x = 2y = 3z$ perpendicular to the plane $5x + 4y - 3z = 8$.

Ans. $17x - 28y - 9z = 0$.

2. Reduce to the normal form the equations of the line of intersection of the planes

$$4x + 4y - 5z = 12, \quad 8x + 12y - 13z = 32,$$

and write down the equations of its projections on the three co-ordinate-planes.

Ans. $\dfrac{x-\mathrm{1}}{2} = \dfrac{y-2}{3} = \dfrac{z}{4}$; $4y - 3z = 8$, $2x - z = 2$, $3x - 2y + \mathrm{1} = 0$.

3. Find the three plane angles at the vertex of the trihedral angle determined by the three planes $x+z=4$, $y-z+5=0$, $x=y+z$, and find the coordinates of the vertex.

Ans. 90°, 90°, 60°; $[1, -2, 3]$.

4. Find the equation of the plane (i) through the origin parallel to each of the lines $(x-y+4z=1, 2x+y-3z=2)$ and $(x+3=2y+1=3z+2)$; (ii) through $[1, 1, 1]$ parallel to each of the lines $(x-3y+2z=0, ax^2+by^2+cz^2=0)$.

Ans. (i) $13x+20y-69z=0$; (ii) $x-3y+2z=0$.

5. Find the equations of the lines through $[1, 2, 3]$ which cut the axes of x and y respectively at right angles, and the equation of the plane containing these lines.

Ans. $(x=1, 3y=2z), (y=2, 3x=z); 6x+3y-2z=6.$

6. Find the equation of the plane through the line

$$(3x-4y+5z=10, 2x+2y-3z=4)$$

and parallel to the line $x=2y=3z$.

Ans. $x-20y+27z=14.$

7. Find the two points on the line $x=2y=3z+6$ at a distance 7 from the plane $2x+y-2z=5$.

Ans. $[12, 6, 2], [-\frac{120}{11}, -\frac{60}{11}, -\frac{62}{11}].$

8. Prove that the lines $2x-y+3z+3=0=x+10y-21$ and $2x-y=0=7x+z-6$ intersect. Find the coordinates of their common point, and the equation of the plane containing them.

Ans. $[1, 2, -1], x+3y+z=6.$

9. Find the condition that the three planes $x=cy+bz$, $y=az+cx$, $z=bx+ay$ may pass through one line.

Ans. $a^2+b^2+c^2+2abc=1.$

10. Find the equations of the two planes through the points $[0, 4, -3], [6, -4, 3]$, other than the plane through the origin, which cut off from the axes intercepts whose sum is zero.

Ans. $2x-3y-6z=6, 6x+3y-2z=18.$ (Math. Trip. I, 1913.)

11. The three lines

$$x=y=\tfrac{1}{4}z, \quad x-2=\tfrac{1}{2}(y+1)=z-1, \quad 2(x+1)=6(y-2)=3(z-3)$$

are three non-intersecting edges of a parallelepiped; find the equations of its six faces.

Ans. $x-5y+z=0$ and -8, $7x-3y-z=0$ and 16,
$\qquad 3x+y-5z=0$ and -16.

12. The coordinates of four points are $[a-b,\ a-c,\ a-d]$, $[b-c,\ b-d,\ b-a]$, $[c-d,\ c-a,\ c-b]$, $[d-a,\ d-b,\ d-c]$. Prove that the straight line joining the mid-points of any two opposite edges of the tetrahedron formed by them passes through the origin.

13. Find the equations of the projection of the line

$$\frac{x-1}{2}=\frac{y+1}{-1}=\frac{z-3}{4}$$

on the plane $x+2y+z=6$.

Ans. $\dfrac{x-3}{4}=\dfrac{y+2}{-7}=\dfrac{z-7}{10}$.

14. Prove that the lines

$$\frac{x-a}{a'}=\frac{y-b}{b'}=\frac{z-c}{c'} \quad \text{and} \quad \frac{x-a'}{a}=\frac{y-b'}{b}=\frac{z-c'}{c}$$

intersect, and find the coordinates of the point of intersection and the equation of the plane in which they lie.

Ans. $[a+a',\ b+b',\ c+c']$, $\Sigma x\,(bc'-b'c)=0$.

15. Prove that the lines

$$\frac{x-a+d}{\alpha-\delta}=\frac{y-a}{\alpha}=\frac{z-a-d}{\alpha+\delta}, \quad \frac{x-b+c}{\beta-\gamma}=\frac{y-b}{\beta}=\frac{z-b-c}{\beta+\gamma}$$

are coplanar, and find the equation of the plane in which they lie. (Wolstenholme.)

Ans. $x-2y+z=0$.

16. Find the equations of the line through $[1,\ 2,\ -1]$ perpendicular to the plane $3x-5y+4z=5$, the length of the perpendicular, and the coordinates of its foot.

Ans. $\dfrac{x-1}{3}=\dfrac{y-2}{-5}=\dfrac{z+1}{4}$, $\dfrac{8\sqrt{2}}{5}$, $\left[\dfrac{49}{25},\ \dfrac{2}{5},\ \dfrac{7}{25}\right]$.

17. Find the distance of the point from the straight line, and the coordinates of its foot:

(i) $[6, 6, -1]$ and $\dfrac{x-2}{1}=\dfrac{y-1}{2}=\dfrac{z+3}{-1}$.

(ii) $[5, 4, 2]$ and $\dfrac{x+1}{2}=\dfrac{y-3}{3}=\dfrac{z-1}{-1}$.

(iii) $[-2, 2, -3]$ and $\dfrac{x-3}{1}=\dfrac{y+1}{2}=\dfrac{z-2}{-4}$.

Ans.

(i) $\sqrt{21}$, $[4, 5, -5]$, (ii) $\sqrt{24}$, $[1, 6, 0]$, (iii) $\sqrt{28}$, $[4, 1, -2]$.

18. Find the length and the equations of the common perpendicular to the two lines:

(i) $\dfrac{x-3}{1}=\dfrac{y-4}{1}=\dfrac{z+1}{-1}$, $\dfrac{x+6}{2}=\dfrac{y+5}{4}=\dfrac{z-1}{-1}$.

(ii) $\dfrac{x-4}{2}=\dfrac{y+2}{1}=\dfrac{z-3}{-1}$, $\dfrac{x+7}{3}=\dfrac{y+2}{2}=\dfrac{z-1}{1}$.

(iii) $\dfrac{x-7}{-1}=\dfrac{y+7}{2}=\dfrac{z+1}{1}$, $\dfrac{x+7}{-5}=\dfrac{y-8}{3}=\dfrac{z+1}{2}$.

(iv) $\dfrac{x}{1}=\dfrac{y-11}{-2}=\dfrac{z-4}{1}$, $\dfrac{x-6}{7}=\dfrac{y+7}{-6}=\dfrac{z}{1}$.

Ans. (i) $\sqrt{14}$, $\dfrac{x-1}{3}=\dfrac{y-2}{-1}=\dfrac{z-1}{2}$.

(ii) $\sqrt{35}$, $\dfrac{x+1}{3}=\dfrac{y-2}{-5}=\dfrac{z-3}{1}$.

(iii) $\sqrt{59}$, $\dfrac{x-4}{1}=\dfrac{y+1}{-3}=\dfrac{z-2}{7}$.

(iv) $2\sqrt{29}$, $\dfrac{x+1}{2}=\dfrac{y+1}{3}=\dfrac{z+1}{4}$.

19. Show that the straight line joining $[a, b, c]$, $[a', b', c']$ passes through the origin if $aa'+bb'+cc'=\rho\rho'$, ρ and ρ' being the distances of the points from the origin.

20. Prove that the equations

$$\frac{a+mz-ny}{m-n}=\frac{b+nx-lz}{n-l}=\frac{c+ly-mx}{l-m}$$

represent the line at infinity on the plane

$$(m-n)\,x+(n-l)\,y+(l-m)\,z=0;$$

unless $al+bm+cn=0$, in which case the line is indeterminate and its locus is the plane

$$(m-n)\,x+(n-l)\,y+(l-m)\,z=a+b+c.$$

21. Find the equations of the planes through the lines of intersection of two of three given planes perpendicular to the third, and show that the three planes pass through one line.

22. If the tetrahedron whose vertices are

$$A_i\equiv[x_i,\,y_j,\,z_i]\quad(i=1,\,2,\,3,\,4)$$

is such that the perpendiculars from the vertices on the opposite faces are concurrent, prove that

$$\Sigma x_2 x_3 + \Sigma x_1 x_4 = \Sigma x_3 x_1 + \Sigma x_2 x_4 = \Sigma x_1 x_2 + \Sigma x_3 x_4,$$

and deduce that the three sums of the squares of opposite edges are equal. (*Orthocentric tetrahedron.*)

23. If two pairs of opposite edges of a tetrahedron are at right angles prove that the third pair are also at right angles. Hence prove that for such a tetrahedron the sums of squares of opposite edges are equal, and that the four altitudes are concurrent.

24. If $ABCD$ is a tetrahedron whose altitudes are concurrent in a point E, show that each of the five points is the orthocentre of the tetrahedron formed by the other four. (*Orthocentric pentad.*)

25. Show that the equation of the plane through the line $x=x_1+l_1t$, etc., perpendicular to the plane $lx+my+nz=0$ is

$$\begin{vmatrix} x & y & z & 1 \\ x_1 & y_1 & z_1 & 1 \\ l_1 & m_1 & n_1 & 0 \\ l & m & n & 0 \end{vmatrix}=0.$$

26. A, B, C, D are four points in space, and P, Q, R, S are points dividing the segments AB, BC, CD, DA in the ratios $p:1$, $q:1$, $r:1$, $s:1$; prove that if $pqrs=1$ the four points P, Q, R, S are coplanar.

27. Four spheres touch in succession, each one touching two others (the number of external contacts being even); prove that the four points of contact lie on a circle.

28. P and Q are two variable points on fixed straight lines, each determined linearly by a parameter, t and u respectively. If the line PQ always cuts a fixed straight line show that t, u are connected by an equation of the form

$$\alpha tu + \beta t + \gamma u + \delta = 0.$$

29. If (ij) denotes the distance between the points P_i and P_j show that the six distances connecting four coplanar points are connected by the relation

$$\Sigma(ij)^2(kl)^4 + \Sigma(ij)^2(jk)^2(ki)^2 - \Sigma(ij)^2(ik)^2(jl)^2 = 0.$$

30. Show that the six angles formed by four concurrent rays $OP_1,\ldots, OP_4$, taken in pairs, are connected by the relation

$$|\cos\theta_{ij}| = 0.$$

31. If (p) and (q) are the line-coordinates of two intersecting lines, show that $(p+\lambda q)$ are the line-coordinates of a straight line belonging to the plane pencil determined by the two given lines.

32. From any point P are drawn PM and PN perpendicular to the planes zx and xy. O is the origin and α, β, γ, θ are the angles which OP makes with the coordinate-planes and with the plane OMN. Prove that

$$\operatorname{cosec}^2\theta = \operatorname{cosec}^2\alpha + \operatorname{cosec}^2\beta + \operatorname{cosec}^2\gamma.$$

33. Deduce from Ex. 32 that if PL is perpendicular to the plane yz, OP makes equal angles with the three planes OMN, ONL, OLM; and that the plane OPL is equally inclined to the planes OLM and OLN.

34. Show that the result in Ex. 33 is equivalent to the following theorem: ABC is a trirectangular spherical triangle and P any point on the sphere; great circles through P perpendicular to the sides meet them in L, M, N; P is the spherical centre of the small circle inscribed in the triangle LMN.

35. If from the point $P \equiv [a, b, c]$ perpendiculars PM, PN are drawn to the planes of zx, xy, find the equation of the plane OMN and the angle which OP makes with it.

Ans. $-bcx + cay + abz = 0$, $\cos^{-1} abc \, (\Sigma a^2 \Sigma b^2 c^2)^{-\frac{1}{2}}$.

36. Find the equation of the plane which bisects the join of $[x_1, y_1, z_1]$ and $[x_2, y_2, z_2]$ perpendicularly.

Ans. $\Sigma x \, (x_1 - x_2) = \frac{1}{2} \Sigma \, (x_1^2 - x_2^2)$.

CHAPTER III

GENERAL HOMOGENEOUS OR PROJECTIVE COORDINATES

3·1. A large section of geometry deals with relations, such as those of incidence, which involve no measurement. This is *projective geometry*. It is more fundamental and primitive than metrical geometry, in the sense that it involves fewer fundamental assumptions or axioms. It is possible to prove theorems of projective geometry by metrical methods, but, without the introduction of additional considerations which do not belong to projective geometry, it is not possible to prove theorems of metrical geometry by projective geometry. Thus we may prove the collinearity of three points by the Theorem of Menelaus, which is a metrical theorem, but if the theorem of collinearity is a purely projective one it should be capable of being proved without any metrical considerations.

In order to deal with projective geometry analytically we have to devise a system of coordinates having no metrical basis. We must distinguish here between metrical and numerical. Naturally a system of coordinates must be numerical.

3·11. In geometry the primitive elements are *points*, *lines*, and *planes*; and the primitive forms or assemblages of elements, arranged according to their dimensions, are as follows:

 1 *a*. *Range of points* (points on a line),
 a'. *Axial pencil of planes* (planes through a line),
 b. *Plane pencil of lines* (lines lying in a plane and passing through a point);
 2 *a*. *Plane field of points* (points in a plane),
 a'. *Bundle of planes* (planes through a point),
 b. *Bundle of lines* (lines through a point),
 b'. *Plane field of lines* (lines in a plane);
 3 *a*. *Space of points* (all points in space),
 a'. *Space of planes* (all planes in space);
 4. *Space of lines* (all lines in space).

Point and plane are *reciprocal* elements, line is self-reciprocal. Forms whose reference-letters are distinguished by an accent are reciprocal; $1b$ and 4 are self-reciprocal.

3·2. One-to-one correspondence. We shall consider first the one-dimensional forms, and, as typical, the range of points. In determining the position of a point on a given line we assume a $(1, 1)$ correspondence between the points and the real numbers, so that with each point is associated a certain number, its coordinate. If the points are renumbered, i.e. subjected to a transformation of coordinates, we assume that the old number x is definitely associated with the new number x' by a $(1, 1)$ correspondence, and that this correspondence is represented by an algebraic relationship linear in both x and x', viz.

$$\gamma xx' - \alpha x + \delta x' - \beta = 0,$$

or
$$x' = \frac{\alpha x + \beta}{\gamma x + \delta}.$$

Instead of x and x' we may write x/y and x'/y', thus introducing homogeneous coordinates, and then the equation of transformation can be replaced by the two equations

$$\rho x' = \alpha x + \beta y,$$

$$\rho y' = \gamma x + \delta y,$$

where ρ is a factor of proportionality. It is essential that $\alpha\delta - \beta\gamma$ should not be zero, for then x'/y' would be a constant ratio. x and y must not be both zero. The parameter x/y has a definite numerical value except when $y = 0$; in the latter case we denote the parameter by the symbol ∞.

3·21. As the transformation is determined by the ratios of the four numbers α, β, γ, δ we can make any three given points have specified numbers, but a fourth point will then have its co-ordinate determined.

We have thus to consider the coordinate of a point determined with reference to three fixed points. A set of four points on a line is associated with a certain function, the *cross-ratio* of the set, which is defined (metrically) as the ratio of the position-ratios of one pair with regard to the other pair, all taken in an assigned order. We denote the cross-ratio of the two pairs P, Q and X, Y by

$$(PQ, XY) = \frac{PX}{QX} \bigg/ \frac{PY}{QY} = \frac{PX}{PY} \bigg/ \frac{QX}{QY}.$$

This function, as is proved in text-books which treat projective geometry metrically, is unaltered by projection; which suggests that we should define the projective coordinate of a point as the value of the cross-ratio of the set consisting of the given point and three fixed points taken in a certain order. But as cross-ratio is here defined metrically we must proceed somewhat differently.

3·22. Coordinate of a point on a line. Take three points on the line, A, B and E, and assign to these respectively the parameters 0, ∞ and 1. Then if P is any other point, the number or parameter x which corresponds to it is called the coordinate of P referred to the *base-points* A, B and *unit-point* E. If we are to avoid actual measurement it is not possible to determine the number corresponding to any point without using a construction which goes outside the line. We shall consider this construction in 3·63. For the present, however, we shall assume that the line has been graduated so that each point has a different number attached to it. If another pair of base-points A', B' and another unit-point E' are chosen, whose parameters are a, b and e, there is a definite linear transformation which changes 0, ∞, 1 into a, b, e respectively, viz.

$$x' = \frac{b(e-a)x + a(b-e)}{(e-a)x + (b-e)}, \qquad x = \frac{e-b}{e-a} \cdot \frac{x'-a}{x'-b}.$$

We can consider this either as a renumbering of the points, so that A, B, E, P take the numbers a, b, e, x'; or as a geometrical transformation by which these points become A', B', E', P'.

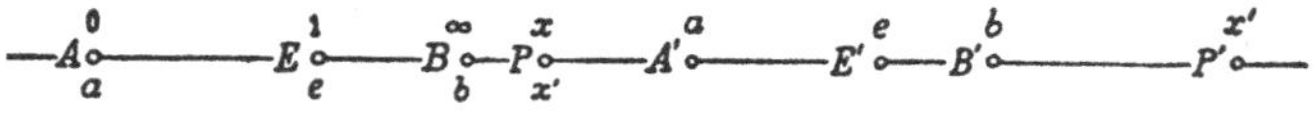

Fig. 11

Then the point P' is related to A', B', E' in exactly the same way as P is to A, B, E, so that when A', B', E' are given the parameters 0, ∞, 1 the coordinate of P' with regard to the system

A', B', E' is also x. The condition which the projective co-ordinate has to satisfy is that the geometrical cross-ratio $(PE, AB) = (P'E', A'B')$.

3·3. Cross-ratio. We now introduce the *cross-ratio of two pairs of numbers* as *a function of the four numbers whose value is not altered by a linear transformation.*

Consider first the cross-ratio $(x, 1; 0, \infty)$ which is a function of the single variable x, and denote it by $f(x)$. Now subject the numbers to the linear transformation which changes 1, 0, ∞ respectively into l', m and m'. The transformation is

$$x = \frac{(l' - m')(x' - m)}{(l' - m)(x' - m')}.$$

Then since the value of the cross-ratio has not been altered, we have

$$(x', l'; m, m') = f(x) = f\left(\frac{(l' - m')(x' - m)}{(l' - m)(x' - m')}\right);$$

or, writing l instead of x',

$$(l, l'; m, m') = f\left(\frac{l - m}{l - m'} \Big/ \frac{l' - m}{l' - m'}\right).$$

The simplest form for $f(x)$ which we can use is x. We therefore define the cross-ratio

$$(l, l'; m, m') = \frac{l - m}{l - m'} \Big/ \frac{l' - m}{l' - m'}.$$

3·31. We may study this function quite apart from geometry. Writing it in terms of four numbers a, b, c, d, we have

$$(ab, cd) = \frac{a - c}{a - d} \Big/ \frac{b - c}{b - d} = \frac{(ab + cd) - (bc + ad)}{(ab + cd) - (ca + bd)} = \frac{C - A}{C - B},$$

where $A = bc + ad, \quad B = ca + bd, \quad C = ab + cd.$

A, B, C are each unaltered if we interchange any two of the numbers and at the same time interchange the other two. Hence the cross-ratio is unaltered by the double interchange. Although there are 24 different orders of the four numbers there are only six different values of the cross-ratio, viz.

$$\frac{A - B}{A - C}, \quad \frac{B - C}{B - A}, \quad \frac{C - A}{C - B}$$

and the reciprocals of these.

The interchange of b and c, or of a and d, leaves A unaltered and interchanges B and C.

Hence $\quad (ab,\, dc) = (ba,\, cd) = \dfrac{C-B}{C-A} = \dfrac{1}{(ab,\, cd)}$,

while $\quad (ac,\, bd) = \dfrac{B-A}{B-C} = 1 - \dfrac{C-A}{C-B} = 1 - (ab,\, cd)$.

If $(ab,\, cd) = k$, the six values are

$$k,\quad k^{-1},\quad 1-k,\quad 1-k^{-1},\quad (1-k)^{-1},\quad (1-k^{-1})^{-1}.$$

3·311. In general the six values are all distinct, but there are certain cases in which two or more of them become equal.

If $k = k^{-1}$, $k^2 = 1$ and $k = \pm 1$.

(1) If $k = +1$, $(ac,\, bd) = 0$ and either $a = b$ or $c = d$. In this case there are three values, 0, 1, ∞.

(2) If $k = -1$, the set of numbers is said to be *harmonic*. The relation in this case can be written $A + B = 2C$, or

$$(a+b)(c+d) = 2(ab+cd),$$

and is symmetrical in both a, b and c, d. In this case there are again three values, -1, 2, $\frac{1}{2}$. The harmonic relation associates the four numbers in pairs; a, b and c, d; and it is only when they are associated in this way that the cross-ratio has the value -1.

(3) If $k = (1-k)^{-1}$, $k^2 - k + 1 = 0$. k is a complex number $= -\omega$, where $\omega^3 = 1$. The set of numbers is then said to be *equianharmonic*. There are only two values, $-\omega$ and $-\omega^2$.

It is easily verified that these are the only cases of equalities.

3·32. When numbers occur in pairs it is often convenient to represent them as the roots of a quadratic equation

$$at^2 + 2ht + b = 0.$$

The discriminant of this quadratic

$$C \equiv ab - h^2$$

determines the nature of the roots, which are real, equal, or imaginary, according as C is negative, zero, or positive.

3·321. Consider two pairs of numbers λ, λ' and μ, μ' represented by the quadratic equations

$$a_1 t^2 + 2h_1 t + b_1 = 0,$$
$$a_2 t^2 + 2h_2 t + b_2 = 0.$$

Then
$$\lambda\lambda' = b_1/a_1, \quad \lambda+\lambda' = -2h_1/a_1,$$
$$\mu\mu' = b_2/a_2, \quad \mu+\mu' = -2h_2/a_2.$$

The condition that $(\lambda\lambda', \mu\mu')$ should be harmonic is
$$\frac{\lambda-\mu}{\lambda-\mu'} \Big/ \frac{\lambda'-\mu}{\lambda'-\mu'} = -1,$$

or
$$2(\lambda\lambda'+\mu\mu') = (\lambda+\lambda')(\mu+\mu').$$

Substituting the values of the symmetric functions we obtain
$$2(b_1/a_1+b_2/a_2) = 4h_1h_2/a_1a_2,$$

i.e.
$$a_1b_2+a_2b_1-2h_1h_2 = 0.$$

This function is the bilinear symmetrical expression which is associated with the quadratic function $ab-h^2$, the discriminant of the quadratic. The term *apolar* is also used to describe this relationship.

3·322. If we write $a_1b_1-h_1{}^2 \equiv C_{11}$ and $a_2b_2-h_2{}^2 \equiv C_{22}$, then we may write $a_1b_2+a_2b_1-2h_1h_2 = 2C_{12}$.

If $C_{11} > 0$ and $C_{22} > 0$, so that both pairs of elements are imaginary, we have
$$(a_1b_2+a_2b_1)^2 - 4h_1{}^2h_2{}^2$$
$$> (a_1b_2+a_2b_1)^2 - 4a_1b_1a_2b_2,$$

i.e.
$$> (a_1b_2-a_2b_1)^2.$$

Hence C_{12} cannot vanish in this case, i.e. *in a harmonic set at least one of the pairs must be real.*

3·323. To find the cross-ratio of the two pairs
$$a_1t^2+2h_1t+b_1 = 0,$$
$$a_2t^2+2h_2t+b_2 = 0.$$

Since the roots of each equation can be taken in either order, there are two values for the cross-ratio, $(\lambda\lambda', \mu\mu')$ and $(\lambda\lambda', \mu'\mu)$. These being reciprocals, their product $= 1$. We proceed to find their sum. Let $C \equiv \lambda\lambda'+\mu\mu'$, $A \equiv \lambda\mu+\lambda'\mu'$, $B \equiv \lambda\mu'+\lambda'\mu$. The two values of the cross-ratio are then
$$(C-A)/(C-B) \quad \text{and} \quad (C-B)/(C-A),$$

and their sum
$$= \frac{(C-A)^2+(C-B)^2}{(C-A)(C-B)} = 2 + \frac{(A+B)^2-4AB}{C^2-C(A+B)+AB}.$$

Now $\quad C = b_1/a_1 + b_2/a_2 = (a_1 b_2 + a_2 b_1)/a_1 a_2,$

$$A + B = (\lambda + \mu)(\lambda' + \mu') = 4h_1 h_2/a_1 a_2,$$

$$AB = \lambda\lambda'(\mu^2 + \mu'^2) + \mu\mu'(\lambda^2 + \lambda'^2) = \lambda\lambda'(\mu + \mu')^2$$
$$+ \mu\mu'(\lambda + \lambda')^2 - 4\lambda\lambda'\mu\mu'$$
$$= 4(h_2^2 a_1 b_1 + h_1^2 a_2 b_2 - a_1 a_2 b_1 b_2)/a_1^2 a_2^2.$$

Hence we find

$$(A + B)^2 - 4AB = 16 C_{11} C_{22}/a_1^2 a_2^2$$

and $\qquad C^2 - C(A + B) + AB = 4(C_{12}^2 - C_{11} C_{22})/a_1^2 a_2^2,$

whence the sum of the cross-ratios is

$$2(C_{12}^2 + C_{11} C_{22})/(C_{12}^2 - C_{11} C_{22}).$$

The two values of the cross-ratio are therefore the roots of the quadratic equation

$$k^2 - 2\frac{C_{12}^2 + C_{11} C_{22}}{C_{12}^2 - C_{11} C_{22}} k + 1 = 0.$$

3·324. *Conditions that the cross-ratio of two pairs should be real, and positive.*

We assume that the coefficients of the two quadratics are all real. The condition that the cross-ratio should be *real* is

$$\left(\frac{C_{12}^2 + C_{11} C_{22}}{C_{12}^2 - C_{11} C_{22}}\right)^2 \geqslant 1,$$

which reduces to $\qquad C_{12}^2 C_{11} C_{22} \geqslant 0,$

or simply $\qquad C_{11} C_{22} > 0,$

i.e. the two pairs are either both real or both imaginary.

3·325. The condition that the cross-ratio should be *positive* is

$$\frac{C_{12}^2 + C_{11} C_{22}}{C_{12}^2 - C_{11} C_{22}} > 0,$$

i.e. $\qquad C_{12}^2 > |C_{11} C_{22}|,$

or, since the cross-ratio must also be real,

$$C_{12}^2 > C_{11} C_{22} > 0.$$

Ex. 1. Show that if $\dfrac{C_{12}^2 + C_{11} C_{22}}{C_{12}^2 - C_{11} C_{22}} = \dfrac{5}{4}$ either $(\lambda\mu, \lambda'\mu')$ or $(\lambda\mu', \lambda'\mu)$ is harmonic.

Ex. 2. Show that if $\dfrac{C_{12}^2 + C_{11} C_{22}}{C_{12}^2 - C_{11} C_{22}} = \dfrac{1}{2}$, the two pairs are equianharmonic.

3·41. We have now to define the geometrical cross-ratio of four points, and to do this we must consider the line as lying in a plane. We think of the cross-ratio in the first place as a property of the four points which is not altered by projection, but we shall have further to endow it with numerical value so that it can be treated as an algebraic quantity capable of addition and multiplication. These operations will be introduced by suitable definitions.

The fundamental property is that if $ABCD$ and $A'B'C'D'$ are two transversals of the same pencil, then the cross-ratio $(AB, CD) = (A'B', C'D')$. We say that $(ABCD)$ is in perspective with $(A'B'C'D')$, with centre of perspective O, and write this

$$(ABCD) \; \overline{\wedge}_0 \, (A'B'C'D').$$

We assume that if $(AB, CD) = (AB, CD')$ the points D and D' coincide, or shortly $D \equiv D'$.

The cross-ratio depends upon the order of the four points. There are 24 different orders, but we can prove that they fall into six sets of four, each set forming equivalent cross-ratios. In Fig. 12 we have

$$(ABCD) \; \overline{\wedge}_0 \, (PBRS)$$

$$\overline{\wedge}_D (PAQO) \; \overline{\wedge}_R (BADC).$$

Thus the cross-ratio is unaltered by the simultaneous interchange of A, B and C, D.

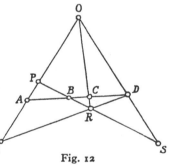

Fig. 12

Thus $(AB, CD) = (BA, DC) = (CD, AB) = (DC, BA).$

3·42. A special case of great importance occurs when

$$(AB, CD) = (BA, CD).$$

Let A, B, C be given. Take any two points O and Q collinear with C. Let OA cut QB in R, and QA cut OB in S. Let RS cut AB in D. Then

$$(AB, CD) \; \overline{\wedge}_0 \, (RS, LD) \; \overline{\wedge}_Q (BA, CD).$$

By this construction D is uniquely determined when A, B, C are given

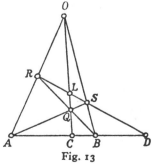

Fig. 13

in this order, and we assume that the same point will be determined if we take any other points O, Q collinear with C*. In this range the relation between A and B is symmetrical, and since also

$$(AB, CD) = (BA, DC),$$

we have

$$(BA, DC) = (BA, CD),$$

so that C, D are also related symmetrically. (AB, CD) is said to be *harmonic*; A, B are harmonic conjugates with regard to C, D, and C, D are harmonic conjugates with regard to A, B.

3·43. If $(AB, CD) = (A'B', C'D')$ the two ranges are not necessarily in perspective, but can be connected by a chain of perspectivities. In general we say that they are projectively related or projective; A and A', B and B', etc., are corresponding points. They are in perspective if three of the four lines AA', BB', CC', DD' are concurrent; for if AA', BB', CC' are concurrent in O, and OD cuts $A'B'$ in D_1, then

$$(A'B', C'D_1) = (AB, CD) = (A'B', C'D'),$$

therefore $D_1 \equiv D'$. In particular if $(AB, CD) = (AB'C'D')$ then BB', CC', DD' are concurrent.

If on each of two straight lines we take three points A, B, C and A', B', C' then if P is any point on the first line there is a definite point P' on the second such that $(ABCP)$ and $(A'B'C'P')$ are projective. A $(\text{1}, \text{1})$ correspondence between the points of the two lines is determined when three pairs of corresponding points are given.

3·44. The last statement is so important that it has been called the fundamental theorem of projective geometry. Assuming the theorem, which implies that the correspondence is unique when three pairs are given, and which requires for its complete proof considerations of continuity, let us consider a construction by which the correspondent of any point P will be determined.

---

* This can be proved with the help of Desargues' Theorem on perspective triangles. Thus if O', Q', R', S' are points determined in the same way as O, Q, R, S, the triangles OQR, $O'Q'R'$ are in perspective since the intersections of QR and $Q'R'$, RO and $R'O'$, OQ and $O'Q'$ are collinear, hence OO', QQ', RR' are concurrent. Similarly the triangles OQS and $O'Q'S'$ are in perspective, hence OO', QQ', SS' are concurrent. Therefore QQ', RR', SS' are concurrent and the intersections of QR and $Q'R'$, QS and $Q'S'$, RS and $R'S'$ are collinear, i.e. RS, $R'S'$ and AB are concurrent.

We have A, B, C on l, and the corresponding points A', B', C' on l'. On AA' take any two points O, Q. Let OB cut QB' in

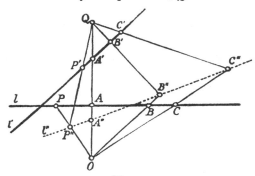

Fig. 14

B'', and OC cut QC' in C''. Let $B''C'' \equiv l''$ cut AA' in A''. Then

$$(ABC) \mathbin{\overline{\wedge}}_O (A''B''C'') \mathbin{\overline{\wedge}}_Q (A'B'C').$$

If P is any point on l its correspondent on l' is found by the construction: let OP cut l'' in P'', then QP'' cuts l' in P'.

3·45. As an example of two projective pencils of lines consider a conic, defined as a plane curve which is cut by an arbitrary line in two points. Let A and B be fixed points on the conic. Then a variable line l through A cuts the conic again in a point P and the line BP or l' uniquely corresponds to l. If P, Q, R, S are four points on the conic we have then

$$A(PQRS) = B(PQRS).$$

3·5. We may have two projective ranges $ABCP\ldots$ and $A'B'C'P'\ldots$ on the same straight line. This is called a *homography*. In this case any point of the line may be regarded as belonging to either of the two ranges, and there will be confusion if the ranges are not kept distinct. As an example of a homography let l be a fixed line, and C a conic, cutting l in D_1 and D_2. Let O_1, O_2 be fixed points on the conic. If P is any point on l, a corresponding

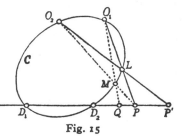

Fig. 15

point P' is uniquely determined by the following construction. Join PO_1 cutting the conic again in L, join O_2L cutting l in P'. Then to P', as a point on the second range, corresponds the unique point P determined by the reverse construction, i.e. join $P'O_2$ cutting the conic in L, then O_1L cuts l in P. But if P is considered as belonging to the second range its correspondent is a different point Q. Thus the homography is not in general symmetrical. The points D_1 and D_2 are *self-corresponding points* or *double-points* of the homography. They may be real, coincident, or imaginary.

3·51. Involution. The case in which the two points corresponding to a point P coincide is important. In this case the relation between P and P' is symmetrical and the homography is said to be *involutory*, or an *involution*. There is no need then to distinguish the two ranges on the line. Points P and P' are connected definitely in pairs. We have then $(PP', D_1D_2) = (P'P, D_1D_2)$ and therefore P, P' *are harmonic conjugates with regard to the double-points*.

In the above construction, when the two pairs O_1P, O_2P' and O_2P, O_1P' both intersect on the conic, we have

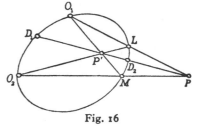

$$L(O_1O_2, D_1D_2)$$
$$= (PP', D_1D_2).$$

The four points O_1O_2, D_1D_2 are said to form a harmonic group on the conic. Since P, P' are harmonic conju-

Fig. 16

gates with regard to D_1, D_2, the polar of P with regard to the conic passes through P'. An involution is thus formed on any line by pairs of conjugate points with regard to a given conic. Similarly pairs of conjugate lines through a fixed point form an involution-pencil.

In an involution the double-points may be real or imaginary; but they cannot coincide unless the involution degenerates, for then either P or P' would coincide with them. When the double-points are real the involution is called *hyperbolic*; when imaginary, *elliptic*.

3·52. An involution is completely determined by its double-points when these are real. More generally, *an involution is determined by two pairs of corresponding points.* For a projectivity is determined by three pairs of corresponding points. Thus taking A, A', X and the corresponding points A', A, X' a projectivity is determined which is an involution since A' corresponds to A, and A to A'.

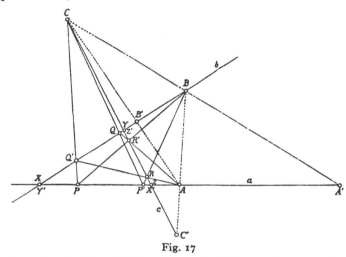

Fig. 17

Let A and A', and X (or Y') and X' (or Z), be two pairs of points on a line a. Through X and X' draw any lines b, c intersecting in Y (or Z'). On c take any point C. Let CA and CA' cut b in B' and B, and let AB cut c in C'.

Then if P is any point on a the corresponding point P' is found by the construction: Let CP cut b in Q', AQ' cut c in R, and BR cut a in P'. To P' should correspond P by a similar construction, i.e. if CP' cuts b in Q, and AQ cuts c in R', then BR' should pass through P. To prove this consider the six points $AQQ'CRP'$ on the two lines $AQ'R$ and $CP'Q$, then by Pappus' Theorem (the particular case of Pascal's Theorem for six points on a conic when the conic degenerates to two straight lines) the three intersections
$$\left(\frac{AQ}{CR}\right) \equiv R', \quad \left(\frac{QQ'}{RP'}\right) \equiv B, \quad \left(\frac{Q'C}{P'A}\right) \equiv P$$
are collinear.

Then
$$(PP'AA'XX')\barwedge_B (R'R\,C'C\,Z'Z)\barwedge_A (QQ'BB'YY')$$
$$\barwedge_O (P'P\,A'A\,X'X),$$

and thus we have involutions on the three lines a, b, c.

3·521. Consider the quadrangle $BCQ'R$. Its opposite sides CQ' and BR, $Q'R$ and BC, BQ' and CR are cut by a in the involution (PP', AA', XX'). This often affords a simple way of recognising that six points are in involution.

3·6. We now return to the definition of geometrical cross-ratio, which up to this point has only been considered as a magnitude in so far as we can recognise when two cross-ratios are equal. We have now to consider cross-ratios as capable of being added or multiplied. We define these operations in two special cases as follows:

(i) $(PQ, XY)\times(QR, XY)=(PR, XY)$,

and   (ii) $(PX, QY)+(PQ, XY)=1$.

In   (i), putting $Q=P$ we have $(PP, XY)=1$,

and in (ii) $(PX, PY)=0$.

Hence

$(PQ, XY)\times(QP, XY)=1$, or $(QP, XY)=\dfrac{1}{(PQ, XY)}$,

and, putting $Q=X$,
$$(PX, XY)=\pm\infty.$$

3·61. If now we take three points O, E, U and attach to them the numbers or parameters 0, 1, $\pm\infty$, we may define the parameter of any point P as the value of the cross-ratio (PE, OU) as this is consistent with the values just determined, for

$$(EE, OU)=1, \quad (OE, OU)=0,$$
and $(UE, OU)=(EU, UO)=\pm\infty.$

We can now identify the numerical value of the geometrical cross-ratio with that of the four parameters as previously defined (3·3).

Let P, P', Q, Q' be four points with parameters λ, λ', μ, μ', so that

$$(PE, OU)=\lambda, \quad (P'E, OU)=\lambda', \quad (QE, OU)=\mu, \quad (Q'E, OU)=\mu'.$$

Then
$$\frac{\lambda}{\mu}=(PE, OU)\times(EQ, OU)=(PQ, OU),$$

whence
$$1-\frac{\lambda}{\mu}=(PO, QU);$$

and similarly
$$1-\frac{\lambda'}{\mu}=(P'O, QU).$$

Hence
$$\frac{\mu-\lambda}{\mu-\lambda'}=(PO, QU)\times(OP', QU)=(PP', QU)=(QU, PP');$$

and similarly
$$\frac{\mu'-\lambda}{\mu'-\lambda'}=(PP', Q'U)=(Q'U, PP');$$

whence finally
$$\frac{\lambda-\mu}{\lambda'-\mu}\Big/\frac{\lambda-\mu'}{\lambda'-\mu'}=(\lambda\lambda', \mu\mu')=(QU, PP')\times(UQ', PP')$$
$$=(QQ', PP')=(PP', QQ').$$

3·62. We have not yet, however, actually determined the parameter of any point except the arbitrarily assumed points O, E, U. The harmonic or quadrilateral construction enables us to determine the point corresponding to the parameter -1. For if

$$(PQ, XY)\ \overline{\wedge}_O\ (P'Q', X'Y)$$
$$\overline{\wedge}_{O'}\ (QP, XY)$$

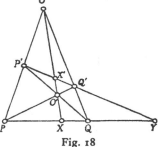

Fig. 18

then since
$$(PQ, XY)\times(QP, XY)=1,$$

and since P and Q, and X and Y, are distinct, it follows that $(PQ, XY)=-1$. Hence if (PE, OU) is harmonic, the parameter of P is -1.

3·63. Now, for the moment altering the notation, if we start with the three points P_0, P_1, P_∞ to which we attach the parameters 0, 1, ∞ we can construct points corresponding to any rational values of the parameter by a repetition of the quadrilateral construction alone. Thus, if $(P_0 P, P_1 P_\infty)$ is harmonic, by 3·6 (ii) $(PP_1, P_0 P_\infty) = 2$, and therefore P corresponds to the parameter 2; if we denote this point by P_2 we have similarly if (P_1, P, P_2, P_∞) is harmonic then $(PP_1, P_0 P_\infty) = 3$, and proceeding in this way we get the points $P_0, P_1, P_2, ..., P_n, ...$ corresponding to all the positive integers, with the general relation

$$(P_{n-1} P_{n+1}, P_n P_\infty) = -1.$$

Again, if $(P_0 P_1, PP_\infty) = -1$, then $(PP_1, P_0 P_\infty) = \frac{1}{2}$, and so P is $P_{\frac{1}{2}}$. Similarly if $(P_0 P_{\frac{1}{2}}, PP_1) = -1$, then P is $P_{\frac{1}{4}}$, and proceeding in this way we obtain the points which correspond to the fractional values $1/n$, with the general iterative relation

$$(P_0 P_{1/n}, P_{1/(n+1)} P_{1/(n-1)}) = -1.$$

Further, $(P_{(n-1)/m} P_{(n+1)/m}, P_{n/m} P_\infty) = -1,$

hence we obtain all the positive fractions n/m; and lastly

$$(P_0 P_\infty, P_n P_{-n}) = -1,$$

hence we obtain the corresponding negative numbers.

From the three points P_0, P_1, P_∞ we thus obtain points corresponding to any rational value of the parameter. This construction is called the Möbius Net or net of rationality.

We shall dismiss the case of irrational values by remarking that any irrational number can be expressed as the limit of a sequence of rational numbers.

3·64. Projective coordinates in a plane and in space.

We have defined the projective coordinate of a point P on a line, with reference to the base-points A, B and unit-point E, as the cross-ratio (PE, AB), and as homogeneous coordinates we take two numbers x, y whose ratio

$$x/y = (PE, AB).$$

The extensions to two and three dimensions are precisely similar to one another.

In a plane the position of a point is determined by the ratios of three numbers x, y, z (not all zero); any multiple of these numbers kx, ky, kz represent the same point, for all values of k (not zero). We refer to this shortly as the point (x). A straight line is determined by two points (x_1), (x_2), and on it a point has one degree of freedom and can be determined by the ratio of two homogeneous parameters λ, μ. We assume that the coordinates of any point on the line are represented by

$$\rho x = \lambda x_1 + \mu x_2,$$
$$\rho y = \lambda y_1 + \mu y_2,$$
$$\rho z = \lambda z_1 + \mu z_2,$$

where ρ is a factor of proportionality. Eliminating ρ, λ, μ between these three equations we obtain an equation in x, y, z, homogeneous and of the first degree. Hence a straight line is represented by a linear homogeneous equation. The three equations $x = 0$, $y = 0$, $z = 0$ represent three straight lines which form a triangle, the *triangle of reference* or *fundamental triangle ABC*. It is essential that these three lines should have no common point, for if they had then the equation $lx + my + nz = 0$ would represent a line passing through this point and could not represent an arbitrary line.

3·65. To complete the determination of the coordinates we attach to one assigned point, the unit-point E, the coordinates $[1, 1, 1]$.

We can show now that every point has a definite set of co-ordinates. Let $P \equiv [x', y', z']$ be any point, and let the line PE cut the sides of the triangle of reference in L, M, N. PE is represented by the parametric equations

$$\rho x = x' + t,$$
$$\rho y = y' + t,$$
$$\rho z = z' + t,$$

and the parameters of P, E, L, M, N are 0, ∞, $-x'$, $-y'$, $-z'$. The cross-ratio

$$(PE, MN) = (0, \infty; -y', -z') = y'/z',$$

and $(PE, LN) = x'/z'.$

Hence the ratios of the coordinates are determined in terms of certain cross-ratios.

3·66. Similarly in three dimensions a point is represented by the ratios of four homogeneous coordinates $[x, y, z, w]$. A plane is determined by three points (x_1), (x_2), (x_3), and is represented by parametric equations

$$\rho x = \lambda x_1 + \mu x_2 + \nu x_3, \text{ etc.,}$$

or by a homogeneous linear equation in x, y, z, w. The four equations $x = 0$, $y = 0$, $z = 0$, $w = 0$ determine the tetrahedron of reference. If $E \equiv [1, 1, 1, 1]$ is the unit-point and EP cuts the fundamental planes in L, M, N, K the ratios of the coordinates are determined by cross-ratios, viz.

$$x'/w' = (PE, LK), \text{ etc.}$$

3·67. Analytically, there is no difficulty in extending this process to a system of five or more homogeneous coordinates. Generally, a point is represented by the ratios of $n+1$ homogeneous coordinates $[x_0, x_1, ..., x_n]$. A line is determined by two points (a), (b), and is represented by parametric equations

$$\rho x_i = \lambda a_i + \mu b_i;$$

a plane is determined by three points (a), (b), (c), and is represented by parametric equations

$$\rho x_i = \lambda a_i + \mu b_i + \nu c_i.$$

Then a region of three dimensions, solid, or flat space, is determined by four points, and is represented by parametric equations involving three parameters or the ratios of four homogeneous parameters. Similarly in the n-dimensional region S_n there are flat regions of all dimensions up to $n-1$. The figure formed of the determining points, triangle, tetrahedron, etc., is called in general a *simplex*. A single linear homogeneous equation in the $n+1$ coordinates represents a flat space of $n-1$ dimensions or $(n-1)$-flat. This, the space of highest dimensions in the containing space, is also called a *prime*. Two equations represent an $(n-2)$-flat, which is therefore the intersection of two $(n-1)$-flats, and so on. We shall have occasion later to make use of relations in higher space in the representation of ordinary spatial relations.

3·71. We have defined the cross-ratio of four collinear points, and shown that it is equal to the cross-ratio of the parameters of the four points. For the other one-dimensional forms we have similar relations. Thus for a plane pencil of lines we define the cross-ratio as equal to that of four collinear points one on each line. If we take the vertex of the pencil as $A \equiv [1, 0, 0]$ and the transversal as the line $x = 0$, the lines are represented by equations $y = \lambda z$, $y = \lambda' z$, $y = \mu z$, $y = \mu' z$, and the points by $[0, \lambda, 1]$, etc. The parameters of the points are λ, λ', μ, μ', and the cross-ratio of the four points, which is equal to that of the four lines, is equal to that of the four parameters $(\lambda\lambda', \mu\mu')$ which are also the parameters of the four lines.

Similarly for four planes through a line the cross-ratio is equal to that of the four lines of intersection with any plane, or of the four points of intersection with any line, and is equal to the cross-ratio of the parameters when the planes are represented by equations of the form $u - \lambda v = 0$.

The fundamental principle at the basis of these determinations is that if two one-dimensional systems are in $(1, 1)$ correspondence, the cross-ratio of four elements of the one system is equal to that of the corresponding four elements of the other system.

3·72. Polar plane of a point with regard to a tetrahedron.

As an example of the use of the projective coordinates we shall prove the following projective theorem.

$ABCD$ is a tetrahedron, and P any point. Let AP cut BCD in P_1, and similarly define the points P_2, P_3, P_4. The plane $P_2 P_3 P_4$ cuts the plane BCD in a line l_1; and similarly we define the lines l_2, l_3, l_4. These four lines lie in one plane which is called the *polar plane* of P with regard to the tetrahedron.

Let $P \equiv [X, Y, Z, W]$, then $P_1 \equiv [0, Y, Z, W]$. The plane $P_2 P_3 P_4$ is

$$\begin{vmatrix} x & y & z & w \\ X & 0 & Z & W \\ X & Y & 0 & W \\ X & Y & Z & 0 \end{vmatrix} = 0,$$

and this cuts $x = 0$ where
$$ZWy + WYz + YZw = 0,$$
i.e. where $\quad\quad y/Y + z/Z + w/W = 0,$

i.e. where $\quad\quad x/X + y/Y + z/Z + w/W = 0.$

The symmetry of this equation shows that the other three lines also lie in it.

3·81. Transition from projective to metrical geometry.

The system of coordinates which has been established is a purely projective one and makes no use of measurement. Distances and angles are metrical functions and are foreign to projective geometry. When the processes of projection are applied in euclidean geometry we find that parallel lines are projected into concurrent lines, which intersect on the "vanishing line". Parallelism and concurrency become merged in the same idea, and this is facilitated by the use of the phrase "point at infinity". The line corresponding to the vanishing line is the "line at infinity". In euclidean geometry there is one point at infinity on each line, one line at infinity in each plane.

Considering now lines through a point O, $y = \mu x$, there are two lines through O such that the "angle" which they determine with any line of the pencil is independent of μ. If one of these lines is $y = \lambda x$ we have by the ordinary formula in rectangular cartesian coordinates
$$\tan\theta = \frac{\lambda - \mu}{1 + \lambda\mu} = \frac{1}{\lambda}\left\{-1 + \frac{\lambda^2 + 1}{1 + \lambda\mu}\right\},$$
and this is independent of μ if $\lambda^2 = -1$ and therefore $\lambda = \pm i$. These two lines are called the *absolute lines* through O. If $\mu = \tan\theta$ and $\mu' = \tan\theta'$ are the parameters of any two lines through O, the cross-ratio of this pair and the two absolute lines is
$$(\mu, \mu'; i, -i) = \frac{\tan\theta - i}{\tan\theta + i} \bigg/ \frac{\tan\theta' - i}{\tan\theta' + i}.$$
Now
$$\frac{\tan\theta - i}{\tan\theta + i} = -\frac{\cos\theta + i\sin\theta}{\cos\theta - i\sin\theta} = -e^{2i\theta}.$$
Hence
$$(\mu, \mu'; i, -i) = e^{2i(\theta - \theta')},$$
and therefore
$$\theta' - \theta = \frac{i}{2}\log(\mu, \mu'; i, -i).$$

This formula, due to Laguerre, expresses the angle between two lines in terms of a cross-ratio. We have therefore a means of reducing angular measurements to projective relations. There is a marked difference between the pencil of lines and the range of points, for in the latter there is only one special element, the point at infinity, while in the former there are two, the two absolute lines. We are led therefore to consider the metrical geometry on a line as specialised by the coalescing of two points. When we replace the single point at infinity by two, we obtain the more general *non-euclidean* geometry of which euclidean geometry is a limiting case.

3·82. Distance in metrical geometry of one dimension.

In the general metrical geometry we begin by assuming that on any line there are two special points, X, Y, the points at infinity. Distance must be defined so that the distance of any finite point from either X or Y is infinite, and, if P, Q, R are three points on a line, $PQ + QR = PR$. We have three cross-ratios (PQ, XY), (QR, XY) and (PR, XY); and

$$(PQ, XY).(QR, XY) = (PR, XY).$$

Hence $\quad \log(PQ, XY) + \log(QR, XY) = \log(PR, XY).$

The function $\qquad\qquad c \log(PQ, XY),$

where c is some constant, therefore satisfies the required conditions, since $(PX, XY) = \infty$, $(XQ, XY) = 0$, and the logarithm of each of these is infinite*.

---

* To show that this is the only expression which can represent the distance we observe first that the distance OP is some function of the cross-ratio $(OP, XY) = x$, say $(OP) = f(x)$. Let Q be another point on the line OP, and let $(OQ, XY) = y$. Then since $(OP, XY).(PQ, XY) = (OQ, XY)$, $(PQ, XY) = y/x$.

Now $\qquad\qquad (OP) + (PQ) = (OQ),$

therefore $\qquad\qquad f(y/x) = f(y) - f(x).$

Differentiating partially with respect to x and y, we have

$$f'(x) = \frac{y}{x^2} f'\left(\frac{y}{x}\right) \quad \text{and} \quad f'(y) = \frac{1}{x} f'\left(\frac{y}{x}\right),$$

hence $\qquad\qquad x f'(x) = y f'(y) = c,$ say.

Therefore $\qquad\qquad f(x) = \int \frac{c}{x} dx = c \log x + C.$

But $(OO) = 0$ and $(OO, XY) = 1$, therefore $f(1) = 0$ and $C = 0$.

In euclidean geometry the two points X, Y coincide; (PQ, XY) becomes $=1$ and its logarithm is zero. The expression for the distance can, however, be obtained as a finite limiting expression by supposing that, as $Y \to X$, $c \to \infty$ in some way.

Let the parameters of P, Q, X and Y be p, q, ∞ and ϵ^{-1} where $\epsilon \to 0$. Then

$$(PQ, XY) = \frac{q - \epsilon^{-1}}{p - \epsilon^{-1}} = (1 - q\epsilon)(1 + p\epsilon) = 1 + (p - q)\epsilon,$$

neglecting ϵ^2. Then

$$\log(PQ, XY) = (p - q)\epsilon.$$

Let $\operatorname*{Lim}_{\epsilon \to 0} c\epsilon = -1$, then we obtain

$$\text{distance } (PQ) = q - p,$$

and, if O is the point corresponding to the parameter 0, the distance $(OP) = p$.

In euclidean geometry then the metrical expression for the cross-ratio of four points

$$(PQ, RS) = (pq, rs) = \frac{p - r}{p - s} \Big/ \frac{q - r}{q - s} = \frac{(PR)}{(PS)} \Big/ \frac{(QR)}{(QS)}.$$

3·83. Distance in two dimensions.

In two dimensions the locus of points at infinity is a curve of the second order or conic, and is represented by an equation of the second degree in the coordinates. In euclidean geometry, since the two points at infinity on any line coincide, the conic degenerates to two coincident lines. Take this line as $z = 0$ and

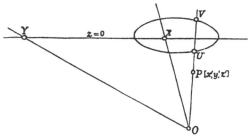

Fig. 19

two conjugate lines through its pole O as the other sides of the triangle of reference, then, with proper choice of unit-point, the equation of the conic can be written

$$-\epsilon^2(x^2 + y^2) + z^2 = 0,$$

where $\epsilon \to 0$. Let $P \equiv [x', y', z']$ be any point and let OP cut the conic in U and V. Then

$$\text{dist. } (OP) = c \log(OP, UV).$$

Parametric equations of OP are

$$\rho x = x', \quad \rho y = y', \quad \rho z = z' + t.$$

Substituting in the equation of the conic we have

$$t^2 + 2z't - \epsilon^2(x'^2 + y'^2) + z'^2 = 0.$$

The parameters of O, P, U, V are $\infty, 0, t_1, t_2$, where t_1, t_2 are the roots of this quadratic equation.

The cross-ratio

$$(OP, UV) = (\infty\ 0, t_1 t_2) = t_2/t_1.$$

Write

$$\epsilon^2(x'^2 + y'^2) = R^2,$$

then

$$\frac{t_2}{t_1} = \frac{z' + R}{z' - R},$$

and $\quad \log(OP, UV) = \log(1 + R/z') - \log(1 - R/z') = 2R/z'.$

Choose c so that $\underset{\epsilon \to 0}{\text{Lim}}\ 2c\epsilon = 1$, then

$$\text{dist. } (OP) = \underset{\epsilon \to 0}{\text{Lim}}\ 2c\epsilon(x'^2 + y'^2)^{\frac{1}{2}}/z'$$

$$= (x^2 + y^2)^{\frac{1}{2}},$$

where

$$x = x'/z', \quad y = y'/z'.$$

3·84 The absolute.

For angles, since there are two special or absolute lines through each point, the assemblage of absolute lines is a curve of the second class or conic-envelope. The angle between two lines is zero only in two cases: either the lines themselves coincide or the two absolute lines through their point of intersection coincide. Identifying the latter case with the case of parallelism, when the point of intersection is a point at infinity, we find that the conic-locus of points at infinity and the conic-envelope of absolute lines form one and the same figure. This is called the *Absolute*. In euclidean geometry when the conic-locus degenerates to two coincident straight lines (the line at infinity) the conic-envelope becomes a pair of imaginary points on this line (the circular points at infinity). In euclidean geometry of three dimensions we shall find that the absolute figure is a pair of coincident planes (the plane at infinity) and a conic in this plane (the circle at infinity).

3·85. Metrical coordinates. For metrical geometry the most convenient coordinates are rectangular cartesians, but a system based, like projective coordinates, on a tetrahedron of reference, is sometimes useful. We may derive from the projective system the metrical meanings to be attached to the coordinates.

Let $P \equiv [x, y, z, w]$, $E \equiv [1, 1, 1, 1]$, and let PE cut the fundamental planes $x = 0$, etc., in L_1, L_2, L_3, L_4. Then (3·66)

$$\frac{x}{w} = (PE, L_1 L_4) = \frac{PL_1}{PL_4} \bigg/ \frac{EL_1}{EL_4},$$

using the metrical value of the cross-ratio.

Draw PP_1 and EE_1 perpendicular to the plane $x = 0$, and PP_4 and EE_4 perpendicular to the plane $w = 0$. Then

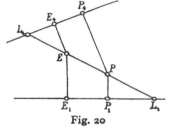

$$\frac{PL_1}{EL_1} = \frac{PP_1}{EE_1}, \quad \frac{PL_4}{EL_4} = \frac{PP_4}{EE_4}.$$

Hence $\quad \dfrac{x}{w} = \dfrac{PP_1}{EE_1} \bigg/ \dfrac{PP_4}{EE_4},$

and therefore

$$x = k \frac{PP_1}{EE_1}, \qquad w = k \frac{PP_4}{EE_4},$$

Fig. 20

where k is some constant; and similarly

$$y = k \frac{PP_2}{EE_2}, \qquad z = k \frac{PP_3}{EE_3}.$$

Hence the coordinates are certain multiples of the distances of P from the fundamental planes.

3·851. When the unit-point E is the centre of the inscribed sphere of the tetrahedron, so that

$$EE_1 = EE_2 = EE_3 = EE_4,$$

the coordinates x, y, z, w are proportional to the distances of P from the fundamental planes. These are analogous to trilinear coordinates in a plane and may be called *quadriplanar coordinates*.

3·852. If E is the centroid of the tetrahedron, and A_1, A_2, A_3, A_4 denote the areas of the faces,

$$A_1 . EE_1 = A_2 . EE_2 = A_3 . EE_3 = A_4 . EE_4.$$

Then $$x = k' . PP_1 . A_1, \text{ etc.,}$$

and the coordinates are proportional to the volumes of the tetrahedra $PDBC$, $PDCA$, $PDAB$, $PABC$. These are analogous to areal coordinates, and may be called *volume-coordinates*.

3·853. If masses m_1, m_2, m_3, m_4 are placed at the vertices and if P is the centre of mass, then if d and p denote the distances of P and A from the plane BCD, we have

$$pm_1 = d\Sigma m, \quad \text{and} \quad \frac{\text{vol. } PDBC}{\text{vol. } ABCD} = \frac{d}{p} = \frac{m_1}{\Sigma m}.$$

Hence m_1, m_2, m_3, m_4 are proportional to the volume-coordinates. From this point of view they are called *barycentric coordinates*.

3·86. In the theorem of 3·72 if P is the centroid of the tetrahedron, the planes $P_2P_3P_4$, etc., are parallel to the corresponding faces of the tetrahedron, hence the polar plane of the centroid is the plane at infinity. Hence in barycentric coordinates, the centroid being [1, 1, 1, 1] the equation of the plane at infinity is

$$x + y + z + w = 0.$$

It is often most convenient to specify the particular system of metrical coordinates by the equation of the plane at infinity.

Ex. Show that if l, m, n, p represent the volumes of the tetrahedra $EDBC$, $EDCA$, $EDAB$, $EABC$, E being the unit-point, the equation of the plane at infinity is

$$lx + my + nz + pw = 0.$$

3·861. General homogeneous coordinates, referred to a fundamental tetrahedron, are not usually the most convenient for metrical geometry, but are used mostly in projective geometry. The simplest metrical system is obtained by taking the plane at infinity itself as one of the fundamental planes $w = 0$, and this leads at once to cartesian coordinates, the opposite vertex of the tetrahedron being the origin, and the edges through that point the coordinate-axes.

3·91. Analytical representation of a homography.

The general homography on a line is represented by a linear transformation. The parameters t, t' of corresponding points P, P' are connected by a bilinear equation

$$att' + bt + ct' + d = 0.$$

The double-points of the homography are then represented by the roots of the quadratic

$$at^2 + (b+c)t + d = 0.$$

The homography is an involution if the equation is symmetrical in t, t', i.e. if $b = c$.

3·92. *An involution is determined by two pairs of points.*

Let the parameters of the two pairs be the roots of the quadratic equations

$$a_1 t^2 + 2h_1 t + b_1 = 0,$$
$$a_2 t^2 + 2h_2 t + b_2 = 0.$$

If the double-points are the roots of the quadratic

$$At^2 + 2Ht + B = 0,$$

then, since the double-points are harmonic conjugates with regard to each pair, we have (3·321)

$$b_1 A - 2h_1 H + a_1 B = 0,$$

and $\qquad b_2 A - 2h_2 H + a_2 B = 0.$

These two equations determine uniquely the ratios

$$A : 2H : B = (a_1 h_2 - a_2 h_1) : (a_1 b_2 - a_2 b_1) : (h_1 b_2 - h_2 b_1).$$

The condition that the double-points should be real is

$$(a_1 b_2 - a_2 b_1)^2 - 4(a_1 h_2 - a_2 h_1)(h_1 b_2 - h_2 b_1) > 0,$$

which reduces to $\qquad C_{12}{}^2 - C_{11} C_{22} > 0.$

If the two pairs are both real this is the condition that their cross-ratio should be positive (3·325). If one or both of the pairs be imaginary the double-points must be real since in a harmonic range one pair at least must be real (3·322).

Hence for an involution determined by two pairs

3·921. *If both pairs are real, the involution is elliptic or hyperbolic according as the cross-ratio of the two pairs is negative or positive,*

3·922. *If one pair at least is imaginary, the involution is hyperbolic.*

An elliptic involution contains no conjugate imaginary pairs.

3·93. *Two involutions on the same line have a unique common pair of elements.*

This is proved in 3·92, considering the first two quadratics as determining the double-points of the two involutions, and the third the pair of common elements.

Hence *the common elements of two involutions are real if one at least of the involutions is elliptic; if both are hyperbolic, the common elements are real or imaginary according as the cross-ratio of the two pairs of double-points is positive or negative.*

3·95. EXAMPLES.

1. $P_1P_2P_3P_4$ is a skew quadrilateral, and a plane cuts the sides P_1P_2, etc., in the ratios $k_{12} : 1$, etc. Prove that

$$k_{12}k_{23}k_{34}k_{41} = 1.$$

2. A plane cuts the six edges of a tetrahedron $A_1A_2A_3A_4$ in six points P_{12}, etc., and Q_{12} is the harmonic conjugate of P_{12} with respect to A_1 and A_2, etc. Show that the six planes $Q_{12}A_3A_4$, etc., have a point in common.

3. A plane cuts the sides P_1P_2, etc., of a skew quadrilateral $P_1P_2P_3P_4$ in points P_{12}, etc., and Q_{12} is the harmonic conjugate of P_{12} with respect to P_1 and P_2, etc. Show that Q_1, Q_2, Q_3, Q_4 are coplanar.

4. Show that the tetrahedron whose faces are $y - z + 2w = 0$, $x - 4z + w = 0$, $-x + 4y + w = 0$, $-2x + y + z = 0$ is both circumscribed and inscribed to the tetrahedron of reference.

5. Show that the tetrahedron whose vertices are $[0, -b, c, p]$, $[a, 0, -c, q]$, $[-a, b, 0, r]$, $[ap, bq, cr, 0]$ is both inscribed and circumscribed to the tetrahedron of reference.

6. Show that the tetrahedron whose faces are

$$-mry + nqz + w = 0, \quad lrx - npz + w = 0,$$
$$-lqx + mpy + w = 0, \quad lx + my + nz = 0$$

is both inscribed and circumscribed to the tetrahedron of reference.

7. Show that the tetrahedron whose faces are $-y+z+pw=0$, $x-z+qw=0$, $-x+y+rw=0$, $px+qy+rz=0$ is both inscribed and circumscribed to the tetrahedron of reference.

8. Show that the product of the cross-ratios of the points in which four lines are cut by their two transversals is

$$m_{13}m_{24}/m_{14}m_{23};$$

where the factors are the mutual moments of the lines taken in pairs.

Consider the equation

$$(m_{23}m_{14})^n + (m_{31}m_{24})^n + (m_{12}m_{34})^n = 0,$$

and show that it is true (i) for $n=\frac{1}{2}$ when the transversals coincide, (ii) for $n=\frac{1}{4}$ when the four lines are tangents of a cubic curve. Give an interpretation for the case $n=1$.

(Math. Trip. II, 1913.)

9. Examine the figure of two mutually inscribed tetrahedra; and show that any one of the eight conditions (making a vertex of one lie on a face of the other) is a consequence of the rest.

What plane theorem is obtained by taking an arbitrary plane section of the tetrahedra? Show that a plane figure can be drawn in which AA', BB', CC' are three lines, and the bisectors at A of the angles subtended by BB' and CC' are the same; and similarly for all the other points. (Math. Trip. II, 1913.)

10. A set of collinear points P_1, P_2, P_3, ..., P_n are obtained by projecting from a point a set of collinear points which are spaced at equal intervals. Find the coordinates of P_n in terms of those of P_1, P_2, P_3, and find the limiting position of P_n when n is large.

(Math. Trip. II, 1914.)

Ans. $x_n = -\dfrac{(n-1)\,x_2x_3 - 2\,n - 2)\,x_3x_1 + (n-3)\,x_1x_2}{(n-1)\,x_1 - 2\,(n-2)\,x_2 + (n-3)\,x_3}$.

11. From a variable point P in one fixed plane a transversal is drawn to two fixed straight lines in space and produced to meet another fixed plane in P'. Find in the simplest form the relation connecting the positions of the points P, P' in their respective planes.

If ABC, $A'B'C'$ be triangles in two different planes such that

BC passes through A', and $B'C'$ passes through A, and variable points P, P' in these planes be related by the fact that PP' intersects BB' and CC', find geometrically the locus of P', (i) when P is on a straight line passing through A or B, (ii) when P is on BC or CA, (iii) when P is on a conic circumscribed to ABC, (iv) when P is on any conic in the plane ABC.

(Math. Trip. II, 1915.)

Ans. Taking ABC and $A'B'C'$ as triangles of reference, $P \equiv [X, Y, Z]$, $P' \equiv [X', Y', Z']$, then $XX' : YY' : ZZ' = p : q : r$. (i) A straight line through A' or B' respectively, (ii) The points A' and B' respectively, (iii) A straight line, (iv) A quartic curve with double-points at A', B', C'.

12. Show that it is possible to choose a tetrahedron of reference so that the vertices of a hexahedron, which is the projection of a parallelepiped, may have the coordinates (i) $[\pm 1, \pm 1, \pm 1, 1]$, (ii) $[1, 0, 0, 0]$, $[0, 1, 0, 0]$, $[0, 0, 1, 0]$, $[0, 0, 0, 1]$, $[-1, 1, 1, 1]$, $[1, -1, 1, 1]$, $[1, 1, -1, 1]$, $[1, 1, 1, -1]$.

CHAPTER IV

THE SPHERE

4·1. A sphere is the locus of a point which is at a constant distance k from a fixed point, the centre. If the rectangular co-ordinates of the centre are $[X, Y, Z]$ the equation of the sphere is

$$(x-X)^2+(y-Y)^2+(z-Z)^2=k^2.$$

This is an equation of the second degree and is characterised by two properties:

(1) The coefficients of x^2, y^2, z^2 are all equal;

(2) The coefficients of yz, zx, xy are all zero, i.e. there are no product terms.

4·11. The general equation of the second degree which satisfies these conditions may be written

$$ax^2+ay^2+az^2+2px+2qy+2rz+d=0.$$

If $a \neq 0$ we may write this in the form

$$(x+p/a)^2+(y+q/a)^2+(z+r/a)^2=(p^2+q^2+r^2-ad)/a^2,$$

and comparing this with (4·1) we see that it represents a sphere with centre $[-p/a, -q/a, -r/a]$ and radius $(p^2+q^2+r^2-ad)^{\frac{1}{2}}/a$. The radius is real provided $p^2+q^2+r^2-ad>0$.

If $$p^2+q^2+r^2-ad<0,$$

the radius is imaginary, and we call it a *virtual sphere*.

If $$p^2+q^2+r^2=ad,$$

the radius is zero, and we call it a *point-sphere*.

4·12. If $a=0$ the equation reduces to

$$2px+2qy+2rz+d=0,$$

which is no longer of the second degree and represents a plane. Now a locus whose equation is of the second degree has the geometrical property that it is cut by an arbitrary straight line in two points, while a plane is cut in only one point. The apparent discrepancy is explained when we use homogeneous coordinates.

If we write x/w, y/w and z/w for x, y and z, the equation of the sphere becomes

$$ax^2 + ay^2 + az^2 + 2pxw + 2qyw + 2rzw + dw^2 = 0,$$

and when $a = 0$ we have

$$w(2px + 2qy + 2rz + dw) = 0.$$

The complete locus therefore consists of the plane

$$2px + 2qy + 2rz + dw = 0,$$

together with the plane at infinity $w = 0$. As $a \to 0$ the radius of the sphere $\to \infty$, and the centre tends to infinity. We may say roughly that the parts of the sphere in the neighbourhood of the plane flatten out on to this plane, while the farther parts of the sphere recede to infinity.

4·2. Power of a point with regard to a sphere.

If P is a fixed point and PUV a variable line cutting a fixed sphere in U, V, it is easily seen by elementary geometry that the product $PU.PV$ is constant; this constant is called the *power* of P with regard to the sphere. If k is the radius of the sphere, and d the distance of P from the centre, the power of P is

$$d^2 - k^2;$$

or if $P \equiv [x, y, z]$ and the equation of the sphere is

$$S \equiv x^2 + y^2 + z^2 + 2px + 2qy + 2rz + c = 0,$$

then the power of P is equal to S.

4·21. S is positive or negative according as P is outside or inside the sphere, zero when P lies on the sphere. For a point-sphere with centre C it reduces to CP^2.

4·22. As the sphere tends to a plane, the radius tending to infinity, the power of a point in general $\to \infty$, but there is another quantity which remains finite in this case. Let PUV pass through the centre C, and let C tend to infinity while U remains fixed. Then

$$S = PU.PV = PU(PU + 2k),$$

and the ratio
$$\frac{S}{k} = PU\left(\frac{PU}{k} + 2\right) \to 2PU.$$

Analytically, if

$$aS \equiv a(x^2+y^2+z^2)+2px+2qy+2rz+d,$$
$$k=(p^2+q^2+r^2-ad)^{\frac{1}{2}}/a.$$

Then
$$\frac{S}{k} = \frac{a(x^2+y^2+z^2)+2px+2qy+2rz+d}{(p^2+q^2+r^2-ad)^{\frac{1}{2}}}$$
$$\rightarrow \frac{2px+2qy+2rz+d}{(p^2+q^2+r^2)^{\frac{1}{2}}} \quad \text{as } a \rightarrow 0$$

= twice the distance of P from the plane.

4·3. A sphere is completely determined by four points. Let

$$x^2+y^2+z^2+2px+2qy+2rz+d=0$$

be the equation of the sphere through the four points (x_1), (x_2), (x_3), (x_4). Then

$$x_1^2+y_1^2+z_1^2+2px_1+2qy_1+2rz_1+d=0$$

and three other similar equations. Eliminating p, q, r and d we have

$$\begin{vmatrix} x^2+y^2+z^2 & x & y & z & 1 \\ x_1^2+y_1^2+z_1^2 & x_1 & y_1 & z_1 & 1 \\ \cdots & & & & \\ x_4^2+y_4^2+z_4^2 & x_4 & y_4 & z_4 & 1 \end{vmatrix} = 0$$

as the equation of the sphere through the four points.

4·31. The condition that the sphere should pass through five given points is found by substituting in this equation the co-ordinates of the fifth point (x_5). This condition may be transformed as follows. We may write it in either of the forms:

$$\begin{vmatrix} x_1^2+y_1^2+z_1^2 & x_1 & y_1 & z_1 & 1 \\ \cdots & & & & \\ x_5^2+y_5^2+z_5^2 & x_5 & y_5 & z_5 & 1 \end{vmatrix} = 0,$$

or
$$\begin{vmatrix} 1 & -2x_1 & -2y_1 & -2z_1 & x_1^2+y_1^2+z_1^2 \\ \cdots & & & & \\ 1 & -2x_5 & -2y_5 & -2z_5 & x_5^2+y_5^2+z_5^2 \end{vmatrix} = 0.$$

Multiply together these two determinants by rows. Taking row i of the first determinant with row j of the second we get the element

$$x_i^2+y_i^2+z_i^2-2x_ix_j-2y_iy_j-2z_iz_j+x_j^2+y_j^2+z_j^2$$
$$=(x_i-x_j)^2+(y_i-y_j)^2+(z_i-z_j)^2=d_{ij}^2,$$

where d_{ij} denotes the distance between the points (x_i) and (x_j), d_{ii} being $= 0$. Hence the condition that five points should lie on a sphere is represented by

$$\begin{vmatrix} 0 & d_{12}^2 & d_{13}^2 & d_{14}^2 & d_{15}^2 \\ d_{21}^2 & 0 & d_{23}^2 & d_{24}^2 & d_{25}^2 \\ d_{31}^2 & d_{32}^2 & 0 & d_{34}^2 & d_{35}^2 \\ d_{41}^2 & d_{42}^2 & d_{43}^2 & 0 & d_{45}^2 \\ d_{51}^2 & d_{52}^2 & d_{53}^2 & d_{54}^2 & 0 \end{vmatrix} = 0.$$

The corresponding relation in two dimensions, or the condition that four points should lie on a circle, viz.

$$\begin{vmatrix} 0 & d_{12}^2 & d_{13}^2 & d_{14}^2 \\ d_{21}^2 & 0 & d_{23}^2 & d_{24}^2 \\ d_{31}^2 & d_{32}^2 & 0 & d_{34}^2 \\ d_{41}^2 & d_{42}^2 & d_{43}^2 & 0 \end{vmatrix} = 0,$$

is equivalent to $\quad d_{23}d_{14} \pm d_{31}d_{24} \pm d_{12}d_{34} = 0,$

and represents the Theorem of Ptolemy.

4·41. A circle is the intersection of a sphere and a plane. The equations of a circle can therefore be given by the equations of a sphere and a plane together.

Ex. **1.** Find the equations of the circle through the points $[a, 0, 0]$, $[0, b, 0]$, $[0, 0, c]$.

The equation of the plane through these points is

$$x/a + y/b + z/c = 1.$$

Let the equation of a sphere through the three points be

$$x^2 + y^2 + z^2 + 2px + 2qy + 2rz + d = 0.$$

Then

$$a^2 + 2pa + d = 0,$$
$$b^2 + 2qb + d = 0,$$
$$c^2 + 2rc + d = 0.$$

We may give d any value, say 0, and then determine

$$2p = -a, \quad 2q = -b, \quad 2r = -c.$$

Hence the equations of the circle are

$$x/a + y/b + z/c = 1,$$
$$x^2 + y^2 + z^2 = ax + by + cz.$$

Ex. 2. Find the equations of the circle with centre $[X, Y, Z]$, radius k, and axis (the line through the centre perpendicular to its plane) in the direction $[l, m, n]$.

Ans.
$$(x-X)^2+(y-Y)^2+(z-Z)^2=k^2,$$
$$l(x-X)+m(y-Y)+n(z-Z)=0.$$

4·42. Intersection of two or more spheres.

4·421. Consider the two spheres
$$S \equiv x^2+y^2+z^2+2px+2qy+2rz+d=0,$$
$$S' \equiv x^2+y^2+z^2+2p'x+2q'y+2r'z+d'=0.$$
The coordinates of all the points of intersection satisfy both of these equations and therefore also satisfy the equation
$$S-S' \equiv 2(p-p')x+2(q-q')y+2(r-r')z+(d-d')=0.$$
The curve of intersection is therefore a circle lying in this plane. This plane is called the *radical plane* of the two spheres, and is the locus of points which have the same power with respect to the two spheres; it is perpendicular to the line joining the centres.

4·422. Three spheres S_1, S_2, S_3 have three radical planes when taken in pairs. Their equations may be written
$$S_2-S_3=0, \quad S_3-S_1=0, \quad S_1-S_2=0.$$
Since
$$(S_2-S_3)+(S_3-S_1)+(S_1-S_2)=0,$$
the three planes have a line in common. This line is called the *radical axis* of the three spheres.

4·423. Similarly four spheres have six radical planes and four radical axes, and these planes and axes all pass through one point which is called the *radical centre* of the four spheres.

4·43. Angle of intersection of two spheres.

The angle of intersection of two spheres at a common point is the angle between the tangent-planes to the spheres at that point, and, as the tangent-planes are perpendicular to the radii, it is equal to the angle between the radii. If C, C' are the centres, and P, P' two common points, the triangles $CC'P$ and $CC'P'$ are congruent, hence at all common points the angle of intersection is the same. If this angle is a right angle the spheres are said to be *orthogonal*.

4·431. Condition that two spheres should be orthogonal.

The geometrical condition is $CP^2 + C'P^2 = CC'^2$. Hence

$$(p-p')^2 + (q-q')^2 + (r-r')^2 = (p^2+q^2+r^2-d)$$
$$+ (p'^2+q'^2+r'^2-d'),$$

i.e. $\qquad\qquad 2pp' + 2qq' + 2rr' - (d+d') = 0.$

The left-hand side of this equation is the bilinear symmetrical expression in p, q, r, d and p', q', r', d' which reduces to

$$2(p^2+q^2+r^2-d)$$

when the spheres coincide.

If θ is the angle of intersection in the general case

$$CC'^2 = CP^2 + C'P^2 - 2CP \cdot C'P \cos\theta.$$

Hence $\qquad 2kk' \cos\theta = 2pp' + 2qq' + 2rr' - (d+d').$

Ex. A point-sphere is self-orthogonal.

4·432. Sphere orthogonal to four given spheres.

Since the relation of orthogonality is linear in the coefficients of each sphere, four such relations connecting the coefficients of a sphere will uniquely determine it. Since the tangents from the centre of the orthogonal sphere to each of the four spheres are also its radii, the centre has the same power with respect to each of the four spheres and is therefore the radical centre. The orthogonal sphere will be real unless the radical centre lies within each of the given spheres.

If $\quad S_i \equiv x^2+y^2+z^2+2p_i x + 2q_i y + 2r_i z + d_i = 0 \quad (i=1, 2, 3, 4)$

represent the given spheres, and

$$S \equiv x^2+y^2+z^2+2px+2qy+2rz+d = 0$$

their orthogonal sphere, we have

$$2p_i p + 2q_i q + 2r_i r - d - d_i = 0 \quad (i=1, 2, 3, 4).$$

The equation of the orthogonal sphere is obtained by eliminating p, q, r, d between these five equations.

4·5. Pole and polar with respect to a sphere.

Let $P \equiv [x', y', z']$ be a fixed point, and draw any line through P cutting the sphere in U, V. Then the locus of $Q \equiv [x, y, z]$, the harmonic conjugate of P with respect to U, V, is a plane, which is called the *polar plane* of P with respect to the sphere. Let the position-ratio of U with respect to P, Q be λ. Then the coordinates of U are $(\lambda x' + x)/(\lambda + 1)$, etc. Substituting in the equation of the sphere,

$$S \equiv x^2 + y^2 + z^2 + 2px + 2qy + 2rz + d = 0,$$

we have

$$(\lambda x' + x)^2 + (\lambda y' + y)^2 + (\lambda z' + z)^2 + 2\{p(\lambda x' + x)$$
$$+ q(\lambda y' + y) + r(\lambda z' + z)\}(\lambda + 1) + d(\lambda + 1)^2 = 0,$$

or, rearranging,

$$S'\lambda^2 + 2\{xx' + yy' + zz' + p(x + x') + q(y + y')$$
$$+ r(z + z') + d\}\lambda + S = 0.$$

This is a quadratic in λ, the roots of which, λ_1 and λ_2, are the position-ratios of U and V. If (UV, PQ) is harmonic, $\lambda_1 = -\lambda_2$, hence

$$xx' + yy' + zz' + p(x + x') + q(y + y') + r(z + z') + d = 0,$$

or $\quad (x' + p)x + (y' + q)y + (z' + r)z + (px' + qy' + rz' + d) = 0.$

This is the equation of the polar plane of $[x', y', z']$, which is therefore perpendicular to the line joining $P \equiv [x', y', z']$ to the centre $[-p, -q, -r]$. When P lies on the surface the polar-plane passes through P and becomes the *tangent-plane* at P.

With homogeneous coordinates the equation of the polar may be written in the form

$$x\frac{\partial S}{\partial x'} + y\frac{\partial S}{\partial y'} + z\frac{\partial S}{\partial z'} + w\frac{\partial S}{\partial w'} = 0.$$

4·51. If the spheres S_1 and S_2 are orthogonal the polar plane of any point P on S_1 with respect to S_2 passes through the other end of the diameter of S_1 through P.

Let $\qquad S_1 \equiv x^2 + y^2 + z^2 - k^2 = 0,$

$$S_2 \equiv x^2 + y^2 + z^2 + 2px + 2qy + 2rz + k^2 = 0.$$

Let $P \equiv [x', y', z']$ be any point on S_1, so that

$$x'^2 + y'^2 + z'^2 = k^2.$$

Its polar plane with respect to S_2 is

$$x'x + y'y + z'z + p(x + x') + q(y + y') + r(z + z') + k^2 = 0,$$

and this equation is satisfied by

$$[x, y, z] = [-x', -y', -z'].$$

If S_1, S_2, S_3, S_4 are four given spheres, their orthogonal sphere is therefore the locus of points P whose polar planes pass through a common point, the other end of the diameter of the orthogonal sphere through P. Hence the equation of the orthogonal sphere is

$$\begin{vmatrix} \dfrac{\partial S_1}{\partial x} & \dfrac{\partial S_1}{\partial y} & \dfrac{\partial S_1}{\partial z} & \dfrac{\partial S_1}{\partial w} \\[2ex] \dfrac{\partial S_2}{\partial x} & \dfrac{\partial S_2}{\partial y} & \dfrac{\partial S_2}{\partial z} & \dfrac{\partial S_2}{\partial w} \\[2ex] \dfrac{\partial S_3}{\partial x} & \dfrac{\partial S_3}{\partial y} & \dfrac{\partial S_3}{\partial z} & \dfrac{\partial S_3}{\partial w} \\[2ex] \dfrac{\partial S_4}{\partial x} & \dfrac{\partial S_4}{\partial y} & \dfrac{\partial S_4}{\partial z} & \dfrac{\partial S_4}{\partial w} \end{vmatrix} = 0.$$

This determinant is called the *Jacobian* of the four homogeneous functions S_1, S_2, S_3, S_4, and the locus is called the Jacobian locus of the four spheres. It appears to be of the fourth degree, but w^2 is a factor*, hence the complete Jacobian consists of the orthogonal sphere together with the plane at infinity taken twice.

4·6. Linear systems of spheres.

4·61. If S_1, S_2 are two spheres,

$$S_1 + \lambda S_2 = 0$$

represents, for all values of λ, a sphere through their circle of intersection. For $\lambda = -1$ it reduces to the radical plane of the two spheres, and any two spheres of the system have the same radical plane. This system is called a *pencil* or *linear one-parameter system* of spheres; it is called also a *coaxial system*.

Writing the equation in full we have

$$(1 + \lambda)(x^2 + y^2 + z^2) + 2(p_1 + \lambda p_2)x + 2(q_1 + \lambda q_2)y$$
$$+ 2(r_1 + \lambda r_2)z + (d_1 + \lambda d_2) = 0.$$

Hence the homogeneous coordinates of the centre are

$$[p_1 + \lambda p_2, \ q_1 + \lambda q_2, \ r_1 + \lambda r_2, \ -(1 + \lambda)],$$

---

* Putting $w = 0$ the first three columns of the determinant become identical, except for a factor, hence w^2 is a factor. Geometrically, the polar of a point at infinity is the diametral plane perpendicular to the given direction. Hence the plane at infinity is part of the Jacobian locus since the polars of any given point at infinity are parallel and have a line at infinity in common.

and the locus of the centre is a straight line perpendicular to the radical plane.

Taking the radical plane as the plane of yz and the line of centres as axis of x we have $q_1 = 0 = q_2$, $r_1 = 0 = r_2$, $d_1 = d_2 = k$, say. Then putting $(p_1 + \lambda p_2)/(1 + \lambda) = -\mu$, the equation of the system becomes

$$x^2 + y^2 + z^2 - 2\mu x + k = 0,$$

or

$$(x - \mu)^2 + y^2 + z^2 = \mu^2 - k.$$

If $k > 0$ the radius of the sphere becomes zero when $\mu = \pm\sqrt{k}$. Hence in this case the system contains two point-spheres; these are called the *limiting-points* of the system. In this case the radical plane cuts the spheres in a virtual circle.

If $k < 0$ there are no real point-spheres, the minimum radius, for $\mu = 0$, being $\sqrt{-k}$; and the common circle of the system is real.

If S is a sphere cutting orthogonally two spheres of the system, so that

$$2pp_1 + 2qq_1 + 2rr_1 = d + d_1,$$

$$2pp_2 + 2qq_2 + 2rr_2 = d + d_2,$$

then

$$2p(p_1 + \lambda p_2) + 2q(q_1 + \lambda q_2) + 2r(r_1 + \lambda r_2) = (1 + \lambda)d + (d_1 + \lambda d_2),$$

and therefore S cuts orthogonally every sphere of the system.

4·62. If S cuts orthogonally the sphere

$$x^2 + y^2 + z^2 - 2\lambda x + k = 0,$$

we have

$$2\lambda p + d + k = 0,$$

which is satisfied identically if $p = 0$ and $d = -k$. Hence the equation of a sphere cutting orthogonally every sphere of the one-parameter system is

$$x^2 + y^2 + z^2 - 2\mu y - 2\nu z - k = 0,$$

which represents a *linear two-parameter system*. The spheres of this system have a common *radical axis* $y = 0$, $z = 0$, and pass through two fixed points $(\pm\sqrt{k}, 0, 0)$ which are real when $k > 0$, i.e. when the common circle of the first system is virtual. If $k < 0$, the common points are imaginary, but there is a real locus of point-spheres. Writing the equation in the form

$$x^2 + (y - \mu)^2 + (z - \nu)^2 = \mu^2 + \nu^2 + k,$$

the radius $=0$ when $\mu^2 + \nu^2 = -k$. Hence the locus of point-spheres is the common circle of the first system.

If S_1, S_2, S_3 are three spheres, the equation

$$S_1 + \lambda S_2 + \mu S_3 = 0$$

represents a linear two-parameter system, which is thus determined by three spheres. Any two spheres of the system determine a pencil of spheres which is contained in the two-parameter system and the limiting-points of the pencil are point-spheres of both systems. Hence in order that the two-parameter system should have imaginary point-spheres every pair of spheres of the system must have real intersection; all the spheres then pass through two real common points. If any one pair of spheres have imaginary intersection the one-parameter system determined by these has real limiting-points and the two-parameter system has a real circle of point-spheres.

4·63. Lastly, if S_1, S_2, S_3, S_4 are four spheres, the equation

$$S_1 + \lambda S_2 + \mu S_3 + \nu S_4 = 0$$

represents a linear three-parameter system. There is one sphere S which cuts the four spheres orthogonally, and then every sphere of the system will cut S orthogonally. A linear three-parameter system is thus a system cutting a fixed sphere orthogonally. The system has a common radical centre, the centre of the orthogonal sphere, and this sphere is the locus of point-spheres of the system. If the equation of the orthogonal sphere is
$$x^2 + y^2 + z^2 = k$$
the equation of a sphere cutting this orthogonally is
$$x^2 + y^2 + z^2 + 2\lambda x + 2\mu y + 2\nu z + k = 0.$$

Ex. 1. Prove that the condition that the sphere S_1 should cut the sphere S_2 in a great circle is
$$2\,(p_1 p_2 + q_1 q_2 + r_1 r_2) - d_1 - d_2 = 2k_2{}^2.$$

Ex. 2. If the sphere S cuts orthogonally the sphere $x^2 + y^2 + z^2 = k$, show that it cuts the sphere $x^2 + y^2 + z^2 + k = 0$ in a great circle.

Ex. 3. If the common orthogonal sphere of a three-parameter system is virtual, show that every sphere of the system cuts a certain fixed sphere in great circles.

4·64. If the coefficients of the equation of a sphere satisfy an equation of the first degree a linear system of spheres is determined. Let the coefficients p, q, r, d be connected by the equation

$$Ap + Bq + Cr + Dd + E = 0.$$

Then, comparing this with the equation

$$2p'p + 2q'q + 2r'r - d - d' = 0,$$

we see that the general linear equation connecting the coefficients expresses that the sphere cuts orthogonally the fixed sphere

$$D(x^2 + y^2 + z^2) - Ax - By - Cz + E = 0.$$

A single equation thus represents a three-parameter system. A sphere in general has four degrees of freedom. Two equations determine a two-parameter system, three equations a one-parameter system, while four equations determine the sphere completely.

4·7. Inversion in a sphere.

Let S be a fixed sphere with centre O and radius k, and P any point. Then if P' lies on OP and $OP . OP' = k^2$, P' is called the *inverse* of P with respect to the given sphere.

Let the equation of the fixed sphere be

$$x^2 + y^2 + z^2 = k^2,$$

and $P \equiv [x, y, z]$, $P' \equiv [x', y', z']$, then

$$\frac{x'}{x} = \frac{y'}{y} = \frac{z'}{z}$$

and

$$(x^2 + y^2 + z^2)(x'^2 + y'^2 + z'^2) = k^4.$$

Hence

$$\frac{x'}{x} = \frac{y'}{y} = \frac{z'}{z} = \frac{k^2}{x^2 + y^2 + z^2} = \frac{x'^2 + y'^2 + z'^2}{k^2}.$$

These are the equations of the transformation of inversion.

4·71. *The inverse of a sphere is in general a sphere.*

The equation

$$x^2 + y^2 + z^2 + 2px + 2qy + 2rz + d = 0$$

becomes

$$\frac{k^4}{x'^2 + y'^2 + z'^2} + 2\frac{px' + qy' + rz'}{x'^2 + y'^2 + z'^2}k^2 + d = 0,$$

i.e.

$$d(x'^2 + y'^2 + z'^2) + 2k^2(px' + qy' + rz') + k^4 = 0,$$

which represents a sphere with centre

$$[-pk^2/d, \ -qk^2/d, \ -rk^2/d]$$

and radius

$$= k^2(p^2 + q^2 + r^2 - d)^{\frac{1}{2}}/d.$$

If $d = 0$, i.e. if the given sphere passes through the origin, the inverse is a plane
$$px' + qy' + rz' + \tfrac{1}{2}k^2 = 0,$$
which is perpendicular to the line joining O to the centre $[-p, -q, -r]$ of the sphere.

The inverse of a circle, which is the intersection of two spheres, is again a circle.

4·72. Invariance of angles under inversion.

Let U_1 and U_2 denote two surfaces and P a point on their curve of intersection, and let L_1 and L_2 denote the tangent-planes at P. The angle between the surfaces is equal to the dihedral angle between these planes. The inverses of the two planes are spheres S_1' and S_2' passing through O and having their tangent-planes at O parallel respectively to L_1 and L_2. Also the inverse surfaces U_1' and U_2' into which U_1 and U_2 are transformed touch S_1' and S_2' respectively at the inverse point P'. Hence the angle between U_1' and U_2' at P' is equal to the angle between the spheres S_1' and S_2' at P' or at O, and is therefore equal to the angle between L_1 and L_2.

4·73. Stereographic projection.

If S is a sphere through O its inverse is a plane parallel to the tangent-plane at O. The inverse of a point P on the sphere is the point P' on the plane such that O, P, P' are collinear. Thus P' is the projection of P on this fixed plane. This transformation between a sphere and a plane is called *stereographic projection*, and, as it is a particular case of inversion, circles on the sphere are transformed into circles on the plane, and angles are unaltered.

Ex. If $P \equiv [x, y, z, w]$ is any point on the sphere $x^2 + y^2 + (z - r)^2 = r^2$ and $[x', y']$ the coordinates of its projection on $z = 0$ from the centre $[0, 0, 2r]$, show that
$$\lambda x = 4r^2 x',$$
$$\lambda y = 4r^2 y',$$
$$\lambda z = 2r\,(x'^2 + y'^2),$$
$$\lambda w = x'^2 + y'^2 + 4r^2.$$

(The stereographic projection can be regarded as a plane representation of the sphere, and these equations represent freedom-equations of the sphere with the parameters x', y'.)

4·74. Inversion is a birational, quadratic, point-transformation, i.e. points are transformed into points, a plane is transformed into a locus of the second order, and the equations of transformation are rational in both sets of coordinates, both when (x, y, z) are expressed in terms of (x', y', z') and *vice versa*. It is also said to be conformal since angles are unaltered; and further it is called a spherical transformation, since spheres are transformed into spheres.

There are two other ways in which the transformation may be defined, which bring out more particularly the two distinct properties (1) that it is birational, (2) that it transforms spheres into spheres.

(1) Let S be a fixed sphere, centre O, and let P be any point. Then the inverse of P is the point of intersection of OP with the polar of P with respect to the sphere.

(2) With the same data, the sphere S together with the point-sphere at P determine a linear one-parameter system of spheres, and in this system there is a second point-sphere P'; P' is the inverse of P.

4·8. The circle at infinity.

The points at infinity on a sphere are of fundamental importance in metrical geometry. Writing the equation of a sphere homogeneously

$$x^2 + y^2 + z^2 + 2pxw + 2qyw + 2rzw + dw^2 = 0,$$

we see that it cuts the plane at infinity $w = 0$ where

$$x^2 + y^2 + z^2 = 0.$$

Hence all spheres cut the plane at infinity in the same curve. This curve, whose equations are

$$w = 0, \quad x^2 + y^2 + z^2 = 0,$$

is called the *circle at infinity*, or *absolute circle*.

Every surface of the second order which contains the circle at infinity is a sphere, for, the general equation of the second degree being

$$ax^2 + by^2 + cz^2 + 2fyz + 2gzx + 2hxy$$
$$+ 2pxw + 2qyw + 2rzw + dw^2 = 0,$$

if this is satisfied identically by $w=0$ and $x^2+y^2+z^2=0$, we must have $a=b=c$ and $f=0=g=h$.

Any plane cuts the circle at infinity in two points, the *circular points* in that plane; and every conic, in this plane, which passes through these two circular points is a circle.

Since every plane section of a sphere passes through the circular points in its plane it is a circle.

Since the pole of the plane at infinity is the centre of the sphere, the circle at infinity is the locus of points of contact of tangents from the centre of the sphere. The assemblage of these tangents is the asymptotic cone of the sphere.

If the points at infinity $P \equiv [l, m, n, 0]$ and $P' \equiv [l', m', n', 0]$ are conjugate with regard to the circle at infinity,

$$ll' + mm' + nn' = 0,$$

hence the lines OP, OP' are at right angles. If the line p is the polar of P with regard to the circle at infinity its equations are

$$w=0, \quad lx+my+nz=0.$$

Hence the plane Op is perpendicular to the line OP.

4·81. Isotropic lines and planes.

A line which cuts the circle at infinity, called an *isotropic* or *absolute line*, has certain peculiar properties. If the line

$$x/l = y/m = z/n$$

cuts the circle at infinity we have

$$l^2 + m^2 + n^2 = 0.$$

Hence the distance from the origin to any point of the line, and hence the distance between any two points of the line, is zero.

If θ is the angle between the (real) line $x/l = y/m = z/n$ and the isotropic line $x/l' = y/m' = z/n'$,

$$\tan^2\theta = \frac{\Sigma l^2 . \Sigma l'^2 - (\Sigma ll')^2}{(\Sigma ll')^2} = -1.$$

The angle is of course unreal, but the fact that the square of its tangent is a real number independent of l, m, n is held to justify the term isotropic.

An absolute line is orthogonal to itself since

$$ll + mm + nn = 0.$$

The assemblage of absolute lines through a point form a virtual cone. If the point is $[X, Y, Z]$ the equation of the cone is

$$(x-X)^2+(y-Y)^2+(z-Z)^2=0.$$

This is called the isotropic or absolute cone at the given point; it is also a point-sphere.

Through every point there is an assemblage of planes tangent to the absolute cone at the point; these are tangent-planes to the circle at infinity and are called *absolute planes*. The plane $lx+my+nz=0$ is an absolute plane if $l^2+m^2+n^2=0$. The line $x/l=y/m=z/n$ is orthogonal to this plane, and also lies in the plane.

Consider two lines $[l_1, m_1, n_1]$ and $[l_2, m_2, n_.]$ both lying in the absolute plane $[l, m, n]$, so that $\Sigma ll_1=0$, $\Sigma ll_2=0$ and $\Sigma l^2=0$. Then we have

$$l:m:n=(m_1 n_2-m_2 n_1):(n_1 l_2-n_2 l_1):(l_1 m_2-l_2 m_1),$$

and therefore

$$\Sigma(m_1 n_2-m_2 n_1)^2=0=\Sigma l_1^2.\Sigma l_2^2-(\Sigma l_1 l_2)^2.$$

Hence if θ is the angle between the two lines

$$\cos\theta=\Sigma l_1 l_2/\sqrt{(\Sigma l_1^2.\Sigma l_2^2)}=\pm 1.$$

The angle between any two lines on an absolute plane is therefore zero or a multiple of π. If, however, one of the lines is absolute, $\Sigma l_1^2=0$ and therefore $\Sigma l_1 l_2=0$ also, so that θ is indeterminate.

The equations of the circle at infinity are not altered by any transformation of coordinates so long as they remain rectangular. This is the fundamental reason why we are able to express metrical relations in terms of this figure.

By Laguerre's theorem (3·81): *The angle between two intersecting straight lines is $i/2$ times the logarithm of the cross-ratio of the pencil formed by the two lines and the two absolute lines through their intersection and lying in their plane.*

With the same algebra, $y=\mu x$ and $y=\mu'x$ represent two planes passing through the axis of z, while $y=\pm ix$ are the two absolute planes through their line of intersection, and we obtain the result that *the angle between two planes is $i/2$ times the logarithm*

*of the cross-ratio of the sheaf of planes formed by the two planes
and the two absolute planes through their line of intersection.*

If the pencil is harmonic, so that the two lines are conjugate with regard to the absolute lines, the cross-ratio $= -1$, the logarithm has the value $i\pi$, and the angle $= \frac{1}{2}\pi$.

If the two lines coincide, the cross-ratio $= 1$, the logarithm $= 0$, and the angle is zero. The same is true if the two absolute lines coincide, i.e. when the two lines lie in an absolute plane.

If one line coincides with an absolute line, the cross-ratio becomes zero or infinite, and the angle becomes infinite (and imaginary). We already saw in this case that $\tan\theta = i$. θ is unreal; let $\theta = \alpha + i\beta$, then

$$i = \tan\theta = \tan(\alpha + i\beta) = \frac{\tan\alpha + i\tanh\beta}{1 + i\tan\alpha\tanh\beta}.$$

If α and β are real we have then

$$\tan\alpha = -\tan\alpha\tanh\beta$$

and

$$\tanh\beta = 1.$$

Hence $\alpha = 0$ and β is infinite.

4·9. EXAMPLES.

1. Show that the equation of the sphere whose diameter is the join of the two points $[x_1, y_1, z_1]$ and $[x_2, y_2, z_2]$ is

$$(x - x_1)(x - x_2) + (y - y_1)(y - y_2) + (z - z_1)(z - z_2) = 0.$$

2. Find the equation of the sphere through the points:

(i) $[1, -1, -1]$, $[3, 3, 1]$, $[-2, 0, 5]$, $[-1, 4, 4]$.

(ii) $[0, 5, -2]$, $[4, 1, 8]$, $[-2, -3, 2]$, $[-4, 1, 0]$.

Ans. (i) $x^2 + (y - 1)^2 + (z - 2)^2 = 14$.

(ii) $(x - 1)^2 + (y - 2)^2 + (z - 3)^2 = 35$.

3. Show that the eight points whose coordinates are $x = a$ or a', $y = b$ or b', $z = c$ or c' lie on one sphere.

4. Find the equation of the sphere circumscribing the tetrahedron whose planes are $x = 0$, $y = 0$, $z = 0$, $lx + my + nz + k = 0$.

Ans. $x^2 + y^2 + z^2 + k(x/l + y/m + z/n) = 0$.

5. Find the equation of the sphere inscribed in the tetra-hedron whose faces are (i) $x=0$, $y=0$, $z=0$, $x+y+z=1$, (ii) $y+z=0$, $z+x=0$, $x+y=0$, $x+y+z=1$.

Ans.

(i) $x^2+y^2+z^2-2a(x+y+z)+2a^2=0$, where $(3+\sqrt{3})a=1$.

(ii) $x^2+y^2+z^2-2a(x+y+z)+a^2=0$, where $(3+\sqrt{6})a=1$.

6. C is a fixed point on OZ and U, V are variable points on OX, OY respectively. Find the locus of a point P when the lines PU, PV, PC are mutually at right angles.

Ans. A sphere with centre C and passing through O.

7. Find the equations of the two spheres which pass through the circle $x^2+y^2+z^2-4x-y+3x+12=0$, $2x+3y-7z=10$ and touch the plane $x-2y+2z=1$.

Ans. $(x-1)^2+(y+1)^2+(z-2)^2=4$

and $(x-3)^2+(y-2)^2+(z+5)^2=16$.

8. Show that the two circles $x-2y+4z-13=0$, $x^2+y^2+z^2=9$ and $x+y+z+2=0$, $x^2+y^2+z^2+6y-6z+21=0$ lie on the same sphere, and verify that the line of intersection of their planes cuts the two circles in the same two points.

9. Find the equation of the sphere for which the circle $2x+3y+4z=8$, $x^2+y^2+z^2+7y-2z+2=0$ is a great circle.

Ans. $(x-1)^2+(y+2)^2+(z-3)^2=4$.

10. If the sphere

$$x^2+y^2+z^2+2ux+2vy+2wz+d=0$$

cuts the sphere

$$x^2+y^2+z^2+2u'x+2v'y+2w'z+d'=0$$

in a great circle,

$$2(uu'+vv'+ww')-(d+d')=2r'^2.$$

11. Find the locus of a point such that the ratio of its distances from two given points is constant.

Ans. A sphere.

12. Find the locus of a point such that the ratio of its distances from three given points are constant.

Ans. A circle.

13. If A_1, A_2, ... are fixed points find the locus of a point P such that $\Sigma k_r PA_r^2$ is constant, k_r being given constants.

Ans. A sphere whose centre is the mean point of the given points.

14. Find the locus of a point such that the feet of the perpendiculars drawn to the faces of a given tetrahedron lie in one plane.

Ans. A cubic surface with conical points at the four vertices.

15. Find the locus of a point such that the feet of the perpendiculars drawn to the faces of a given tetrahedron form a tetrahedron of constant volume.

16. Find by inversion the loci of the points of contact of a variable sphere with three fixed spheres which it touches.

Ans. The three circles in which one of the spheres is cut orthogonally by a sphere through the intersection of the other two.

17. Find the locus of the centre of a variable sphere which cuts each of (i) two, (ii) three given spheres in great circles.

Ans. (i) A plane perpendicular to the line of centres, (ii) a line perpendicular to the plane of centres.

18. Show that the following five spheres are mutually orthogonal:
$$x^2+y^2+z^2=a^2,$$
$$x^2+y^2+z^2-2ay-2az+a^2=0,$$
$$x^2+y^2+z^2-2ax-2az+a^2=0,$$
$$x^2+y^2+z^2-2ax-2ay+a^2=0,$$
$$x^2+y^2+z^2-ax-ay-az+a^2=0.$$

19. Write down the equation of a system of spheres passing through the circle $(z=0, x^2+y^2+2px+2qy+d=0)$, and prove that the spheres of the system cut the plane $y=0$ in a system of coaxal circles.

Ans. $x^2+y^2+z^2+2px+2qy+2\lambda z+d=0.$

20. Show that every sphere through the circle
$$(z=0,\ x^2+y^2-2ax+r^2=0)$$
cuts orthogonally every sphere through the circle
$$(y=0,\ x^2+z^2=r^2).$$

21. Show that every sphere through the circle
$$(z=0,\ x^2+y^2=a^2)$$
cuts orthogonally every sphere through the circle
$$\{y=0,\ (x-\lambda)^2+z^2=\lambda^2-a^2\}.$$

22. Find the limiting-points of the coaxal system of spheres determined by the two spheres
$$(x-a)^2+y^2+z^2=c^2 \text{ and } (x-a')^2+y^2+z^2=c'^2.$$
Ans. $[(a+\lambda a')/(1+\lambda),\ 0,\ 0]$, where λ is a root of the equation
$$c'^2\lambda^2+\{c^2+c'^2-(a-a')^2\}\lambda+c^2=0.$$

23. Show that all the spheres, that can be drawn through the origin and each set of points where planes parallel to the plane $x/a+y/b+z/c=0$ cut the coordinate-axes, form a system of spheres which are cut orthogonally by the sphere
$$x^2+y^2+z^2+2fx+2gy+2hz=0$$
if $af+bg+ch=0$.   (Math. Trip. I, 1914.)

24. If a tetrahedron is self-polar with respect to a sphere show that it has an orthocentre which is the centre of the sphere.

25. Find the equation of the sphere through the origin O and three points A, B, C whose coordinates are $[a,\ 0,\ 0]$, $[0,\ b,\ 0]$, $[0,\ 0,\ c]$.
Show that, if O' is the centre of this sphere, the sphere on OO' as diameter passes through the mid-points of the six edges of the tetrahedron $OABC$, and through the feet of the perpendiculars from O on the sides of the triangle ABC.
                    (Math. Trip. I, 1915.)
Ans. $x^2+y^2+z^2-ax-by-cz=0.$

26. If $ABCDE$ is an orthocentric pentad show that the five spheres for which the tetrahedra $BCDE$, etc., are self-polar are mutually orthogonal.

27. If a tetrahedron has an orthocentre show that this divides the line joining the circumcentre and the centroid externally in the ratio 2 to 1.

28. If the mid-points of the six edges of a tetrahedron lie on a sphere show that the centre of this sphere is the centroid of the tetrahedron. Show also that the tetrahedron has in this case an orthocentre and that the sphere through the mid-points of the edges passes also through the feet of the altitudes.

CHAPTER V

THE CONE AND CYLINDER

5·1. A cone is a surface generated by a line which passes through a fixed point, the *vertex*, and through the points of a fixed curve.

There is no loss of generality in taking the guiding curve as a plane curve since any arbitrary plane section of the surface can be taken as guiding curve. Take the vertex as origin, and as guiding curve a curve in the plane $z = c$, with equations $z = c$, $f(x, y) = 0$. Let $P \equiv [x', y', z']$ be any point on the surface. OP cuts the plane $z = c$ where $f(x, y) = 0$, hence, t being the parameter of this point on the line OP, $tz' = c$ and $f(tx', ty') = 0$. Eliminating t we have $f(cx'/z', cy'/z') = 0$, or, dropping the accents,
$$f(cx/z, cy/z) = 0.$$
This equation is *homogeneous* in x, y, z.

Conversely a homogeneous equation in x, y, z represents a cone whose vertex is at the origin, for if it is satisfied by
$$[x, y, z] = [l, m, n],$$
it is satisfied by the coordinates of all points on the line
$$x/l = y/m = z/n.$$

If the guiding curve is a plane curve of degree n, the equation of the cone is also of degree n, and we call it a cone of order n. A cone of order n is cut by an arbitrary plane through its vertex in n generating lines.

If the guiding curve is determined by two equations
$$f_1(x, y, z) = 0, \ f_2(x, y, z) = 0$$
the equation of the cone with vertex at the origin is found by making these equations homogeneous by writing x/w, y/w, z/w instead of x, y, z, and then eliminating w; for the resulting equation is homogeneous in x, y, z and is satisfied by the co-ordinates of any point on the guiding curve.

If the vertex is at the point $[X, Y, Z]$ we may transform first to this point as origin and proceed as before.

We may also determine the equation by the following method.

Ex. To find the equation of the cone with vertex $[X, Y, Z]$ and guiding curve the conic

$$z=0, \quad f(x, y) \equiv ax^2 + by^2 + 2hxy + 2gx + 2fy + c = 0.$$

Using homogeneous coordinates, the coordinates of any point on the cone are given by

$$\rho x = X + \lambda x',$$
$$\rho y = Y + \lambda y',$$
$$\rho z = Z + \lambda z',$$
$$\rho w = W + \lambda w',$$

where $z' = 0$, and x', y', w' satisfy the equation

$$F(x', y', w') = ax'^2 + by'^2 + 2hx'y' + 2gx'w' + 2fy'w' + cw'^2 = 0.$$

Now $Z = \rho z$, $\lambda x' = \rho x - X$, therefore, eliminating ρ,

$$\lambda z x' = Zx - Xz.$$

Similarly $$\lambda z y' = Zy - Yz,$$

and $$\lambda z w' = Zw - Wz.$$

Hence the required equation is

$$F(Zx - Xz, \ Zy - Yz, \ Zw - Wz) = 0,$$

i.e. by Taylor's theorem

$$Z^2 F(x, y, w) - Zz \left(X \frac{\partial F}{\partial x} + Y \frac{\partial F}{\partial y} + W \frac{\partial F}{\partial w} \right) + z^2 F(X, Y, W) = 0,$$

or in non-homogeneous coordinates

$$Z^2 f(x, y) - Zz \left(x \frac{\partial F}{\partial X} + y \frac{\partial F}{\partial Y} + \frac{\partial F}{\partial W} \right) + z^2 f(X, Y) = 0.$$

5·11. The general equation of a cone of the second order with vertex at the origin is

$$f(x, y, z) \equiv ax^2 + by^2 + cz^2 + 2fyz + 2gzx + 2hxy = 0.$$

5·12. Cone of revolution or circular cone.

Let $[l, m, n]$ be the direction-cosines of the axis, 2α the vertical angle. Then if $P \equiv [x, y, z]$ is any point on the cone we have

$$\cos\alpha = \frac{lx + my + nz}{(x^2 + y^2 + z^2)^{\frac{1}{2}} (l^2 + m^2 + n^2)^{\frac{1}{2}}}.$$

Hence the equation of the cone is

$$(x^2 + y^2 + z^2)(l^2 + m^2 + n^2)\cos^2\alpha - (lx + my + nz)^2 = 0.$$

5·121. Conversely the equation

$$x^2 + y^2 + z^2 - (lx + my + nz)^2 = 0$$

represents a cone of revolution or circular cone whose axis is $[l, m, n]$ and vertical semi-angle α given by

$$\cos^2\alpha = (l^2 + m^2 + n^2)^{-1}.$$

5·122. *Condition that the general homogeneous equation of the second degree in x, y, z should represent a circular cone.*

Comparing with 5·121 we have

$$\lambda a = 1 - l^2, \quad \lambda f = -mn,$$
$$\lambda b = 1 - m^2, \quad \lambda g = -nl,$$
$$\lambda c = 1 - n^2, \quad \lambda h = -lm.$$

Therefore $\qquad\qquad l^2 = -\lambda gh/f$, etc.

Eliminating l, $\qquad\qquad \lambda(af - gh) = f.$

Equating this value of λ to two other values obtained similarly by eliminating m and n we have, provided none of the coefficients f, g, h vanishes,

$$\frac{af - gh}{f} = \frac{bg - hf}{g} = \frac{ch - fg}{h},$$

or $\qquad\qquad F/f = G/g = H/h,$

where $\qquad F \equiv gh - af, \quad G \equiv hf - bg, \quad H \equiv fg - ch.$

Conversely, if f, g and h are all finite, these conditions secure that the cone is circular.

If $f = 0$, then either m or $n = 0$, and hence either h or $g = 0$.

If $g = 0$ and $h = 0$, then $l = 0$, and we find the condition

$$f^2 = (a - b)(a - c).$$

If f, g and h all vanish, then two of a, b, c must be equal.

5·123. From an examination of the equation of a circular cone it is seen that it cuts the plane at infinity in a conic which has double contact with the absolute circle $x^2 + y^2 + z^2 = 0$, the points of contact being on the line $lx + my + nz = 0$, $w = 0$. The conditions for a circular cone can therefore be obtained from those for double contact of two conics.

Let $S_1 = 0$ and $S_2 = 0$ denote two conics, and consider the equation $S_1 + \lambda S_2 = 0$, which represents a pencil of conics through their common points. A conic of the pencil will degenerate to two straight lines if λ has a value which makes the matrix

$$\begin{bmatrix} a_1+\lambda a_2 & h_1+\lambda h_2 & g_1+\lambda g_2 \\ h_1+\lambda h_2 & b_1+\lambda b_2 & f_1+\lambda f_2 \\ g_1+\lambda g_2 & f_1+\lambda f_2 & c_1+\lambda c_2 \end{bmatrix}$$

of rank 2, i.e. for which the determinant vanishes. This gives a cubic equation in λ and the three roots correspond to the three pairs of common chords. If the conics have double contact two pairs of common chords coincide with the chord of contact while the other pair are the tangents at the points of contact. The cubic has then two equal roots, and the matrix is of rank 1. Or thus:

$$(a_1+\lambda a_2)x+(h_1+\lambda h_2)y+(g_1+\lambda g_2)z=0$$

represents the polar of the point [1, 0, 0] with respect to the conic of the system with the parameter λ. When this conic breaks up into two straight lines the polars of all points are concurrent at their point of intersection, and in particular the polars of [1, 0, 0], [0, 1, 0], and [0, 0, 1] are concurrent. The condition for this is that the matrix should be of rank 2. When the conic degenerates to two coincident lines, the polars of all points coincide, and the condition for this is that the matrix should be of rank 1.

Hence the conditions that the equation

$$ax^2+by^2+cz^2+2fyz+2gzx+2hxy=0$$

should represent a circular cone are that for some value of λ the matrix

$$\begin{bmatrix} a+\lambda & h & g \\ h & b+\lambda & f \\ g & f & c+\lambda \end{bmatrix}$$

should be of rank 1.

Ex. If the general homogeneous equation in x, y, z represents a circular cone, prove that the direction of the axis is $[f^{-1}, g^{-1}, h^{-1}]$.

5·2. *Intersection of the cone*

$$f(x, y, z)=ax^2+by^2+cz^2+2fyz+2gzx+2hxy=0$$

with the plane $\qquad lx+my+nz=0$

which passes through the vertex.

l, m, n are not all zero. If n is not zero, eliminate z, and we obtain the quadratic equation

$$n^2(ax^2+by^2+2hxy) - 2n(gx+fy)(lx+my) + c(lx+my)^2 = 0,$$

i.e.

$$x^2(cl^2 - 2gnl + an^2) + 2xy(hn^2 - gmn - fnl + clm) \\ + y^2(bn^2 - 2fmn + cm^2) = 0.$$

This equation determines two values for the ratio y/x, say y_1/x_1 and y_2/x_2, and the equation of the plane then gives corresponding values for z/x. We thus get two sets of values of the ratios $x:y:z$. The plane thus cuts the cone in two generating lines, with direction-cosines proportional to $[x_1, y_1, z_1]$ and $[x_2, y_2, z_2]$.

There are two particular cases of importance.

5·21. If the two generating lines coincide, the plane is a *tangent-plane* to the cone, touching at all points of this generating line. The condition for this is that the equation in y/x should have equal roots, hence

$$(hn^2 - gmn - fnl + clm)^2 - (bn^2 - 2fmn + cm^2)(cl^2 - 2gnl + an^2) = 0.$$

We find that n^2 is a factor of the left-hand side. Rejecting this factor, which is not zero, we obtain the equation

$$(bc-f^2)l^2 + (ca-g^2)m^2 + (ab-h^2)n^2 + 2(gh-af)mn \\ + 2(hf-bg)nl + 2(fg-ch)lm = 0.$$

If capital letters denote the cofactors of the corresponding small letters in the determinant

$$D \equiv \begin{vmatrix} a & h & g \\ h & b & f \\ g & f & c \end{vmatrix}$$

the equation can be written

$$\phi(l, m, n) \equiv Al^2 + Bm^2 + Cn^2 + 2Fmn + 2Gnl + 2Hlm = 0.$$

More generally the conditions that the plane

$$lx + my + nz + p = 0$$

should be a tangent-plane to the cone are $p = 0$ together with the equation $\phi(l, m, n) = 0$. These two equations taken together are called the *tangential equations* of the cone.

5·22. If the two generating lines are at right angles,

$$x_1 x_2 + y_1 y_2 + z_1 z_2 = 0.$$

Now from the equation in y/x

$$\frac{x_1 x_2}{y_1 y_2} = \frac{bn^2 - 2fmn + cm^2}{cl^2 - 2gnl + an^2}.$$

Hence if we put $\lambda x_1 x_2 = bn^2 - 2fmn + cm^2,$

then $\lambda y_1 y_2 = cl^2 - 2gnl + an^2,$

and from symmetry

$$\lambda z_1 z_2 = am^2 - 2hlm + bl^2.$$

Hence the condition that the two generating lines should be at right angles is

$$(b+c)l^2 + (c+a)m^2 + (a+b)n^2 - 2fmn - 2gnl - 2hlm = 0$$

or $$(a+b+c)(l^2 + m^2 + n^2) - f(l, m, n) = 0.$$

5·23. Angle between two generating lines.

If θ is the angle between the two lines, then

$$\cos\theta = (x_1 x_2 + y_1 y_2 + z_1 z_2)(\Sigma x_1^2 \cdot \Sigma x_2^2)^{-\frac{1}{2}}.$$

Now $$\Sigma x_1^2 \cdot \Sigma x_2^2 = \Sigma(x_1 y_2 - x_2 y_1)^2 + (\Sigma x_1 x_2)^2.$$

Also the sum of the roots of the equation in y/x is

$$\frac{x_1}{y_1} + \frac{x_2}{y_2} = -2\frac{hn^2 - gmn - fnl + clm}{an^2 - 2gnl + cl^2},$$

therefore

$$\lambda(x_1 y_2 + x_2 y_1) = -2(hn^2 - gmn - fnl + clm)$$

and $$\lambda^2(x_1 y_2 - x_2 y_1)^2 = \lambda^2\{(x_1 y_2 + x_2 y_1)^2 - 4x_1 x_2 y_1 y_2\}$$

$$= 4(hn^2 - gmn - fnl + clm)^2 - 4(bn^2 - 2fmn + cm^2)(cl^2 - 2gnl + an^2)$$

$$= -4n^2\phi(l, m, n).$$

Hence

$$\lambda^2\Sigma x_1^2 \cdot \Sigma x_2^2 = -4\Sigma l^2 \cdot \phi(l, m, n) + \{\Sigma(b+c)l^2 - 2\Sigma fmn\}^2.$$

Therefore finally

$$\cos\theta = \frac{(a+b+c)(l^2 + m^2 + n^2) - f(l, m, n)}{[\{(a+b+c)(l^2 + m^2 + n^2) - f(l, m, n)\}^2 - 4\Sigma l^2 \cdot \phi(l, m, n)]^{\frac{1}{2}}},$$

or $$\tan\theta = \frac{2\{-\Sigma l^2 \cdot \phi(l, m, n)\}^{\frac{1}{2}}}{\Sigma a \cdot \Sigma l^2 - f(l, m, n)}.$$

5·231. θ is real or imaginary, and therefore the lines of intersection are real or imaginary, according as

$$\phi(l, m, n) < \text{ or } > 0.$$

5·3. Polar of a point with regard to a cone.

The polar of a point $P \equiv [x', y', z']$ is defined as the locus of harmonic conjugates of P with respect to the pairs of points in which a variable line through P cuts the surface.

Let U, V be the points of intersection of any line through P, and $Q \equiv [x, y, z]$ any point on this line. If X divides PQ in the ratio $k : 1$ the coordinates of U are

$$\frac{kx + x'}{k + 1}, \text{ etc.}$$

Substituting in the equation of the cone we have

$$a(kx + x')^2 + \dots + 2f(ky + y')(kz + z') + \dots = 0.$$

Hence

$$k^2 f(x, y, z) + 2k\{axx' + byy' + czz' + f(yz' + y'z)$$
$$+ g(zx' + z'x) + h(xy' + x'y)\} + f(x', y', z') = 0.$$

The roots of this equation correspond to the two points U, V. If (PQ, UV) is harmonic, the two values of k are equal and of opposite sign, therefore

$$axx' + \dots + f(yz' + y'z) + \dots = 0.$$

Hence the equation of the polar, the locus of Q, is

$$x(ax' + hy' + gz') + y(hx' + by' + fz') + z(gx' + fy' + cz') = 0,$$

which may also be written

$$x\frac{\partial f}{\partial x'} + y\frac{\partial f}{\partial y'} + z\frac{\partial f}{\partial z'} = 0.$$

Hence the polar of any point is a plane passing through the vertex, and the polar-planes of all points on a given line through the vertex coincide. If the point P lies on the cone, the polar-plane becomes the tangent-plane at that point.

5·4. Reciprocal cones.

The condition that the plane $lx + my + nz = 0$ should be a tangent-plane to the cone $f(x, y, z) = 0$ is

$$\phi(l, m, n) = 0.$$

The line $x/l = y/m = z/n$ is normal to this plane. Hence the normals at O to the tangent-planes are generators of a cone whose equation is

$$\phi(x, y, z) \equiv Ax^2 + By^2 + Cz^2 + 2Fyz + 2Gzx + 2Hxy = 0.$$

This cone is called the *reciprocal cone*. The relation between the two cones is a mutual one, for $BC - F^2 = aD$, $GH - AF = fD$, etc., where D stands for the determinant

$$\begin{vmatrix} a & h & g \\ h & b & f \\ g & f & c \end{vmatrix}.$$

The intersection of the cone $f(x, y, z) = 0$ with the plane at infinity $w = 0$ is a conic, the conic at infinity on the cone. The two conics f and ϕ are reciprocals with respect to the virtual conic $\Omega \equiv x^2 + y^2 + z^2 = 0$, the circle at infinity, in the usual sense, viz. that the polar with respect to Ω of any point on f is a tangent to ϕ, and *vice versa*.

5·41.   The reciprocal of a surface F in the corresponding sense is defined as that surface F' which is such that the polar of any point on F with respect to the virtual sphere

$$x^2 + y^2 + z^2 + w^2 = 0$$

is a tangent-plane to F', and *vice versa*. In this sense we find that the reciprocal of the cone $f(x, y, z) = 0$ is the assemblage of planes $[l, m, n, p]$ such that $f(l, m, n) = 0$, for the polar-plane of $[x', y', z', w']$ is

$$x'x + y'y + z'z + w'w = 0,$$

i.e. the plane

$$[l, m, n, p] = [x', y', z', w'],$$

and since $f(x', y', z') = 0$ we have the equation $f(l, m, n) = 0$.

But this is the condition that the plane $[l, m, n, 0]$ should be a tangent-plane to the cone $\phi(x, y, z) = 0$; hence the assemblage of planes $[l, m, n, p]$ which satisfy this condition are these tangent-planes and all planes parallel to them. These planes do not envelop a surface but are tangents to the conic in which $\phi(x, y, z) = 0$ cuts the plane at infinity. Thus the reciprocal of a cone is not a surface but a plane curve.

5·5. Rectangular generators.

The plane $lx + my + nz = 0$

cuts the cone $f(x, y, z) = 0$ ......(1)

in rectangular generators if

$$\Sigma a . \Sigma l^2 - f(l, m, n) = 0.$$

The normal to the plane at O is therefore a generator of the cone

$$\Sigma a . \Sigma x^2 - f(x, y, z) = 0. \qquad \ldots\ldots(2)$$

Hence all the planes through O which cut the given cone in rectangular generators touch the cone which is the reciprocal of (2). To find the reciprocal of (2) we have

$$(c+a)(a+b) - f^2 = A + a\Sigma a,$$

$$gh + (b+c)f = F + f\Sigma a.$$

Hence the reciprocal of (2) is

$$\Sigma a . f(x, y, z) + \phi(x, y, z) = 0. \qquad \ldots\ldots(3)$$

A tangent-plane to (3) thus cuts (1) in two rectangular generators.

5·51. If we attempt to get a set of three mutually rectangular generators we must choose the tangent-plane to (3) so that its normal is a generator of the given cone (1). But this normal is a generator of (2), and the generators which are common to (1) and (2) also belong to the absolute cone $\Sigma x^2 = 0$. Every generator of the absolute cone is orthogonal to itself and lies in its own normal plane (just as a point on a conic lies on its own polar). Hence we fail in general to obtain a set of three distinct mutually rectangular generators. Not only are they imaginary, but in any such set two coincide.

5·52. If, however, $a + b + c = 0$, the two cones (1) and (2) coincide, and the normal is always a third generator, i.e. if we take any generator, the plane perpendicular to it through the vertex cuts the cone in rectangular generators. The cone is then said to be a *rectangular cone*. We have thus the poristic theorem: *a cone possesses either no set of three mutually rectangular generators, or an unlimited number.*

5·53. If the reciprocal cone is rectangular, the given cone has the property of possessing an unlimited number of sets of three mutually orthogonal tangent-planes. The condition for this is $A+B+C=0$, and we shall say that in this case the cone is an *orthogonal cone*.

The equation of a rectangular cone referred to a set of three mutually rectangular generators is

$$fyz+gzx+hxy=0;$$

and the equation of an orthogonal cone referred to a set of three mutually orthogonal tangent-planes is

$$p^2x^2+q^2y^2+r^2z^2-2qryz-2rpzx-2pqxy=0.$$

Ex. 1. If OX, OY, OZ and OP, OQ, OR are two sets of three mutually perpendicular lines, prove that they are all generators of the same rectangular cone.

Take OX, OY, OZ as coordinate-axes, and let the direction-cosines of OP, etc., be $[l_1, m_1, n_1]$, etc. Then the direction-cosines of OX, OY, OZ referred to OP, OQ, OR are $[l_1, l_2, l_3]$, etc. Hence

$$\left.\begin{aligned}m_1n_1+m_2n_2+m_3n_3&=0\\n_1l_1+n_2l_2+n_3l_3&=0\\l_1m_1+l_2m_2+l_3m_3&=0\end{aligned}\right\}\qquad \ldots\ldots(1)$$

The equation of a rectangular cone containing OX, OY, OZ is

$$fyz+gzx+hxy=0,$$

and if it contains OP, OQ, OR, we have

$$fm_1n_1+gn_1l_1+hl_1m_1=0,$$
$$fm_2n_2+gn_2l_2+hl_2m_2=0,$$
$$fm_3n_3+gn_3l_3+hl_3m_3=0.$$

But these equations can be simultaneously satisfied since when we add the left-hand sides together the result vanishes in virtue of (1); hence the ratios $f:g:h$ are uniquely determined.

The reciprocal theorem is that two sets of three mutually orthogonal planes through the same point are tangent-planes to one cone.

Ex. 2. If a circular cone is rectangular prove that its vertical semi-angle $=\cos^{-1}\dfrac{1}{\sqrt{3}}$, and if it is orthogonal that the vertical semi-angle $=\sin^{-1}\dfrac{1}{\sqrt{3}}$.

5·6. The geometry of cones with a common vertex is projectively equivalent to the geometry of conics in a plane. To every cone corresponds its conic at infinity C. To reciprocal cones correspond two conics at infinity which are reciprocal with regard to the circle at infinity Ω. A triangle which is self-conjugate with respect to Ω corresponds to a set of three mutually rectangular lines through O. Two conics have in general a unique common self-conjugate triangle. If C and Ω are referred to their common self-conjugate triangle their equations are of the form

$$a'x^2 + b'y^2 + c'z^2 = 0,$$

$$x^2 + y^2 + z^2 = 0.$$

The equation of a cone can therefore always be reduced to the form

$$ax^2 + by^2 + cz^2 = 0.$$

The envelope of lines which cut two conics

$$f(x, y, z) \equiv ax^2 + \ldots + 2fyz + \ldots = 0,$$

$$f'(x, y, z) \equiv a'x^2 + \ldots + 2f'yz + \ldots = 0,$$

in a harmonic range has for its tangential equation

$$(bc' + b'c - 2ff')l^2 + \ldots + 2(gh' + g'h - af' - a'f)mn + \ldots = 0.$$

This is a conic called the *harmonic conic-envelope* of the given conics*. When the conic f' is the circle at infinity

$$\Omega \equiv x^2 + y^2 + z^2 = 0,$$

the equation of the harmonic conic-envelope becomes

$$(b + c)l^2 + \ldots - 2fmn - \ldots = 0,$$

the equation found above, as the condition for rectangular generators.

To a rectangular cone corresponds a conic C which has an infinity of inscribed triangles each self-conjugate with respect to the circle at infinity Ω. The conic-locus C and the conic-envelope Ω are then said to be *apolar*; C is also said to be *outpolar* to Ω and Ω *inpolar* to C. The condition that the conic-locus f and the conic-envelope ϕ' should be apolar is

$$aA' + \ldots + 2fF' + \ldots = 0.$$

---

* See Sommerville's *Analytical Conics*, chap. xx, § 18.

For $A'=B'=C'$ and $F'=G'=H'=0$, this reduces to $a+b+c=0$, the condition for a rectangular cone. Similarly to an orthogonal cone corresponds a conic which has an infinity of circumscribed triangles each self-conjugate with respect to the circle at infinity.

5·7. Cylinders.

A cylinder is a surface generated by a line which passes through points of a fixed curve and is parallel to a fixed direction, the axis. It can therefore be regarded as a cone whose vertex is a point at infinity. Let the given axis be taken as the axis of z, and the guiding curve the plane curve $f(x, y)=0$ in the plane of xy. Let $P\equiv[x', y', z']$ be any point on the surface, then the line $x=x'$, $y=y'$ through P parallel to the axis cuts $z=0$ where $f(x, y)=0$. Hence $f(x', y')=0$, or, dropping the accents,

$$f(x, y)=0.$$

Hence the equation of a cylinder whose axis is the axis of z does not contain z. If we express the equation in homogeneous coordinates it will therefore be homogeneous in x, y, w. We may say that a homogeneous equation in three variables x_0, x_1, x_2 represents a cone with vertex $x_0=0$, $x_1=0$, $x_2=0$; if this is a point at infinity the surface is a cylinder.

5·71. As a quadric cone is cut in two straight lines by any plane through its vertex, so a quadric cylinder is cut in two straight lines by any plane which contains the point at infinity on its axis. One such plane is the plane at infinity. Thus a quadric cylinder cuts the plane at infinity in two straight lines, and the nature of the cylinder will depend upon whether these two lines are real, coincident, or imaginary. Thus, while there is only one type of real cone of the second order, there are three types of cylinders.

When the two lines at infinity are real and distinct, every plane (not parallel to the axis) cuts the surface in a conic which has two real and distinct points at infinity and is therefore a hyperbola. The cylinder in this case is a *hyperbolic cylinder*. The tangent-planes along the lines at infinity are *asymptotic planes*.

When the two lines at infinity are imaginary, the sections are ellipses, and the cylinder is an *elliptic cylinder*. In this case the cylinder has only one real point at infinity, the point at infinity on its axis.

When the two lines at infinity coincide, every section is a conic which meets the line at infinity in its plane in two coincident points, and is therefore a parabola; the cylinder is a *parabolic cylinder*. In this case as the plane at infinity meets the surface in two coincident generating lines it is a tangent-plane, i.e. a parabolic cylinder touches the plane at infinity along a line at infinity.

If the two lines at infinity (necessarily real in this case) are conjugate with respect to the circle at infinity, the asymptotic planes are at right angles, and we have a *rectangular hyperbolic cylinder*.

If the two lines at infinity (necessarily imaginary in this case) are tangents to the circle at infinity—for a real cylinder, if one is a tangent, both must be tangents—we have a *circular cylinder*. In this case a plane through the chord of contact (the polar of the point at infinity on the axis) is perpendicular to the axis and cuts the cylinder in a conic whose points at infinity are the circular points, i.e. a circle.

5·72. The general equation of a quadric cylinder with the axis of z for axis is

$$f(x, y) \equiv ax^2 + 2hxy + by^2 + 2gx + 2fy + c = 0.$$

The polar-plane of $[x', y', z']$ (or the tangent-plane at this point if it is a point of the surface) is

$$x(ax' + hy' + g) + y(hx' + by' + f) + (gx' + fy' + c) = 0.$$

The conditions that the general plane $lx + my + nz + p = 0$ should be a tangent-plane are
$$n = 0,$$

and
$$Al^2 + 2Hlm + Bm^2 + 2Glp + 2Fmp + Cp^2 = 0,$$

where $A = bc - f^2$, etc., and $F = gh - af$, etc.

Thus, like the cone, the cylinder requires two equations to express the tangential condition. The equation $n = 0$ expresses that the plane is parallel to the axis. The other equation is the condition that the line of intersection $lx + my + p = 0$ with the plane $z = 0$ should touch the conic $f(x, y) = 0$, $z = 0$. This equation by itself then is the tangential equation of this conic.

5·73. The reciprocal of the cylinder with respect to the virtual sphere $x^2+y^2+z^2+w^2=0$ is the assemblage of planes $[l, m, n, p]$ such that
$$al^2 + 2hlm + bm^2 + 2glp + 2fmp + cp^2 = 0.$$
But this is the condition that the plane should envelop the conic
$$Ax^2 + 2Hxy + By^2 + 2Gx + 2Fy + C = 0$$
since $Da \equiv BC - F^2$, etc.

Hence, as in the case of a cone, the reciprocal of a cylinder is a conic.

5·74. Both the cone and the cylinder, while they are two-dimensional assemblages of points, or loci, are only one-dimensional assemblages of tangent-planes. The reciprocals are two-dimensional envelopes but only one-dimensional loci, i.e. they are curves. They are, as we shall see later, particular cases of *developables*, or surfaces which can be developed or laid flat on a plane without stretching or tearing.

5·9. EXAMPLES.

1. Find the equation of the cone with vertex $[x', y', z']$ and base $z=0$, $x^2/a^2+y^2/b^2=1$.

Ans. $(z-z')^2 = (xz'-x'z)^2/a^2 + (yz'-y'z)^2/b^2$.

2. Find the equation of the circular cone with vertex $[x', y', z']$, vertical semi-angle α, and direction-cosines of axis $[l, m, n]$.

Ans. $\{\Sigma l(x-x')\}^2 = \Sigma l^2 . \Sigma (x-x')^2 \cos^2\alpha$.

3. Find the equation of the cone which contains the three coordinate-axes and the lines through the origin having direction-ratios $[l_1, m_1, n_1]$ and $[l_2, m_2, n_2]$.

Ans. $\Sigma l_1 l_2 (m_1 n_2 - m_2 n_1) yz = 0$.

4. Find the equations of the lines of intersection of the plane $x+y-z=0$ with the cone $yz+6zx-12xy=0$.

Ans. $x=y/3=z/4$, $x=y/2=z/3$.

5. Find the direction-ratios of the lines of intersection of the plane (i) $x-5y+3z=0$ with the cone $7x^2+5y^2-3z^2=0$, (ii) $5x-11y-7z=0$ and $25yz-38zx+66xy=0$.

Ans. (i) $[1, 2, 3]$, $[-1, 1, 2]$, (ii) $[5, 1, 2]$, $[77, 266, -363]$.

6. Find the angle between the lines of intersection of

$$\text{(i)} \qquad x - 3y + z = 0 \quad \text{and} \qquad x^2 - 5y^2 + z^2 = 0,$$

$$\text{(ii)} \quad x - (a+1)y + z = 0 \quad \text{and} \quad x^2 - (a^2+1)y^2 + z^2 = 0,$$

$$\text{(iii)} \qquad 6x + 3y - 2z = 0 \quad \text{and} \qquad yz + zx + xy = 0,$$

$$\text{(iv)} \qquad x + y + z = 0 \quad \text{and} \qquad 3yz - 2zx - xy = 0,$$

$$\text{(v)} \qquad x + y + z = 0 \quad \text{and} \qquad 6yz + 3zx - 2xy = 0,$$

$$\text{(vi)} \qquad lx + my + nz = 0 \quad \text{and} \qquad yz + zx + xy = 0,$$

$$\text{(vii)} \qquad x + y + z = 0 \quad \text{and} \quad (\lambda+1)(\lambda x + y)z = \lambda xy.$$

Ans. (i) $\cos^{-1}\frac{5}{8}$, (ii) $\cos^{-1}(2a+1)/(a^2+2)$, (iii) $90°$, (iv) $90°$, (v) $60°$, (vi) $\cos^{-1}\Sigma mn/(\Sigma l^2 - \Sigma mn)$, (vii) $60°$.

7. Show that the cone $x^2 - z^2 + 3xy = 0$ cuts the sphere $x^2 + y^2 + z^2 - 2x - 4 = 0$ in two circles.

8. Show that the cone $yz + zx + xy = 0$ cuts the sphere $x^2 + y^2 + z^2 = r^2$ in two equal circles, and find their radius.
Ans. $\frac{1}{3}\sqrt{6}r$.

9. Show that
$$x^2 + 2y^2 + z^2 - 4yz - 6zx - 2x + 8y - 2z + 9 = 0$$
represents a cone and find its vertex.
Ans. $[1, -2, 0]$.

10. Show that any plane through the vertex perpendicular to a generating line of the cone $6yz - 2zx + 5xy = 0$ cuts it in two lines which are at right angles.

11. Find the equation of a cone with vertex at the origin and base a circle in the plane $z = 12$, with centre $[13, 0, 12]$ and radius $= 5$; and show that the section by any plane parallel to $x = 0$ is a circle.
Prove that a sphere can be drawn through the sections of the cone made by the two planes $z = 12$ and $x = 6$, and find its equation.
Ans. Cone $6(x^2 + y^2 + z^2) = 13xz$,
sphere $x^2 + y^2 + z^2 - 26x - 13z + 156 = 0$.

12. Prove that the straight lines which cut two given non-intersecting straight lines, such that the length intercepted is constant, are parallel to the generators of a circular cone.

(Math. Trip. I, 1914.)

13. Prove that the locus of a point whose distance from a fixed line is in a fixed ratio to its distance from a fixed plane is a cone. Examine the cases when the line (i) is parallel to the plane, (ii) lies in the plane.

Ans. (i) A cylinder, (ii) two planes.

14. Show that the locus of vertices of circular cones which contain the ellipse $z=0$, $x^2/a^2+y^2/b^2=1$ $(a>b)$ consists of the virtual conic $x=0$, $y^2/(a^2-b^2)+z^2/a^2+1=0$ and the hyperbola $y=0$, $x^2/(a^2-b^2)-z^2/b^2=1$.

15. Find the locus of the vertex of a rectangular cone which passes through a given conic and through a given point.

Ans. A conic.

16. Find the locus of the vertex of an orthogonal cone which passes through a given conic.

Ans. A sphere.

CHAPTER VI

TYPES OF SURFACES OF THE SECOND ORDER

6·1. Surfaces of revolution.

A surface of revolution is generated by a plane curve rotating about a straight line in its plane. The curve is called the *generating curve* or *meridian curve*. We shall assume that it is symmetrical about the axis of rotation.

Take the axis of z as axis of rotation, and the generating curve originally in the plane of yz. Let its equation in y, z be

$$y^2 = f(z).$$

Let $P \equiv [x, y, z]$ be any point on the surface, and let Q be the corresponding point on the generating curve; draw $QN \perp Oz$. The points generated by Q all lie on a circle with centre N, and the plane of this circle is perpendicular to Oz. Hence P and Q have the same z, and the coordinates of Q are $[0, y', z]$ where $NQ = y' = NP$. But $NP^2 = x^2 + y^2$ and $y'^2 = f(z)$. Hence

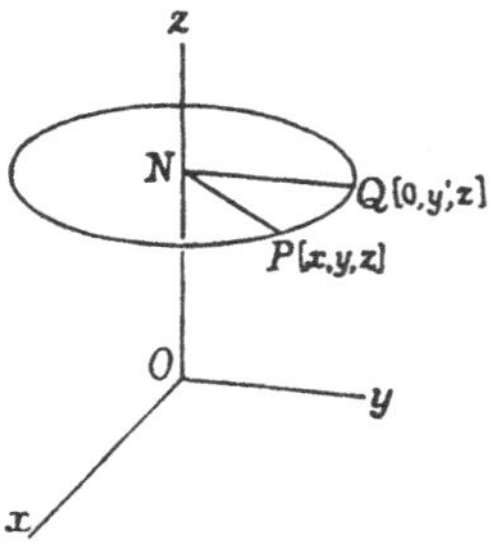

$$x^2 + y^2 = f(z).$$

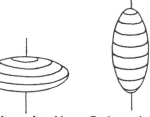

Fig. 21

We are particularly interested in the surfaces of revolution generated by conics.

6·11. *Surfaces of revolution of an ellipse about an axis.*

The ellipse $y^2/b^2 + z^2/c^2 = 1$, by revolution about the axis of z, generates the surface

$$\frac{x^2}{b^2} + \frac{y^2}{b^2} + \frac{z^2}{c^2} = 1,$$

which is called an *ellipsoid of revolution* or a *spheroid*: an *oblate spheroid* if $b < c$, a *prolate spheroid* if $b > c$.

Oblate spheroid   Prolate spheroid

Fig. 22

6·12. *Surfaces of revolution of a hyperbola about an axis.*

6·121. Axis of revolution the transverse axis.

The generating hyperbola is $-y^2/b^2 + z^2/c^2 = 1$, and it generates the surface

$$-\frac{x^2}{b^2} - \frac{y^2}{b^2} + \frac{z^2}{c^2} = 1.$$

The surface consists of two distinct parts and is called a *hyperboloid of revolution of two sheets*.

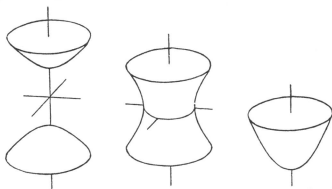

Fig. 23. Hyperboloid of two sheets.   Fig. 24. Hyperboloid of one sheet.   Fig. 25. Paraboloid of revolution

6·122. Axis of revolution the conjugate axis.

The generating hyperbola is $y^2/b^2 - z^2/c^2 = 1$, and it generates the surface

$$\frac{x^2}{b^2} + \frac{y^2}{b^2} - \frac{z^2}{c^2} = 1.$$

The two branches of the hyperbola generate the same surface, which consists of one continuous sheet, and is called a *hyperboloid of revolution of one sheet*.

6·13. *Surface of revolution of a parabola about its axis.*

The equation of the parabola is $y^2 = 4pz$, and it generates the surface $x^2 + y^2 = 4pz$, which consists of one infinite sheet, and is called a *paraboloid of revolution*.

6·14. The equations of these surfaces are all of the second degree. Any meridian plane cuts the surface in a conic congruent to the generating conic, and any plane perpendicular to the axis of revolution cuts it in a circle. If a conic is rotated about an axis which is not an axis of symmetry the generating conic will not come to coincide with itself again after a rotation through two right angles, and the meridian plane will cut the surface in two conics. These together form a curve of the fourth order, and the surface will be of the fourth order, i.e. its equation will be of the fourth degree.

Suppose, for example, that a parabola is rotated about the tangent at the vertex. Its equation is $z^2 = 4py$. The equation of the surface of revolution is formed by replacing y^2 by $x^2 + y^2$, hence it is

$$z^4 = 16p^2(x^2 + y^2),$$

an equation of the fourth degree.

Fig. 26. Surface of revolution of parabola about vertical tangent

As another example let the circle

$$(y - b)^2 + z^2 = a^2$$

be rotated about the axis of z. The equation of the surface of revolution is

$$\{\sqrt{(x^2 + y^2)} - b\}^2 + z^2 = a^2,$$

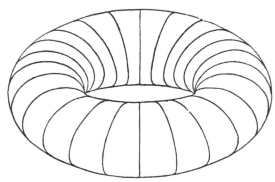

Fig. 27. Anchor-ring

which, on rationalising, becomes

$$(x^2 + y^2 + z^2 + b^2 - a^2)^2 = 4b^2(x^2 + y^2),$$

again an equation of the fourth degree. This surface is called an *anchor-ring* or *tore*.

6·15. A straight line rotated about an axis in the same plane will generate a surface of the second order.

If the equation of the line is $y = \mu z$, the equation of the surface of revolution is $x^2 + y^2 = \mu^2 z^2$. This is a *circular cone*.

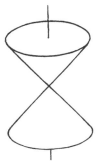

Fig. 28. Circular cone   Fig. 29. Circular cylinder

If the line is parallel to the axis, $y = b$, the equation of the surface of revolution is $x^2 + y^2 = b^2$, a *circular cylinder*.

6·2. These surfaces can now be generalised.

6·21. The equation

$$\frac{x^2}{a^2} + \frac{y^2}{b^2} + \frac{z^2}{c^2} = 1,$$

which represents a spheroid when two of the quantities a, b, c are equal, and a sphere when all three are equal, represents the general *ellipsoid*. Sections by planes parallel to any of the co-ordinate-planes are ellipses which are real only within distances a, b, c respectively from the origin. The surface is closed and finite.

6·22. The equation $-\dfrac{x^2}{a^2} - \dfrac{y^2}{b^2} + \dfrac{z^2}{c^2} = 1$

represents a *hyperboloid of two sheets*. Sections parallel to the yz or xz plane are hyperbolas, sections parallel to the xy plane are ellipses, which are real only at distances greater than c.

6·23. The equation $\dfrac{x^2}{a^2} + \dfrac{y^2}{b^2} - \dfrac{z^2}{c^2} = 1$

represents a *hyperboloid of one sheet*. Sections parallel to the xy plane are always real ellipses.

6·24. The equation of the paraboloid may be generalised to $ax^2+by^2=4pz$. When a, b are of the same sign the surface resembles the paraboloid of revolution, but sections perpendicular to the z-axis are now ellipses when $a \neq b$. It is called an *elliptic paraboloid*.

6·25. If a, b are of opposite sign we have a new type of surface, called the *hyperbolic paraboloid*. Let us write the equation

$$\frac{x^2}{a^2} - \frac{y^2}{b^2} = 2\frac{z}{c},$$

and suppose c to be positive. Sections parallel to the yz or zx plane are parabolas, as before, but turned in opposite directions. Sections parallel to the xy plane are hyperbolas; those on the positive side of the origin have their transverse axes parallel to the axis of x, while those on the negative side have their transverse axes parallel to the axis of y.

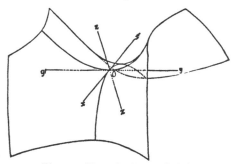

Fig. 30. Hyperbolic paraboloid

6·251. By a change of axes the equation of the hyperbolic paraboloid may be further simplified. Taking the planes

$$x/a - y/b = 0 \text{ and } x/a + y/b = 0$$

as coordinate-planes $x' = 0$ and $y' = 0$ (in general oblique), the equation assumes the form

$$x'y' = kz'.$$

6·3. Ruled surfaces. A surface may be generated also by the motion of a line or a plane. In the former case we have a *ruled surface*, in the latter an *envelope*. We shall consider at present certain simple cases of ruled surfaces of the second order.

6·31. *The surface generated by a straight line which rotates about an axis which it does not cut.*

Take the axis of rotation as axis of z, and let ON be the com-

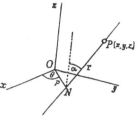

mon perpendicular of the two lines in any position. Then O is a fixed point and ON is of constant length $= p$. Take O as origin and the plane through O perpendicular to Oz as plane of xy. In this plane take two fixed rectangular lines as axes of x and y. The angle between the revolving line and the axis of rotation is also a constant $= \alpha$.

Fig. 31

The position of the line is determined by the variable angle $xON = \theta$. Let $P \equiv [x, y, z]$ be any point on the revolving line and let $NP = r$. Then

$$x = p\cos\theta - r\sin\alpha\sin\theta,$$
$$y = p\sin\theta + r\sin\alpha\cos\theta,$$
$$z = \qquad r\cos\alpha.$$

These are freedom-equations of the locus of P, in terms of the two parameters r, θ. Eliminating r and θ, we have

$$x^2 + y^2 = p^2 + r^2\sin^2\alpha = p^2 + z^2\tan^2\alpha,$$

i.e. $\qquad x^2 + y^2 - z^2\tan^2\alpha = p^2.$

The surface is therefore a hyperboloid of revolution of one sheet.

6·32. The general hyperboloid of one sheet may be generated by a moving line in the following way, as

The surface generated by a line which joins pairs of points with constant difference of eccentric angle on two equal and similarly placed ellipses in parallel planes.

Let the line joining the centres of the ellipses be perpendicular to their planes, and take this line as axis of z, the midpoint of the join of the centres as origin, and axes of x and y parallel to the principal axes.

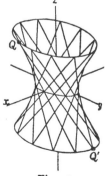

Fig. 32

The coordinates of the corresponding points Q, Q' are

$$Q : [a\cos(\phi-\alpha),\ b\sin(\phi-\alpha),\ c],$$
$$Q' : [a\cos(\phi+\alpha),\ b\sin(\phi+\alpha),\ -c].$$

The direction-cosines of QQ' are proportional to

$$a\sin\alpha\sin\phi,\ -b\sin\alpha\cos\phi,\ c.$$

If $P\equiv[x,\ y,\ z]$ is any point on QQ',

$$x/a = \cos(\phi-\alpha)+t\sin\alpha\sin\phi = \cos\alpha\cos\phi+(1+t)\sin\alpha\sin\phi,$$
$$y/b = \sin(\phi-\alpha)-t\sin\alpha\cos\phi = \cos\alpha\sin\phi-(1+t)\sin\alpha\cos\phi,$$
$$z/c = 1+t.$$

These are freedom-equations of the locus in terms of the two parameters ϕ, t. Eliminating ϕ and t, we have

$$x^2/a^2+y^2/b^2 = \cos^2\alpha+(1+t)^2\sin^2\alpha = \cos^2\alpha+z^2/c^2.\sin^2\alpha,$$

i.e.
$$\frac{x^2}{a^2}+\frac{y^2}{b^2}-\frac{z^2}{c^2\operatorname{cosec}^2\alpha} = \cos^2\alpha.$$

If the sign of α is changed evidently the equation is unaltered, hence the surface can be generated in two different ways; or, from another point of view, the surface is covered by two sets of straight lines, or systems of generating lines. Each system is called a *regulus*.

An effective model of this surface can be made by fixing two elliptic disks together rigidly in parallel planes and passing a continuous thread alternately through holes pierced on the two ellipses at equal intervals of eccentric angle.

6·33. The hyperbolic paraboloid also can be generated by a moving line.

$ABA'B'$ is a regular tetrahedron; Q and Q' are variable points on AB and $A'B'$ such that $AQ=A'Q'$. QQ' generates a hyperbolic paraboloid.

The lines joining the mid-points x, x'; y, y'; z, z' of opposite edges intersect at right angles. Take these lines as coordinate-axes. Let $Ox=Ox'=Oy=Oy'=Oz=Oz'=c$, $Az=$ etc. $=a\sqrt{2}$, $AQ=A'Q'=r\sqrt{2}$. Then the coordinates of Q and Q' are

$$Q : [a-r,\quad a-r,\quad c],$$
$$Q' : [a-r,\quad -a+r,\quad -c].$$

The direction-cosines of QQ' are proportional to $[0, a-r, c]$. Hence the coordinates of any point on QQ' are

$$x = a-r,$$
$$y = (a-r)(1+t),$$
$$z = c(1+t).$$

Eliminating r and t, we obtain

$$\frac{y}{z} = \frac{a-r}{c} = \frac{x}{c},$$

i.e.
$$zx = cy,$$

which is the equation of a hyperbolic paraboloid in the second form.

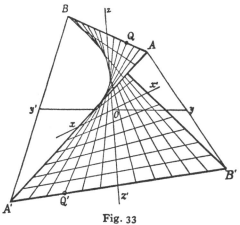

Fig. 33

This may be generated also in a second way. If Q and Q' are variable points on AA' and BB' such that $AQ = BQ'$, the equation of the surface generated by QQ' can be written down from the last result by interchanging x and z, but this does not alter the equation.

6·4. It remains now to examine whether the other surfaces can be generated by a moving line, or if straight lines exist upon them.

Consider first the equation

$$ax^2 + by^2 + cz^2 = k, \qquad \qquad \dots\dots(1)$$

which can represent, for suitable values of the coefficients, an ellipsoid, one of the two hyperboloids, a cone, or a cylinder.

Let
$$\frac{x-X}{l}=\frac{y-Y}{m}=\frac{z-Z}{n} \qquad \ldots\ldots(2)$$

be any straight line. Equating each of these ratios to t and substituting for x, y, z in equation (1), we obtain the quadratic in t:
$$a(lt+X)^2+b(mt+Y)^2+c(nt+Z)^2=k.$$

The line in general cuts the surface in two points. But if it lies entirely in the surface this equation must be true for all values of t and becomes an identity. Equating to zero the coefficients of the different powers of t, we obtain the three equations:

$$al^2+bm^2+cn^2=0, \qquad \ldots\ldots(3)$$
$$alX+bmY+cnZ=0, \qquad \ldots\ldots(4)$$
$$aX^2+bY^2+cZ^2=k. \qquad \ldots\ldots(5)$$

Equation (5) expresses that the point $[X, Y, Z]$ lies on the surface. If we choose any set of values of X, Y, Z satisfying this equation, the other two equations, being homogeneous in l, m, n, determine two sets of values for the ratios $l : m : n$, i.e. two directions through the point. Hence *through every point on the surface there pass two generating lines*. These may not, however, be real. Eliminating n between (3) and (4),

$$cZ^2(al^2+bm^2)+(alX+bmY)^2=0,$$

i.e. $a(aX^2+cZ^2)l^2+2abXYlm+b(bY^2+cZ^2)m^2=0.$

The roots of this equation will be real if

$$a^2b^2X^2Y^2-ab(aX^2+cZ^2)(bY^2+cZ^2)>0,$$

i.e. $-Z^2abc(aX^2+bY^2+cZ^2)>0,$

i.e. (by (5)) $abck<0.$

Taking k positive the generators will be real only if a, b, c are all negative, or one negative and two positive. The latter gives a hyperboloid of one sheet, the former a virtual quadric which has no real points. The other surfaces, ellipsoid and hyperboloid of two sheets, have imaginary generators. The generating lines through a real point of the ellipsoid and the hyperboloid of two sheets are imaginary lines of the first species since each has one real point upon it; those of the virtual quadric are imaginary lines of the second species.

If a, b, c or k is zero, the generators coincide. In the last case the surface is a cone, in the other cases a cylinder.

6·41. The paraboloids are represented by the equation

$$ax^2 + by^2 = 2cz,$$

and a similar investigation shows that the elliptic paraboloid, for which $ab > 0$, has imaginary generators.

Thus all the quadric surfaces are ruled surfaces, but the generating lines are real only in the cases of the hyperboloid of one sheet, the hyperbolic paraboloid, the cone, and the cylinder.

6·5. EXAMPLES.

1. The sum of the squares of the perpendiculars from a point to the lines $y = x\tan\theta$, $z = c$ and $y = -x\tan\theta$, $z = -c$ is $2k^2$. Prove that the locus of the point is an ellipsoid, and state the lengths of its principal axes.   (Math. Trip. I, 1915.)

Ans. $a\operatorname{cosec}\theta$, $a\sec\theta$, a, where $a^2 = k^2 - c^2$. A circular cylinder if the two lines are parallel.

2. Find the locus of the position of the eye at which two given non-intersecting lines will appear to cut at right angles.

Ans. A hyperboloid of one sheet with the given lines as generators and its centre at the midpoint of their common perpendicular. Reduces to two planes, one through each line perpendicular to the other, when the lines are at right angles.

3. Find the locus of a luminous point such that the shadows cast on the plane of xy by the lines $x = 0$, $z = ny + b$ and $y = 0$, $z = mx + a$ should be at right angles. Examine the case when $a = b$.

Ans. $mnx^2 + mny^2 - myz - nzx + anx + bmy = 0$, a hyperboloid having the axis of z and the two given lines as generators. A cone when $a = b$, i.e. when the two lines intersect.

4. Find the locus of a luminous point if the ellipsoid

$$x^2/a^2 + y^2/b^2 + z^2/c^2 = 1$$

casts a circular shadow on the plane of xy.

Ans. An ellipse $y = 0$, $z^2/c^2 + x^2/(a^2 - b^2) = 1$, and a hyperbola $x = 0$, $z^2/c^2 - y^2/(a^2 - b^2) = 1$.

5. If $y^2 = ax$ is the equation of a parabola in the plane of xy, and $y + z = 0$ that of a straight line in the plane of yz, find the locus of the perpendicular drawn from any point of the parabola to the straight line. (St Andrews, 1906.)

Ans. $y^2 - z^2 - ax = 0$, hyperbolic paraboloid.

6. Show that the locus of a line which moves parallel to the plane $y = z$, and intersects the two conics $y = 0$, $z^2 = cx$ and $z = 0$, $y^2 = bx$ is $x = (y - z)(y/b - z/c)$.

7. Find the equation of the surface generated by a line which cuts the three lines

$$\begin{matrix} x = a \\ y = z \end{matrix} \Big\} , \quad \begin{matrix} x = -a \\ y = -z \end{matrix} \Big\} , \quad \frac{x - 2a}{-3} = \frac{y + a}{4} = \frac{z + 2a}{5} .$$

Ans. $x^2 + y^2 - z^2 = a^2$.

8. Find the equation of the surface generated by a line which meets the three lines $x = 2$, $4y = 3z$; $x + 2 = 0$, $4y + 3z = 0$; $y = 3$, $2x + z = 0$.

Ans. $x^2/4 + y^2/9 - z^2/16 = 1$.

9. A line of constant length moves with its extremities on two fixed skew lines; find the locus of its mid-point.

Ans. An ellipse.

10. A circle of constant radius cuts an equal fixed circle in two points and has its plane always parallel to a fixed plane which is perpendicular to that of the fixed circle. Show that the moving circle generates two cylinders.

CHAPTER VII

ELEMENTARY PROPERTIES OF QUADRIC SURFACES DERIVED FROM THEIR SIMPLEST EQUATIONS

7·1. The surfaces of the second order, or quadric surfaces, which we have recognised can be grouped as follows:

$$\pm\frac{x^2}{a^2}\pm\frac{y^2}{b^2}\pm\frac{z^2}{c^2}=1$$
Ellipsoid $(+ + +)$.
Hyperboloid of one sheet (one minus).
Hyperboloid of two sheets (two minuses).
Virtual quadric $(- - -)$.

$$ax^2+by^2+2cz=0$$
Elliptic paraboloid (a and b of same sign).
Hyperbolic paraboloid (a and b of opposite sign).

7·11. Symmetry.

We shall consider these first from the point of view of geometrical symmetry.

A *centre of symmetry* or centre of a figure is a point C such that every line through C cuts the figure in pairs of points which are equidistant from C.

An *axis of symmetry* is a line such that every line which cuts this axis at right angles cuts the figure in pairs of points equidistant from the axis.

A *plane of symmetry* is a plane such that every line perpendicular to this plane cuts the figure in pairs of points equidistant from the plane.

If the plane of xy is a plane of symmetry, any line $x=a$, $y=b$ cuts the figure in points for which z has values equal and opposite in sign. Hence the equation of the surface must contain only even powers of z. If the equation is $f(x, y, z)=0$ we have $f(x, y, z)=f(x, y, -z)$.

If the axis of z is an axis of symmetry, to any point $[x, y, z]$ on the surface corresponds the point $[-x, -y, z]$, i.e.

$$f(x, y, z)=f(-x, -y, z).$$

If the origin is a centre of symmetry, to any point $[x, y, z]$ corresponds the point $[-x, -y, -z]$, i.e.

$$f(x, y, z) = f(-x, -y, -z).$$

If the two planes $x = 0$ and $y = 0$ are both planes of symmetry,

$$f(x, y, z) = f(-x, y, z) = f(-x, -y, z),$$

hence the intersection of these planes is an axis of symmetry.

If the axes of x and y are both axes of symmetry,

$$f(x, y, z) = f(x, -y, -z) = f(-x, -y, z),$$

hence the axis of z is also an axis of symmetry. In this case there may be no plane of symmetry. For example, the locus $xyz + c = 0$ has the coordinate-axes as axes of symmetry, but it has no planes of symmetry.

Lastly, if the three planes $x = 0$, $y = 0$, $z = 0$ are all planes of symmetry,

$$f(x, y, z) = f(-x, y, z) = f(-x, -y, z) = f(-x, -y, -z), \text{ etc.},$$

hence the coordinate-axes are all axes of symmetry and the origin is a centre of symmetry.

Applying these results, we see that the surfaces of the first group, ellipsoid and hyperboloids, have the coordinate-planes as planes of symmetry, the coordinate-axes as axes of symmetry, and the origin as centre of symmetry. On the other hand, the paraboloids have only the planes $x = 0$ and $y = 0$ as planes of symmetry and the axis of z as axis of symmetry.

7·2. We shall investigate first the elementary tangential properties, taking the central quadric

$$ax^2 + by^2 + cz^2 = 1$$

as the typical surface, noting the modifications which are required in the case of the paraboloids.

7·21. Intersection of a straight line with a quadric.

Let the equations of the straight line be

$$x = x' + lt,$$
$$y = y' + mt,$$
$$z = z' + nt.$$

Substituting for x, y, z in the equation of the surface, we obtain the equation

$$(al^2+bm^2+cn^2)t^2+2(alx'+bmy'+cnz')t$$
$$+(ax'^2+by'^2+cz'^2-1)=0,$$

a quadratic in t. Hence the surface is cut by an arbitrary straight line in two points, real, coincident, or imaginary.

If the point $[x', y', z']$ lies on the surface,

$$ax'^2+by'^2+cz'^2=1,$$

and one root of the quadratic is $t=0$.

7·22. Tangent at a point.

If also $\qquad alx'+bmy'+cnz'=0,$ $\qquad\qquad$ ......(1)

the other root of the quadratic in t also vanishes, and the line therefore meets the surface in two coincident points. It is said to be a *tangent* at $[x', y', z']$.

If the direction-cosines of the line are allowed to vary, consistent with equation (1), the line is always perpendicular to the direction $[ax', by', cz']$. Hence all the tangents at a given point lie in one plane. This plane is called the *tangent-plane* at the point, and the line through the point of contact perpendicular to the tangent-plane is called the *normal*.

7·221. The direction-cosines of the normal at $[x', y', z']$ are therefore proportional to

$$[ax', by', cz'],$$

and the equation of the tangent-plane at $[x', y', z']$ is

$$ax'(x-x')+by'(y-y')+cz'(z-z')=0.$$

Since also $\qquad ax'^2+by'^2+cz'^2=1,$

the equation of the tangent-plane can be written

7·222. $\qquad\qquad ax'x+by'y+cz'z=1.$

7·223. Exactly the same method applied to the equation of the paraboloid

$$ax^2+by^2+2cz=0$$

gives for the equation in t to determine the points of intersection with a line

$$(al^2 + bm^2 + 2cn)t^2 + 2(ax'l + by'm + cz')t$$
$$+ (ax'^2 + by'^2 + 2cz') = 0.$$

The direction-cosines of the normal are $[ax', by', c]$, and the equation of the tangent-plane at $[x', y', z']$ is

$$ax'x + by'y + c(z + z') = 0.$$

7·23. Tangential equation.

The equation connecting the point-coordinates $[x, y, z]$ represents the surface as a locus of points. Reciprocally we may consider the surface as an envelope of planes and investigate the equation connecting the coordinates $[l, m, n, p]$ of a variable plane $lx + my + nz + p = 0$ which touches the surface.

Let the point of contact of the plane be $[x', y', z']$. The equation of the tangent-plane at this point is

$$ax'x + by'y + cz'z = 1.$$

Identifying this with the equation of the given plane we have

$$\frac{ax'}{l} = \frac{by'}{m} = \frac{cz'}{n} = -\frac{1}{p}.$$

But $$ax'^2 + by'^2 + cz'^2 = 1,$$

hence $$l^2/a + m^2/b + n^2/c = p^2.$$

This is called the *tangential equation* of the quadric.

7·231. In a similar way it may be shown that the tangential equation of the paraboloid

$$ax^2 + by^2 + 2cz = 0$$

is $$l^2/a + m^2/b + 2np/c = 0.$$

7·3. Pole and polar.

The equation $ax'x + by'y + cz'z = 1$ always represents a plane whether the point $[x', y', z']$ lies on the surface or not. When it lies on the surface it is the tangent-plane. In general it is called the *polar-plane* of $[x', y', z']$ with regard to the given surface.

7·31. *The polar-plane of a point* $P \equiv [x', y', z']$ *is the locus of harmonic conjugates of P with regard to the surface.*

Take any line through P and on it take a point $Q \equiv [x, y, z]$. The line cuts the surface in two points X, Y. Let one of these X divide PQ in the ratio $k : 1$. The coordinates of X are then, by Joachimsthal's formulae,

$$\frac{kx+x'}{k+1}, \quad \frac{ky+y'}{k+1}, \quad \frac{kz+z'}{k+1}.$$

But X lies on the surface, therefore

$$a(kx+x')^2 + b(ky+y')^2 + c(kz+z')^2 = (k+1)^2,$$

i.e. $\quad k^2(ax^2+by^2+cz^2-1) + 2k(ax'x+by'y+cz'z-1)$
$$+ (ax'^2+by'^2+cz'^2-1) = 0.$$

This quadratic, which we call Joachimsthal's ratio-equation, has two roots k_1, k_2 which correspond to the two points X, Y.

If (PQ, XY) is a harmonic range, X and Y divide PQ internally and externally in equal ratios, hence $k_1+k_2=0$, and therefore

$$ax'x+by'y+cz'z=1.$$

This is the equation of the locus of harmonic conjugates Q, and represents the polar of P.

7·32. If the join of two points $[x_1, y_1, z_1]$ and $[x_2, y_2, z_2]$ is cut harmonically by the surface, each point lies on the polar of the other, and
$$ax_1x_2+by_1y_2+cz_1z_2=1.$$
Two such points are called *conjugate points.*

7·33. Pole of a plane. Conjugate planes.

The polar of the point $[x', y', z']$ with respect to the quadric
$$ax^2+by^2+cz^2=1$$
is
$$ax'x+by'y+cz'z=1.$$

Identifying this equation with the equation of the general plane
$$lx+my+nz+p=0,$$
we have
$$\frac{ax'}{l}=\frac{by'}{m}=\frac{cz'}{n}=-\frac{1}{p}.$$

Hence the pole of the plane $lx+my+nz+p=0$ is
$$\left[-\frac{l}{ap}, \quad -\frac{m}{bp}, \quad -\frac{n}{cp}\right].$$

7·34. Two planes which have the property that one passes through the pole of the other are said to be conjugate. If the plane $l'x + m'y + n'z + p' = 0$ passes through the pole of the plane

$$lx + my + nz + p = 0$$

we have $ll'/a + mm'/b + nn'/c = pp'.$

As this equation is symmetrical in the two sets of coefficients it follows that each plane contains the pole of the other. This equation bears the same relation to the tangential equation that the equation connecting the coordinates of the conjugate points bears to the point-equation, the left-hand side being in each case the bilinear symmetrical expression corresponding to the quadratic expression in the coordinates.

The planes through the line of intersection of the two planes $[l_1, m_1, n_1, p_1]$ and $[l_2, m_2, n_2, p_2]$ form a pencil of planes whose equation is

$$(l_1 x + m_1 y + n_1 z + p_1) + \lambda(l_2 x + m_2 y + n_2 z + p_2) = 0.$$

Its coordinates may be represented by

$$\rho l = l_1 + \lambda l_2,$$
$$\rho m = m_1 + \lambda m_2,$$
$$\rho n = n_1 + \lambda n_2,$$
$$\rho p = p_1 + \lambda p_2.$$

If this variable plane is tangent to the quadric,

$$(l_1 + \lambda l_2)^2/a + (m_1 + \lambda m_2)^2/b + (n_1 + \lambda n_2)^2/c = (p_1 + \lambda p_2)^2,$$

i.e.

$$(l_1{}^2/a + m_1{}^2/b + n_1{}^2/c - p_1{}^2) + 2\lambda(l_1 l_2/a + m_1 m_2/b + n_1 n_2/c - p_1 p_2)$$
$$+ \lambda^2(l_2{}^2/a + m_2{}^2/b + n_2{}^2/c - p_2{}^2) = 0.$$

This quadratic equation determines two values of λ, λ_1 and λ_2, which correspond to the two tangent-planes passing through the given line. The condition

$$l_1 l_2/a + m_1 m_2/b + n_1 n_2/c = p_1 p_2$$

is that the two planes should be harmonic conjugates with regard to the two tangent-planes through their line of intersection.

7·35. Tangent-cone.

If a plane be drawn through P touching the surface in T, the polar of T, which is the tangent-plane at T, passes through P, and therefore the polar-plane of P passes through T. Hence the points of contact of all tangent-planes through P lie in the polar of P. These are also the points of contact of tangent-lines from P to the surface.

The assemblage of all the tangent-lines through a point $P[x', y', z']$ is a cone with vertex P, called the *tangent-cone*. If $Q[x, y, z]$ is any point on the tangent-cone, PQ is a tangent and the two points X, Y in which PQ meets the surface coincide. Hence Joachimsthal's equation has equal roots, and therefore

$$(ax'x + by'y + cz'z - 1)^2$$
$$= (ax'^2 + by'^2 + cz'^2 - 1)(ax^2 + by^2 + cz^2 - 1).$$

This is the equation of the tangent-cone with vertex $[x', y', z']$.

7·351. Rectangular hyperboloid.

Ex. 1. Show that the locus of points from which three mutually rectangular tangent-lines can be drawn to the surface $ax^2 + by^2 + cz^2 = 1$ is

$$a(b+c)x^2 + b(c+a)y^2 + c(a+b)z^2 = a+b+c.$$

The locus in question is the locus of vertices of rectangular tangent-cones. The tangent-cone with vertex $[x', y', z']$ is

$$(ax'x + by'y + cz'z - 1)^2 = (ax'^2 + by'^2 + cz'^2 - 1)(ax^2 + by^2 + cz^2 - 1),$$

or, say,

$$a_1 x^2 + b_1 y^2 + c_1 z^2 + 2f_1 yz + 2g_1 zx + 2h_1 xy + \text{etc.} = 0.$$

The condition that it should be rectangular is $a_1 + b_1 + c_1 = 0$, and this gives the required equation in x', y', z'.

If $a + b + c = 0$ the equation reduces to

$$a^2 x^2 + b^2 y^2 + c^2 z^2 = 0,$$

which represents a virtual cone. The only real point on the locus is the centre. In this case the asymptotic cone is rectangular, and is the only real rectangular tangent-cone. The surface in this case is called a *rectangular hyperboloid*.

Ex. 2. Discuss the nature of the locus in Ex. 1 according to the values of a, b, c.

7·352. *Locus of points through which three mutually orthogonal tangent-planes can be drawn to the quadric*

$$ax^2 + by^2 + cz^2 = 1.$$

The locus in question is the locus of vertices of orthogonal tangent-cones. Denoting for shortness

$$ax'^2 + by'^2 + cz'^2 - 1$$

by F, we have

$$a_1 = a^2 x'^2 - aF, \text{ etc.}, \quad f_1 = bcy'z', \text{ etc.}$$

The condition that the tangent-cone should be orthogonal is

$$\Sigma(b_1 c_1 - f_1^2) = 0.$$

Now $\quad b_1 c_1 - f_1^2 = (b^2 y'^2 - bF)(c^2 z'^2 - cF) - b^2 c^2 y'^2 z'^2$
$$= -Fbc(by'^2 + cz'^2 - F) = Fbc(ax'^2 - 1).$$

Hence the required condition is

$$abc(x'^2 + y'^2 + z'^2) = \Sigma bc.$$

The locus is therefore a sphere, concentric with the quadric. This is called the *orthoptic sphere**. In the case of the ellipsoid the orthoptic sphere is always real and encloses the ellipsoid, but in the case of the hyperboloids it may be virtual or a point-sphere. In the last case $\Sigma bc = 0$ and the quadric is an *orthogonal hyperboloid*, its asymptotic cone being an orthogonal cone.

For the paraboloid $ax^2 + by^2 + 2cz = 0$ the equation of the locus is found to be $2abz = c(a + b)$, which represents a plane.

7·353. The enveloping cylinder. If the vertex of the cone $[x', y', z']$ becomes a point at infinity the cone becomes a cylinder. Let the direction of the axis of the cylinder be $[l, m, n]$. Then $P \equiv [l, m, n, 0]$. Using homogeneous coordinates x, y, z, w, the equation of the cone becomes

$$(ax'x + by'y + cz'z - w'w)^2$$
$$= (ax'^2 + by'^2 + cz'^2 - w'^2)(ax^2 + by^2 + cz^2 - w^2).$$

Substituting $[x', y', z', w'] = [l, m, n, 0]$ this reduces to

$$(alx + bmy + cnz)^2 = (al^2 + bm^2 + cn^2)(ax^2 + by^2 + cz^2 - 1).$$

---

* It is also called the *director sphere* on the analogy of the director circle of a conic, which reduces to the directrix in the case of the parabola.

7·36. Polar of a line. If A and B are two points on a line l, the polar-planes α, β of A, B intersect in a line l'. Let C, D be any two points on l'. Since the polars of A and B both pass through C and D, the polars of C and D both pass through A and B. Let P be any other point on l. Then the polars of C and D both pass through P, hence the polar of P passes through C and D. Hence the polars of all points on l pass through l', and, reciprocally, the polars of all points on l' pass through l. l and l' are called mutual polars.

The polar of the line joining the points $[x_1, y_1, z_1]$ and $[x_2, y_2, z_2]$ is the line of intersection of the two planes

$$ax_1x + by_1y + cz_1z = 1,$$
$$ax_2x + by_2y + cz_2z = 1.$$

The line l cuts the quadric in two points P, Q. The tangent-planes at P and Q are the polars of P and Q. Hence the polar of l is the line of intersection of the tangent-planes at P and Q.

If a line intersects its polar the point of intersection, being on its polar, lies on the surface; the polar-plane of this point thus contains both the lines, which are therefore tangents to the surface. If a line coincides with its polar it lies entirely on the surface.

Ex. 1. Show that the polar of the line $[l, m, n; l', m', n']$ with respect to $ax^2 + by^2 + cz^2 = 1$ is $[-bcl', -cam', -abn'; al, bm, cn]$.

Ex. 2. Show that the condition that the line $[l, m, n; l', m', n']$ should be a tangent to the quadric $ax^2 + by^2 + cz^2 = 1$ (the "line-equation" of the quadric) is

$$al^2 + bm^2 + cn^2 - bcl'^2 - cam'^2 - abn'^2 = 0.$$

Ex. 3. Show that the lines through $[x', y', z']$ which are perpendicular to their polars form a cone $\Sigma a (b-c) x' (y-y') (z-z') = 0$.

7·37. Polar tetrahedra.

If $ABCD$ is any tetrahedron the polar-planes of the vertices form another tetrahedron $A'B'C'D'$, and the polar-planes of A', B', C', D' are the faces of the former tetrahedron. Each tetrahedron is the polar of the other. If a tetrahedron coincides with its polar it is said to be *self-polar*. A self-polar tetrahedron may be chosen in an infinite variety of ways. The first vertex A

has three degrees of freedom; the second vertex B may be any point on the polar-plane of A and has therefore two degrees of freedom; the third vertex C lies on the polar-line of AB and has one degree of freedom; and then the fourth vertex is determined. We may say therefore that there are ∞^6 self-polar tetrahedra.

7·38. Conjugate lines.

If l and m are two lines such that the polar of l cuts m then the polar of m cuts l. Let l', the polar of l, cut m in P. Then since P lies on the polar of l, the polar-plane of P contains l. But since P lies on m, the polar of m also lies in this plane and therefore cuts l. The two lines l, m are said to be *conjugate* with respect to the given quadric.

Ex. If the quadric is $x^2 + y^2 + z^2 + w^2 = 0$, show that the condition that the two lines (p), (q) should be conjugate is

$$p_{01}q_{01} + p_{02}q_{02} + p_{03}q_{03} + p_{23}q_{23} + p_{31}q_{31} + p_{12}q_{12} = 0.$$

7·381. *If A, B, C, D are four points such that AB is conjugate to CD and AC is conjugate to BD, then BC is conjugate to AD.*

Let the four points be $[x_i, y_i, z_i, w_i]$ $(i = 1, 2, 3, 4)$, and the quadric $S \equiv x^2 + y^2 + z^2 + w^2 = 0$. The polar of AB is the intersection of the planes $\Sigma x_1 x = 0$ and $\Sigma x_2 x = 0$; these planes both contain a point of CD, say $x = x_3 + \lambda x_4$, etc. Hence λ satisfies the two equations

$$\Sigma x_1(x_3 + \lambda x_4) = 0, \quad \Sigma x_2(x_3 + \lambda x_4) = 0.$$

Eliminating λ we have

$$\Sigma x_1 x_3 . \Sigma x_2 x_4 = \Sigma x_1 x_4 . \Sigma x_2 x_3.$$

Similarly the condition that AC should be conjugate to BD is

$$\Sigma x_1 x_2 . \Sigma x_3 x_4 = \Sigma x_1 x_4 . \Sigma x_2 x_3.$$

Therefore $\quad \Sigma x_1 x_2 . \Sigma x_3 x_4 = \Sigma x_1 x_3 . \Sigma x_2 x_4,$

which is the condition that BC should be conjugate to AD.

The tetrahedron $ABCD$ is said to be *self-conjugate* with respect to the given quadric. Since the twelve coordinates of the four vertices are connected by only two relations there are ∞^{10} self-conjugate tetrahedra. Every self-polar tetrahedron is self-conjugate, but not *vice versa*.

7·382. If $A'B'C'D'$ is the polar tetrahedron corresponding to a self-conjugate tetrahedron $ABCD$ each edge of the one tetrahedron intersects the similarly named edge of the other. Consider any two tetrahedra, without reference to any quadric, having these relationships of incidence. Denote the point of intersection of BC and $B'C'$ by (23), that of AD and $A'D'$ by (14), and so on. Then the three points (23), (31) and (12) all lie in both of the planes ABC and $A'B'C'$, and are therefore collinear. Hence the triangles ABC and $A'B'C'$, in different planes, are in perspective, and AA', BB', CC' are concurrent in a point O. Similarly AA', BB', DD' are concurrent in O. Hence the two tetrahedra are in perspective, i.e. *if two tetrahedra are such that each edge of one cuts the corresponding edge of the other, they are in perspective.*

One condition is required in order that two given lines should intersect, and two conditions in order that a given line should pass through a given point, hence in order that two tetrahedra should be in perspective five conditions are required. It follows that *if two tetrahedra have five pairs of corresponding edges incident the sixth pair also are incident.*

Ex. If two tetrahedra are mutually polar with respect to a given quadric and are also in perspective, show that each is self-conjugate with respect to the quadric.

7·4. Diametral planes.

A line through the centre is a *diameter*, and is bisected at the centre. A plane through the centre is a *diametral plane*.

Since the harmonic conjugate of the mid-point of a segment, with respect to the ends of the segment, is a point at infinity, the polar of the centre is the plane at infinity; and reciprocally the polar of a point at infinity passes through the centre and is a diametral plane.

Hence the locus of the mid-points of a system of chords parallel to $[l, m, n]$ is a diametral plane, the polar-plane of the point at infinity $[l, m, n, o]$.

Since a diameter passes through the centre its polar is a line at infinity. Hence the polar-planes of all points on a diameter have a common line at infinity and are therefore parallel.

Let A be any point, and α the polar-plane of A, and let the diameter OA cut α in C. The polar of C is parallel to α, i.e. it meets α in a line at infinity. Hence C is the centre of the conic in which α cuts the quadric. In particular if OA cuts the quadric in X, the tangent-plane at X is parallel to α. Hence the diameter OX cuts all planes, which are parallel to the tangent-plane at X, in the centres of the sections.

The diameter OX is said to be *conjugate* to the diametral plane which is parallel to the sections through whose centres OX passes.

Taking the equation of the surface

$$ax^2 + by^2 + cz^2 = 1,$$

consider a diameter

$$x = lt, \quad y = mt, \quad z = nt.$$

The polar-plane of the point t on this diameter is

$$alx + bmy + cnz = t^{-1}.$$

For all values of t these planes are parallel, and as $t \to \infty$ we get the diametral plane

$$alx + bmy + cnz = 0$$

conjugate to $[l, m, n]$.

7·41. To find the centre of the section by the plane

$$lx + my + nz + p = 0$$

we may find the pole P of this plane, then OP cuts the plane in the centre C. The coordinates of P are found by identifying the equation of the plane with the equation of the polar of $[x', y', z']$

$$ax'x + by'y + cz'z - 1 = 0,$$

hence $\quad x' = -l/ap, \quad y' = -m/bp, \quad z' = -n/cp.$

The freedom-equations of OP are

$$ax = lt, \quad by = mt, \quad cz = nt.$$

This line cuts the given plane where

$$t(l^2/a + m^2/b + n^2/c) + p = 0;$$

whence the coordinates of the centre are given by

$$\frac{ax}{l} = \frac{by}{m} = \frac{cz}{n} = -\frac{p}{l^2/a + m^2/b + n^2/c}.$$

7·411. The following alternative method is also useful.

Take any point $P \equiv [x', y', z']$. The freedom-equations of a line through P are

$$x = x' + lr, \quad y = y' + mr, \quad z = z' + nr.$$

This line cuts the surface where

$$(al^2 + bm^2 + cn^2)r^2 + 2(alx' + bmy' + cnz')r \\ + (ax'^2 + by'^2 + cz'^2 - 1) = 0.$$

If P is the mid-point of the chord through P we have

$$alx' + bmy' + cnz' = 0.$$

7·412. If P is a fixed point this equation determines the directions of the chords which are bisected at P, and we find that all these lines lie in the plane

$$ax'(x - x') + by'(y - y') + cz'(z - z') = 0.$$

This represents the section which has its centre at P.

7·413. If P is a variable point and $[l, m, n]$ a fixed direction the equation

$$alx + bmy + cnz = 0$$

represents the locus of mid-points of the system of chords parallel to $[l, m, n]$, i.e. the diametral plane conjugate to this direction.

7·414. The equation

$$ax_1(x_2 - x_1) + by_1(y_2 - y_1) + cz_1(z_2 - z_1) = 0$$

connects the two points (x_1) and (x_2) which have the unsymmetrical property that the chord joining them is bisected at (x_1). If (x_2) is fixed and (x_1) variable the locus of (x_1) is that of the mid-point of chords through the fixed point (x_2). The equation of this locus is therefore

$$ax(x_2 - x) + by(y_2 - y) + cz(z_2 - z) = 0,$$

i.e.

$$a(x - \tfrac{1}{2}x_2)^2 + b(y - \tfrac{1}{2}y_2)^2 + c(z - \tfrac{1}{2}z_2)^2 \\ = \tfrac{1}{4}(ax_2^2 + by_2^2 + cz_2^2),$$

which represents a similar quadric with centre at the mid-point of the radius-vector to (x_2).

7·42. Conjugate diameters.

The diametral plane $alx + bmy + cnz = 0$ is conjugate to the direction $[l, m, n]$ and we say in particular that it is conjugate to the diameter in that direction. Denote the point at infinity in the direction $[l, m, n]$ by P, and let $P' \equiv [l', m', n', 0]$ be any point at infinity in the conjugate diametral plane. Then the polar of P' passes through P. Hence the diametral plane conjugate to $[l', m', n']$ contains P and therefore contains the diameter OP. Let these two diametral planes intersect in OP'', where

$$P'' \equiv [l'', m'', n'', 0]$$

is the point at infinity in both planes. Then the diametral plane conjugate to OP'' contains both OP and OP''. We have then a set of three diametral planes, each conjugate to the line of intersection of the other two.

Take the ellipsoid $\dfrac{x^2}{a^2} + \dfrac{y^2}{b^2} + \dfrac{z^2}{c^2} = 1,$

and let $P_1 \equiv [x_1, y_1, z_1]$ be any point on the surface, so that

$$x_1^2/a^2 + y_1^2/b^2 + z_1^2/c^2 = 1.$$

The diametral plane conjugate to OP_1 is

$$xx_1/a^2 + yy_1/b^2 + zz_1/c^2 = 0.$$

Choose $P_2 \equiv [x_2, y_2, z_2]$ any point on the section by this plane, so that

$$x_2^2/a^2 + y_2^2/b^2 + z_2^2/c^2 = 1$$

and

$$x_1 x_2/a^2 + y_1 y_2/b^2 + z_1 z_2/c^2 = 0.$$

The diametral plane conjugate to OP_2 is

$$xx_2/a^2 + yy_2/b^2 + zz_2/c^2 = 0,$$

which passes through P_1.

These two diametral planes cut in a diameter which cuts the surface in a point $P_3 \equiv [x_3, y_3, z_3]$. We have then the two sets of equations

$$\frac{x_2 x_3}{a^2} + \frac{y_2 y_3}{b^2} + \frac{z_2 z_3}{c^2} = 0, \qquad \frac{x_1^2}{a^2} + \frac{y_1^2}{b^2} + \frac{z_1^2}{c^2} = 1,$$

$$\frac{x_3 x_1}{a^2} + \frac{y_3 y_1}{b^2} + \frac{z_3 z_1}{c^2} = 0, \qquad \frac{x_2^2}{a^2} + \frac{y_2^2}{b^2} + \frac{z_2^2}{c^2} = 1,$$

$$\frac{x_1 x_2}{a^2} + \frac{y_1 y_2}{b^2} + \frac{z_1 z_2}{c^2} = 0, \qquad \frac{x_3^2}{a^2} + \frac{y_3^2}{b^2} + \frac{z_3^2}{c^2} = 1.$$

If we put $x_1/a = \lambda_1$, $y_1/b = \mu_1$, $z_1/c = \nu_1$, etc., these equations show that λ_1, μ_1, ν_1, etc., are the direction-cosines of three mutually perpendicular lines, and we have the derived relations (see 2·9)

$$\lambda_1{}^2 + \lambda_2{}^2 + \lambda_3{}^2 = 1, \qquad \mu_1\nu_1 + \mu_2\nu_2 + \mu_3\nu_3 = 0, \text{ etc.,}$$

hence

$$x_1{}^2 + x_2{}^2 + x_3{}^2 = a^2, \qquad y_1 z_1 + y_2 z_2 + y_3 z_3 = 0,$$
$$y_1{}^2 + y_2{}^2 + y_3{}^2 = b^2, \qquad z_1 x_1 + z_2 x_2 + z_3 x_3 = 0,$$
$$z_1{}^2 + z_2{}^2 + z_3{}^2 = c^2, \qquad x_1 y_1 + x_2 y_2 + x_3 y_3 = 0.$$

7·421. Lengths of conjugate diameters of an ellipsoid.

$$OP_1{}^2 = x_1{}^2 + y_1{}^2 + z_1{}^2, \text{ etc.,}$$

hence

$$OP_1{}^2 + OP_2{}^2 + OP_3{}^2 = a^2 + b^2 + c^2.$$

7·422. Volume of the parallelepiped whose edges are three conjugate semi-diameters.

The volume of the parallelepiped whose corners are $[0, 0, 0]$, $[x_1, y_1, z_1]$, etc.,

$$V = \begin{vmatrix} x_1 & y_1 & z_1 \\ x_2 & y_2 & z_2 \\ x_3 & y_3 & z_3 \end{vmatrix}.$$

Squaring this determinant we have

$$\begin{vmatrix} x_1 & x_2 & x_3 \\ y_1 & y_2 & y_3 \\ z_1 & z_2 & z_3 \end{vmatrix}^2 = \begin{vmatrix} \Sigma x^2 & \Sigma xy & \Sigma xz \\ \Sigma yx & \Sigma y^2 & \Sigma yz \\ \Sigma zx & \Sigma zy & \Sigma z^2 \end{vmatrix} = \begin{vmatrix} a^2 & 0 & 0 \\ 0 & b^2 & 0 \\ 0 & 0 & c^2 \end{vmatrix} = a^2 b^2 c^2.$$

Hence

$$V = abc.$$

7·43.

If the diametral plane $alx + bmy + cnz = 0$ of the surface $ax^2 + by^2 + cz^2 = 1$ is perpendicular to the conjugate direction $[l, m, n]$ we have

$$\frac{al}{l} = \frac{bm}{m} = \frac{cn}{n}.$$

If $l \neq 0$, either $m = 0$ or $a = b$, and either $n = 0$ or $a = c$. Hence if a, b, c are all unequal, two of the quantities l, m, n must vanish, and the only diametral planes which are perpendicular to their conjugate directions are the coordinate-planes. These are called the *principal diametral planes* and the *principal axes*. They are the only planes and axes of symmetry.

If $a = b \neq c$, then $n = 0$. The surface is a surface of revolution, the axis of revolution being the axis of z, a principal axis. Any diameter perpendicular to this is also a principal axis.

If $a = b = c$, the surface is a sphere, and any diameter is a principal axis.

7·5. In the case of the hyperboloids the metrical relations require modification, and it is convenient to take together the two associated hyperboloids

$$\frac{x^2}{a^2} + \frac{y^2}{b^2} - \frac{z^2}{c^2} = \pm 1.$$

These are separated from one another by the asymptotic cone

$$\frac{x^2}{a^2} + \frac{y^2}{b^2} - \frac{z^2}{c^2} = 0.$$

Diameters which lie within this cone cut the hyperboloid of two sheets in real points, but cut the hyperboloid of one sheet in imaginary points. Diameters lying outside this cone are intersectors of the hyperboloid of one sheet, and non-intersectors of the hyperboloid of two sheets.

Let $[l, m, n]$ be the direction-cosines of a diameter within the asymptotic cone. Then

$$l^2/a^2 + m^2/b^2 - n^2/c^2 < 0, \quad = -k^{-2}, \text{ say.}$$

The conjugate diametral plane for either of the hyperboloids is

$$\frac{lx}{a^2} + \frac{my}{b^2} - \frac{nz}{c^2} = 0.$$

This cuts the hyperboloid of one sheet in a real ellipse. The diameter $[l, m, n]$ cuts the hyperboloid of two sheets where

$$x/l = y/m = z/n = \pm(-l^2/a^2 - m^2/b^2 + n^2/c^2)^{-\frac{1}{2}} = \pm k,$$

and the equation of the tangent-plane at one of these points is

$$lx/a^2 + my/b^2 - nz/c^2 = k^{-1}.$$

The enveloping cylinder with axis $[l, m, n]$ of the hyperboloid of one sheet is

$$(-x^2/a^2 - y^2/b^2 + z^2/c^2 + 1) = (lx/a^2 + my/b^2 - nz/c^2)^2 k^2,$$

and this is cut by the tangent-plane in the same section as the asymptotic cone (Fig. 34).

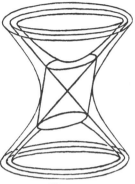

Fig. 34

7·6. When a central quadric is referred to a set of conjugate diameters as coordinate-axes (oblique) its equation is of the same form as when referred to its principal axes. This follows because each diametral plane bisects all chords in the conjugate direction.

The centre and the points at infinity on the three conjugate diameters form a self-polar tetrahedron.

Ex. The ellipsoid with conjugate diameters the lines which bisect pairs of opposite edges of a tetrahedron touches the edges at their mid-points.

Let A, A'; B, B'; C, C' be the mid-points of the edges. Then $BCB'C'$ is a parallelogram, and therefore BB' and CC' intersect at their mid-points. Hence AA', BB', CC' are all concurrent at their mid-points. Let the coordinates of A, A' be $[\pm a, 0, 0]$, B, B' $[0, \pm b, 0]$, and C, C' $[0, 0, \pm c]$. Then, referred to axes OA, OB, OC, the equation of the ellipsoid is

$$x^2/a^2 + y^2/b^2 + z^2/c^2 = 1.$$

The edge through A is parallel to BC whose direction-ratios are $[0, b, -c]$. Hence the equations of this edge are $x = a, y/b + z/c = 0$. But $x = a$ is the tangent-plane to the ellipsoid at A, hence the edge is a tangent to the surface.

7·7. Normals.

The tangent-plane at $[x', y', z']$ is

$$ax'x + by'y + cz'z = 1.$$

Hence the normal at this point is

$$\frac{x - x'}{ax'} = \frac{y - y'}{by'} = \frac{z - z'}{cz'}.$$

7·71. *Every normal is perpendicular to its polar*, for the polar of the normal at P lies in the tangent-plane at P. Hence all the normals which can be drawn from a given point $P[X, Y, Z]$, not in general on the surface, are generators of the cone which consists of the lines through P which are perpendicular to their polars, i.e. the cone (7·36, Ex. 3)

$$\Sigma \frac{X}{x-X}\left(\frac{1}{b}-\frac{1}{c}\right)=0.$$

7·72. Number of normals which can be drawn from a given point.

If the normal at $[x', y', z']$ passes through $[X, Y, Z]$,

$$\frac{X-x'}{ax'}=\frac{Y-y'}{by'}=\frac{Z-z'}{cz'}=t, \text{ say,}$$

therefore $x'=\dfrac{X}{1+at}$, $y'=\dfrac{Y}{1+bt}$, $z'=\dfrac{Z}{1+ct}$.

But $ax'^2+by'^2+cz'^2=1$,

therefore $\dfrac{aX^2}{(1+at)^2}+\dfrac{bY^2}{(1+bt)^2}+\dfrac{cZ^2}{(1+ct)^2}=1$,

i.e. $(at+1)^2(bt+1)^2(ct+1)^2-\Sigma aX^2(bt+1)^2(ct+1)^2=0.$

This is an equation of the sixth degree in t. Hence six normals may be drawn through a given point.

7·73. The normals at two points P, Q do not in general intersect. Suppose they intersect in R. Let α, β be the tangent-planes at P, Q, and γ the polar of R. The polar of P is α, and the polar of Q is β, hence the polar of PQ is the line $(\alpha\beta)$. Now the plane PQR is perpendicular to both α and β and therefore perpendicular to $(\alpha\beta)$. This is therefore a condition that the normals should intersect. All the lines through P which are perpendicular to their polars form a cone, and this cone cuts the surface in a curve (of the fourth order).

The tangent-plane at P cuts the cone in two generating lines which are such that each is perpendicular to its polar, but their polars are also in the tangent-plane and have this same property. Hence each is the polar of the other, and the two generators are perpendicular. Hence at any point P on the surface there are two directions in the surface (principal directions) such that the normals at points near P cut the normal at P, and these two directions are at right angles. The curves on the surface which have the property that normals to the surface at neighbouring points on the curve intersect are called *lines of curvature*. Through every point on the surface there pass two lines of curvature, and these are at right angles.

7·8. The paraboloids.

7·81. The diametral properties of the paraboloids are somewhat different from those of the central quadrics.

Let the equation of the surface be

$$ax^2 + by^2 + 2cz = 0.$$

The line $\qquad x = x' + lr, \; y = y' + mr, \; z = z' + nr$

cuts the surface where

$$a(lr + x')^2 + b(mr + y')^2 + 2c(nr + z') = 0,$$

i.e. $\quad (al^2 + bm^2)r^2 + 2(ax'l + by'm + cn)r$
$$+ (ax'^2 + by'^2 + 2cz') = 0.$$

If $l = 0$ and $m = 0$, one root of this quadratic is infinite, hence all lines parallel to the axis of z meet the surface in a point at infinity.

The point $[x', y', z']$ will be the mid-point of the chord if

$$alx' + bmy' + cn = 0.$$

Hence if l, m, n are given, i.e. for a system of parallel chords, the locus of the mid-points is the plane

$$alx + bmy + cn = 0,$$

and this plane is parallel to the z-axis. We have therefore to consider all planes parallel to the z-axis as diametral planes, and all lines parallel to the z-axis as diameters, and hence the centre must be regarded as a point at infinity.

If the diametral plane $alx + bmy + cn = 0$ is perpendicular to the conjugate direction $[l, m, n]$, we have

$$\frac{al}{l} = \frac{bm}{m} = \frac{0}{n}.$$

Hence, provided that $a \neq b$, $n = 0$, and either $l = 0$ or $m = 0$. Hence the two planes $x = 0$, $y = 0$ are principal planes. If $a = b$ we have a paraboloid of revolution and all planes through the axis of z are principal planes.

There is no diametral plane conjugate to the direction of the z-axis [o, o, 1] except the plane at infinity, but all sections parallel to the plane of xy have their centres on the z-axis; in particular the plane of xy is the tangent-plane at O. O is called the *vertex*.

7·82. Normals to the paraboloids.

The tangent-plane at $[x', y', z']$ is

$$ax'x + by'y + c(z + z') = 0,$$

and freedom-equations of the normal are

$$x = x'(1 + at),$$
$$y = y'(1 + bt),$$
$$z = z' + ct.$$

If the normal passes through the fixed point $[X, Y, Z]$,

$$X = x'(1 + at),$$
$$Y = y'(1 + bt),$$
$$Z = z' + ct.$$

But $$ax'^2 + by'^2 + 2cz' = 0,$$

hence $$\frac{aX^2}{(1 + at)^2} + \frac{bY^2}{(1 + bt)^2} + 2c(Z - ct) = 0.$$

This equation is of the fifth degree in t, hence five normals pass through a given point. The sixth normal must be the diameter through the given point. This counts as a normal since the plane at infinity is a tangent-plane. [o, o, 1, o] is the point at infinity, and the tangent there is $w = 0$.

7·9. EXAMPLES.

1. Find the equations of the tangent-planes to

$$2x^2 - 6y^2 + 3z^2 = 5$$

which pass through the line

$$x + 9y - 3z = 0 = 3x - 3y + 6z - 5.$$

Ans. $4x + 6y + 3z = 5$, $2x - 12y + 9z = 5$.

2. Show that $2x - 2y - 3z + 8 = 0$ is a tangent-plane to $4x^2 + y^2 - 9z^2 = 16$, and find the point of contact.

Ans. $[-1, 4, -\frac{2}{3}]$.

3. If l and l' are polar-lines with respect to a quadric and the common transversal through the centre meets them in P and P', prove that the polar-planes of P and P' are parallel to both l and l'.

4. Prove that the product of the lengths of a pair of opposite edges of a tetrahedron, self-polar with respect to a sphere, is inversely proportional to the shortest distance between that pair.
(Math. Trip. I, 1914.)

5. Prove that, if the chord which joins two points of the ellipsoid

$$x^2/a^2 + y^2/b^2 + z^2/c^2 = 1$$

touches the ellipsoid

$$x^2/a^2 + y^2/b^2 + z^2/c^2 = \tfrac{1}{2},$$

the two points must lie at the extremities of conjugate diameters of the former, and the point of contact must bisect the chord.
(Math. Trip. I, 1915.)

6. Show that any set of three equal conjugate diameters of the ellipsoid whose equation is $x^2/a^2 + (y^2 + z^2)/b^2 = 1$ lie on a circular cone and that the cosine of the angle between any two is $(a^2 - b^2)/(a^2 + 2b^2)$.
(Math. Trip. I, 1913.)

7. Show that the plane through the extremities P, Q, R of three conjugate diameters of the ellipsoid

$$x^2/a^2 + y^2/b^2 + z^2/c^2 = 1$$

touches the ellipsoid

$$x^2/a^2 + y^2/b^2 + z^2/c^2 = \tfrac{1}{3}$$

at the centroid of the triangle PQR.

8. If a point P be chosen on a line l such that the polar-plane of P with respect to the quadric $ax^2 + by^2 + cz^2 = 1$ is parallel to l, and l' is the polar-line of l, prove that the plane through l' and the origin O cuts l in P. If the coordinates of P are $[X, Y, Z]$, show that the plane through l and O cuts l' in the point $X/\Sigma aX^2$, $Y/\Sigma aX^2$, $Z/\Sigma aX^2$, and that the polar of this point is parallel to l'.

9. Show that the locus of points on the quadric

$$ax^2 + by^2 + cz^2 = 1$$

the normals at which all intersect the straight line

$$(x - X)/l = (y - Y)/m = (z - Z)/n$$

is the curve of intersection with the quadric

$$\Sigma l(b - c)yz + \Sigma(mZ - nY)ax = 0.$$

10. Prove that the condition that all the normals to the ellipsoid $x^2/a^2 + y^2/b^2 + z^2/c^2 = 1$ at the points of intersection with the plane $lx/a + my/b + nz/c = 1$ should intersect one straight line is $\Sigma(m^2n^2 - l^2)(b^2 - c^2)^2 = 0$, and that if this condition is satisfied all the normals at the points of intersection with the plane $x/al + y/bm + z/cn + 1 = 0$ will intersect the same straight line.

Prove also that when $l = m = n = 1$ the normals all intersect the straight line $a(b^2 - c^2)x = b(c^2 - a^2)y = c(a^2 - b^2)z$.

11. Prove that the six normals to the quadric

$$ax^2 + by^2 + cz^2 = 1$$

which pass through the point $P \equiv [X, Y, Z]$ are generators of the cone

$$\Sigma aX(b - c)(y - Y)(z - Z) = 0.$$

12. Show that the cone in Ex. 11 contains also as generators the line OP and the lines through P parallel to the axes; also the normal through P to its polar-plane.

13. Show that the feet of the six normals from $[X, Y, Z]$ to the quadric $ax^2 + by^2 + cz^2 = 1$ are the intersections of the surface with the cubic curve whose parametric equations are

$$\rho x = X(1 + bt)(1 + ct), \ldots, \rho w = (1 + at)(1 + bt)(1 + ct).$$

14. Show that the cubic curve in Ex. 13 lies on the cone

$$\Sigma \frac{X}{x - X}\left(\frac{1}{b} - \frac{1}{c}\right) = 0.$$

15. If the quadric is a surface of revolution, show that there are just four normals through a point and that they lie in a meridian plane.

The other two normals have become isotropic lines through the point in a plane perpendicular to the axis.

16. If the feet of three of the six normals drawn from a given point to the quadric $ax^2 + by^2 + cz^2 = 1$ lie on the plane

$$lx + my + nz + p = 0,$$

show that the feet of the other three will lie on the plane

$$ax/l + by/m + cz/n - 1/p = 0.$$

17. Find the radius of a sphere concentric with an ellipsoid of semi-axes a, b, c, such that an octahedron can be inscribed in the ellipsoid and circumscribed about the sphere, and show that if one such octahedron exists, an infinite number exist, all with their diagonals intersecting at right angles at the centre.

Ans. $r = abc/\sqrt{\Sigma b^2 c^2}$.

18. Show that, for the same radius as in Ex. 17, an infinity of parallelepipeds can be inscribed in the ellipsoid and circumscribed about the sphere.

19. Show that the condition that the plane

$$lx + my + nz = 0$$

should cut the cone

$$ax^2 + by^2 + cz^2 + 2fyz + 2gzx + 2hxy = 0$$

in two conjugate diameters of the quadric

$$Ax^2 + By^2 + Cz^2 = 1$$

is $$(Bc + Cb)l^2 + \ldots - 2Afmn - \ldots = 0.$$

20. Prove that the condition that the cone

$$ax^2 + by^2 + cz^2 + 2fyz + 2gzx + 2hxy = 0$$

should have three generating lines coinciding with three conjugate diameters of the quadric

$$Ax^2 + By^2 + Cz^2 = 1$$

is $$a/A + b/B + c/C = 0,$$

and that if this condition is satisfied there are an infinite number of such triads of generating lines.

21. If $[l, m, n; l', m', n']$ are line-coordinates of a normal to the quadric $ax^2 + by^2 + cz^2 = 1$ show that

$$\frac{ll'}{a(b-c)} = \frac{mm'}{b(c-a)} = \frac{nn'}{c(a-b)} = \frac{\{\Sigma a(b-c)^2 m'^2 n'^2\}^{\frac{1}{2}}}{(b-c)(c-a)(a-b)}.$$

CHAPTER VIII

THE REDUCTION OF THE GENERAL EQUATION OF THE SECOND DEGREE

8·1. We consider now the general equation of the second degree in x, y, z, or the homogeneous equation in x, y, z, w. The surfaces represented must include the ellipsoid, hyperboloids, and paraboloids, since their equations are of the second degree. The geometrical property which they have in common is that an arbitrary straight line cuts the surface in two points, real, coincident, or imaginary. We call the general surface a *quadric surface* or quadric.

The general equation of the second degree in x, y, z is

$$F(x, y, z) \equiv ax^2 + by^2 + cz^2 + 2fyz + 2gzx + 2hxy$$
$$+ 2px + 2qy + 2rz + d = 0.$$

The homogeneous equation is written more symmetrically, using x_0, x_1, x_2, x_3 for the coordinates,

$$a_{00}x_0^2 + \dots + 2a_{01}x_0x_1 + \dots = 0,$$

or shortly $\quad \Sigma\Sigma a_{rs}x_rx_s = 0 \quad (r, s = 0, 1, 2, 3),$

where* $\quad a_{rs} = a_{sr}.$

8·11. The equation contains ten terms; the ratios of these give nine independent constants whose values determine the equation. We say that the *constant-number* of a quadric is 9. The condition that the surface should pass through a given point gives an equation which is linear and homogeneous in the ten constants; this is therefore called a *linear condition*. Nine such equations in general determine the ratios of the constants uniquely. Hence in general *one quadric can be drawn to pass through nine given points.*

If three of the points lie on a straight line all points of this straight line must lie on the surface (otherwise the line would cut the surface in three points), and the line is a generating line of the surface. If a second set of three points lie on another line

---

* There is no loss of generality in this assumption, for the sum of the two terms $a_{rs}x_rx_s + a_{sr}x_sx_r$ can always be replaced by $2a_{rs}'x_rx_s$.

which does not cut the first one, we have a second generating line. Three mutually skew lines therefore uniquely determine a quadric. If two sets of three points lie on two intersecting lines the six points may, so far as they serve to determine the quadric, be replaced by five: the point of intersection, and any other points, two on each line. We would then have an insufficient number of points to determine the quadric.

8·12. Any plane section of a quadric is a conic, for any line in the plane of section cuts the surface, and therefore the curve of section, in two points.

Five coplanar points determine a conic. Any quadric which passes through five given points in a plane α will contain the conic C which is determined by the five points; for the plane α cuts the quadric in a conic, and since this conic contains the five given points it must be the conic C. Hence if a quadric has to pass through a given conic it has still 4 degrees of freedom. This may be shown also as follows. Let the conic be determined by the two equations $F_2 = 0$ and $F_1 = 0$, the latter an equation of the first degree representing the plane of the conic, the former an equation of the second degree representing any quadric which contains the conic. Then the equation

$$F_2 + F_1 L = 0,$$

where L is an expression of the first degree, represents any quadric which contains the conic, and L contains four disposable constants.

If six of the points which determine a quadric lie in one plane, either they all lie on one conic, in which case the quadric is not fully determined; or the plane itself must form part of the quadric; the other three points then determine a second plane, and the quadric consists of these two planes.

In order that the quadric may be determinate no four of the nine points may be collinear, not more than six may lie in one plane, and not more than five on one conic.

8·13. A proper quadric cannot in general be made to pass through two conics, for each conic requires five data and therefore ten are given. But the planes of the two conics together form a degenerate quadric which contains the two conics, and this is in general the only quadric containing the two conics.

But if we cut a quadric by two planes we get two conics through which a proper quadric passes. These two planes intersect in a straight line, and this cuts the quadric in two points which are common to the two conics. Hence the condition that two conics should be capable of having a proper quadric surface passed through them is that they should have two points (real, coincident, or imaginary) in common. Each of these requires three other conditions to determine it, hence we have only $3+3+2=8$ conditions. Hence *if one quadric, other than two planes, can be drawn to pass through two conics, an infinite number can be drawn.* All these quadrics touch one another in the two common points, for the tangents to the two conics are tangents to each quadric, and the plane determined by them is a tangent-plane to each quadric.

Ex. 1. Prove that any two spheres have double contact with one another.

The two spheres intersect in a circle. They have also in common the circle at infinity. Therefore they touch in two points, viz. the points I, J in which the circle cuts the plane at infinity.

Ex. 2. Show that an infinity of quadrics can be drawn to pass through two circles in parallel planes.

Ex. 3. Given a conic S in a plane α, and three points A, B, C in another plane β, to construct the conic which passes through A, B, C and the two points H, K (real or imaginary) in which S cuts the plane β.

When two conics have two common points the two involutions, determined on their common chord by pairs of conjugate points with respect to the two conics, coincide, for they each have the two common points as double-points. Thus, even when the conics intersect in imaginary points, the polars of any point on their common chord intersect on that line.

Hence a solution of the problem is afforded as follows.

Let l be the line of intersection of α and β. Projecting l to infinity and at the same time S into a circle, the required conic becomes a circle also, and its centre may be constructed in the usual way by bisecting the joins of A, B, C perpendicularly. Hence we can construct the pole of l with regard to the required conic in the following way. Let BC cut l in P and determine D the harmonic conjugate of P with respect to B, C. Let the polar of P with respect to S cut l in P'. Similarly determine E, F and Q', R'. Then DP', EQ', FR' are concurrent in a point O, the pole of l with respect to the required conic. Let AO cut l in L and determine A', the harmonic conjugate of A with respect to O, L. Then A' is also a point on the required conic. Similarly we find B' on OB and C' on OC, and thus six real points have been determined on the conic, and any number of other points can be found by Pascal's theorem.

8·2. Conjugate points.

Let $F(x, y, z, w) = 0$ be the equation of the quadric in homogeneous coordinates, and let $P \equiv [x, y, z, w]$ and $P' \equiv [x', y', z', w']$ be any two points. The coordinates of any point Q on the line PP' are $[x' + \lambda x, y' + \lambda y, z' + \lambda z, w' + \lambda w]$. If this point lies on the quadric we have

$$F(x' + \lambda x, y' + \lambda y, z' + \lambda z, w' + \lambda w) = 0,$$

i.e. $\quad \lambda^2 F(x, y, z, w) + \lambda \left(x \dfrac{\partial F}{\partial x'} + y \dfrac{\partial F}{\partial y'} + z \dfrac{\partial F}{\partial z'} + w \dfrac{\partial F}{\partial w'} \right)$

$$+ F(x', y', z', w') = 0, \quad \ldots\ldots(1)$$

a quadratic giving the two points of intersection Q_1 and Q_2 of the line PP' with the quadric, corresponding to the roots λ_1 and λ_2.

If (PP', Q_1Q_2) is harmonic, $\lambda_1 + \lambda_2 = 0$, hence the condition that P, P' be conjugate points with respect to the quadric is

$$x \frac{\partial F}{\partial x'} + y \frac{\partial F}{\partial y'} + z \frac{\partial F}{\partial z'} + w \frac{\partial F}{\partial w'} = 0. \quad \ldots\ldots(2)$$

This relation is bilinear and symmetrical in x, y, z, w and x', y', z', w'.

8·21. If P' is fixed equation (2) represents the *polar-plane* of P'.

If P, P' are conjugate points, and P' lies on the surface, equation (1) becomes

$$\lambda^2 F(x, y, z, w) = 0.$$

Hence either $F(x, y, z, w) = 0$ or $\lambda = 0$.

8·22. In the first case equation (1) is identically satisfied. Hence *if P and P' are conjugate points and both lie on the surface, the line PP' lies entirely in the surface and is therefore a generating line.*

8·23. In the second case, when the equation reduces to $\lambda^2 = 0$, PP' meets the surface in two coincident points at P' and is therefore a *tangent*. The locus of points P such that PP' is a tangent is represented by equation (2), which is therefore the equation of the *tangent-plane* at P'.

8·231. The tangent-plane at P', like any plane, cuts the quadric in a conic. Let P be any point on this conic. Then since P, P' are conjugate points, both lying on the conic, every point of the line PP' also lies on the surface. The conic therefore breaks up into the line PP' and another line through P'. Hence *the tangent-plane at any point cuts the surface in two generating lines.*

8·24. The two lines l_1, l_2 through P' may be real and distinct, coincident, or imaginary, and the point P' is called accordingly a *hyperbolic, parabolic,* or *elliptic* point of the surface.

If l_1 coincides with l_2 every line in the tangent-plane π at P' meets the surface in two coincident points, and is therefore a tangent. π is then tangent at all points of l_1. Let Q be any other point on the surface, not on l_1. The tangent-plane at Q cannot contain l_1 but meets it in a point C, and CQ is a generating line. There can be no other generating line through Q, for this would meet π in a point lying on the surface and therefore on l_1. Hence Q is also a parabolic point. For any other point the tangent-plane must meet CQ and CP' in the same point, i.e. C. Hence all the generating lines pass through one point C, and the surface is a cone with vertex C.

If l_1 and l_2 are real and distinct the tangent-plane at any other point Q meets them in real and distinct points R_1, R_2, both conjugate to Q and lying in the surface. Hence QR_1 and QR_2 are real and distinct generating lines. If l_1 and l_2 are imaginary, the generating lines through any other point are also imaginary. Hence if one point on the surface is a hyperbolic (elliptic) point, every point on the surface is hyperbolic (elliptic).

8·25. Since the relation between conjugate points P and P' is symmetrical it follows that each point lies on the polar-plane of the other. If π is a given plane and P its pole, the polar-planes of all points on π pass through P.

If l is a given line and P, Q two points on it, the polar-planes α and β of P and Q determine a line l'. If P' is any point on l' the polar-planes of P and Q both pass through P' and therefore the polar-plane of P' passes through P and Q and therefore contains l. The two lines l and l' are therefore such that the polar-plane of any point on one of them contains the other. These are called *polar-lines*.

If a line l cuts its polar l' in P, P is conjugate to itself and therefore lies on the surface. P is conjugate to every point in l and in l', hence l and l' are tangents to the surface. If t and t' are the generating lines through P, (tt', ll') is harmonic.

If l cuts the surface in P and Q the polar of l is the intersection of the tangent-planes at P and Q since these are the polar-planes of these points. Let l, l' be the generating lines through P, and m, m' those through Q. Then l, m' meet in a point P', and l', m in a point Q'. $P'Q'$ is the polar of PQ and cuts the surface in P' and Q'. Hence when the generating lines are real two polar-lines either both cut the surface in real points or both in imaginary points. If the surface has imaginary generators, and one line cuts the surface in real points P, Q, the polar-line, being the intersection of the tangent-planes at P and Q, cannot meet the surface in real points. Hence, in this case, of two polar-lines one cuts the surface in real points and the other in imaginary points.

8·31. The equations $\dfrac{\partial F}{\partial x} = 0$, $\dfrac{\partial F}{\partial y} = 0$, $\dfrac{\partial F}{\partial z} = 0$, $\dfrac{\partial F}{\partial w} = 0$ represent the polar-planes of the vertices of the tetrahedron of reference. In general these form a tetrahedron. In a special case they may pass through one point. Eliminating x, y, z, w between these equations we obtain the condition

$$\Delta \equiv \begin{vmatrix} a & h & g & p \\ h & b & f & q \\ g & f & c & r \\ p & q & r & d \end{vmatrix} = 0.$$

This determinant is called the *discriminant* of the quadric. If $C \equiv [x', y', z', w']$ is the point through which the polar-planes all pass, we have

$$0 = \Sigma x' \frac{\partial F}{\partial x'} = 2F(x', y', z', w'),$$

therefore C lies on the surface. If $P \equiv [x, y, z, w]$ is any other point on the surface the equation (1) of 8·2 is identically satisfied, and hence the line CP lies entirely in the surface. The surface therefore consists of lines passing through C, and is a cone with vertex C.

8·311. The quadric may be further specialised. If the polar-planes all have a line in common the quadric degenerates to two planes passing through this line. The condition for this is that the matrix [Δ] should be of rank 2. If the matrix [Δ] is of rank 1, all polar-planes coincide, and the quadric degenerates to two coincident planes.

8·32. Invariants.

An invariant property of a geometrical figure is a property of the figure which remains true when the figure is subjected to some geometrical transformation. Analytically, the figure is determined by certain equations in the coordinates, and a geometrical transformation consists of a set of equations by which the coordinates are changed into a new system. An invariant is then a function of the coefficients of the equations of the figure which is unaltered by the transformation, or alternatively the invariant property is represented by an equation in the coefficients of the figure which is unaltered by the transformation. These are not quite the same thing. In the first case the functions in question are *absolute* invariants; in the second we have to deal with functions of the coefficients whose ratios only remain unaltered; these are *relative* invariants.

Thus when a pair of points $P_1 \equiv [x_1, y_1]$, $P_2 \equiv [x_2, y_2]$ in a plane is subjected to a linear transformation

$$x = l_1 x' + m_1 y',$$
$$y = l_2 x' + m_2 y',$$

the function $x_1 y_2 - x_2 y_1$ becomes

$$(l_1 x_1' + m_1 y_1')(l_2 x_2' + m_2 y_2') - (l_1 x_2' + m_1 y_2')(l_2 x_1' + m_2 y_1')$$
$$= (l_1 m_2 - l_2 m_1)(x_1' y_2' - x_2' y_1').$$

If $x_1 y_2 - x_2 y_1 = 0$, then $x_1' y_2' - x_2' y_1' = 0$ also, i.e. the property that the line joining the two points passes through the origin is an invariant property for this transformation. If the transformation is orthogonal, so that $l_1 = \cos \alpha$, $m_1 = \sin \alpha$, $l_2 = -\sin \alpha$, $m_2 = \cos \alpha$, then $l_1 m_2 - l_2 m_1 = 1$ and $x_1 y_2 - x_2 y_1$ is invariant; it represents in fact double the area of the triangle $OP_1 P_2$. $x_1 y_2 - x_2 y_1$ is an *absolute invariant* for the orthogonal transformation, but only a relative invariant, or simply an invariant, for the general linear transformation. The factor

$$l_1 m_2 - l_2 m_1 \equiv \begin{vmatrix} l_1 & m_1 \\ l_2 & m_2 \end{vmatrix}$$

is called the *modulus* of the transformation.

Similarly in three dimensions if the coordinates are subjected to a linear transformation

$$x = l_1 x' + m_1 y' + n_1 z',$$
$$y = l_2 x' + m_2 y' + n_2 z',$$
$$z = l_3 x' + m_3 y' + n_3 z',$$

the function
$$\Delta \equiv \begin{vmatrix} x_1 & y_1 & z_1 \\ x_2 & y_2 & z_2 \\ x_3 & y_3 & z_3 \end{vmatrix}$$

of the coordinates of three points becomes

$$\begin{vmatrix} l_1 & m_1 & n_1 \\ l_2 & m_2 & n_2 \\ l_3 & m_3 & n_3 \end{vmatrix} \begin{vmatrix} x_1' & y_1' & z_1' \\ x_2' & y_2' & z_2' \\ x_3' & y_3' & z_3' \end{vmatrix} = L\Delta'.$$

Δ is then an invariant and L is the modulus of the transformation.

Ex. Interpret the equation $\Delta = 0$, and the value of the invariant when the transformation is an orthogonal one.

8·33. We now define more precisely an invariant as a function $I(a)$ of the coefficients, represented collectively by a, in the equations which determine a figure, such that if a' represent the new coefficients after a linear transformation $x_r = \sum_s l_{rs} x_s'$

$$I(a') = \phi(l_{rs}) I(a),$$

the factor $\phi(l_{rs})$ being a function of the constants l_{rs} belonging to the equations of transformation. We shall prove that *if L is the determinant $|l_{rs}|$ and $I(a)$ is a polynomial function of a, then $\phi(l_{rs})$ is a power of L.* To prove this we shall require the following:

8·331. LEMMA. *A determinant cannot be expressed as a product of factors rational and integral as regards the elements l_{rs}.* For suppose $\Delta = \phi\psi$. Since Δ is linear in the elements of any row or column, no element can be contained in both factors. Suppose that ϕ contains l_{pq} and that ψ contains l_{rs}. Then ϕ cannot contain any of the elements l_{pv} or l_{uq} except l_{pq}, and ψ cannot contain any of the elements l_{rv} or l_{us} except l_{rs}. Hence neither ϕ nor ψ can contain the elements l_{ps} and l_{rq}, which are both involved in Δ. The factorisation is therefore impossible.

8·332. Now the inverse transformation, which expresses x' in terms of x, is

$$Lx_s' = \Sigma L_{rs} x_r,$$

where L_{rs} is the cofactor of l_{rs} in the determinant; hence

$$I(a) = \phi(L_{rs}/L) I(a').$$

We have therefore

$$\phi(l_{rs}) . \phi(L_{rs}/L) = 1.$$

Now $\phi(l_{rs})$ is rational and integral in l_{rs}, and $\phi(L_{rs}/L)$ is rational and integral in L_{rs}/L. The latter can be made integral in L_{rs} by multiplying by a power of L, say L^k, and then, L_{rs} being expressed in terms of l_{rs}, we have

$$\phi(l_{rs}) . \psi(l_{rs}) = L^k.$$

But by the lemma the determinant L has no rational factors in l_{rs}, hence both $\phi(l_{rs})$ and $\psi(l_{rs})$ must be powers of L.

If $\phi(l_{rs}) = L^w$, w is called the *weight* of the invariant I.

8·34. Projective invariant of the general quadric.

The general quadratic expression $\Sigma\Sigma a_{rs} x_r x_s$ is transformed by the general linear equations

$$x_r = \Sigma l_{ri} x_i' \quad (r = 0, 1, 2, 3)$$

into a similar expression

$$\Sigma\Sigma a_{rs}' x_r' x_s'.$$

The condition that the quadric should be specialised as a cone is that the discriminant $\Delta \equiv |a_{rs}|$ should vanish. This is clearly an invariant property.

Now
$$\Sigma\Sigma_{r\,s} a_{rs} x_r x_s = \Sigma\Sigma_{i} (a_{rs} \Sigma_{i} l_{ri} x_i' \Sigma_{j} l_{sj} x_j')$$
$$= \Sigma\Sigma_{i\,j} (a_{ij} \Sigma_{r} l_{ir} x_r' \Sigma_{s} l_{js} x_s')$$
$$= \Sigma\Sigma\Sigma\Sigma_{i\,j\,r\,s} a_{ij} l_{ir} l_{js} x_r' x_s'.$$

Hence
$$a_{rs}' = \Sigma\Sigma_{i\,j} a_{ij} l_{ir} l_{js} = \Sigma_{i} (l_{ir} \Sigma_{j} l_{js} a_{ij}).$$

Write $$\sum_j l_{js} a_{ij} = c_{is},$$

then $$a_{rs}' = \sum_i l_{ir} c_{is}.$$

Hence the determinant

$$\Delta' \equiv |\, a_{rs}' \,| = |\, c_{rs} \,| \, |\, l_{rs} \,|.$$

But $$|\, c_{rs} \,| = |\, a_{rs} \,| \, |\, l_{rs} \,|.$$

Therefore $$|\, a_{rs}' \,| = |\, a_{rs} \,| \, |\, l_{rs} \,|^2,$$

i.e. $$\Delta' = L^2 \Delta,$$

where $$L = |\, l_{rs} \,|.$$

Hence the discriminant $\Delta \equiv |\, a_{rs} \,|$ is an invariant, and the modulus of the transformation is $L \equiv |\, l_{rs} \,|$. As the general linear equations represent the general projective transformation, Δ is called a *projective invariant*.

8·341. *A quadric cannot have an absolute projective invariant* for the equations of the transformation contain sixteen constants l_{ij} while the equation of the quadric contains only ten. It is therefore possible in an infinite variety of ways to transform the given quadratic expression (not necessarily by a real transformation) into any other quadratic expression, provided only it is not specialised; projectively, any unspecialised quadric is equivalent to any other unspecialised quadric.

8·342. From this it can be deduced that *a quadric cannot have more than one projective invariant*. For if I_1 and I_2 were two invariants, so that

$$I_1' = L^\alpha I_1 \quad \text{and} \quad I_2' = L^\beta I_2,$$

then $$\frac{I_1'^{\beta}}{I_2'^{\alpha}} = \frac{I_1^{\beta}}{I_2^{\alpha}},$$

and I_1^{β}/I_2^{α} would be an absolute invariant.

8·35. Condition for real generating lines.

Since Δ is of the fourth degree in the coefficients, its sign will remain always the same for any real transformation. The sign of Δ is therefore an invariant for real projective transformations.

Let P be any point on the quadric. Take a tetrahedron of reference with one vertex $P \equiv [1, 0, 0, 0]$, and one plane the tangent-plane at $P(w=0)$. Then the equation of the quadric becomes

$$by^2 + cz^2 + dw^2 + 2fyz + 2pxw + 2qyw + 2rzw = 0.$$

$w = 0$ cuts the surface where

$$by^2 + cz^2 + 2fyz = 0.$$

This gives real lines if $f^2 - bc > 0$. Now the discriminant of the quadric is

$$\Delta \equiv \begin{vmatrix} 0 & 0 & 0 & p \\ 0 & b & f & q \\ 0 & f & c & r \\ p & q & r & d \end{vmatrix} = p^2(f^2 - bc).$$

Hence *the condition for real generating lines is that the discriminant should be positive.* Conversely, if the discriminant is positive, and the quadric has real points, it has also real lines.

8·4. Polarity.

The relation of pole and polar with regard to a quadric establishes a $(1, 1)$ correspondence or correlation of a dualistic kind between points and planes, lines and lines. To every point corresponds a plane, to every plane a point, and to every line another line.

If the equation of the quadric is written

$$\Sigma\Sigma a_{rs}x_r x_s = 0 \quad (r, s = 0, 1, 2, 3),$$

where $a_{rs} = a_{sr}$, the equation of the polar of the point

$$(x_0', x_1', x_2', x_3'),$$

or simply (x'), is $\qquad \Sigma\Sigma a_{rs}x_r' x_s = 0.$

We may represent a plane by its own coordinates, which are proportional to the coefficients of x_0, x_1, x_2, x_3 in its equation. Denoting the coordinates of the polar-plane of (x) by

$$\xi_0', \xi_1', \xi_2', \xi_3',$$

we have then

$$\xi_0' = a_{00}x_0 + a_{01}x_1 + a_{02}x_2 + a_{03}x_3, \text{ etc.,}$$

or $\qquad \xi_r' = \Sigma a_{rs}x_s \quad (r = 0, 1, 2, 3).$

If a point lies on its own polar it lies on the quadric, and if a plane passes through its own polar it is a tangent-plane to the quadric.

8·41. Correlations.

We shall consider now the general dual correlation between points and planes determined by the equations

$$\xi_r' = \sum_s a_{rs} x_s \quad (r=0,1,2,3), \qquad \ldots\ldots(1)$$

without making the assumption $a_{rs} = a_{sr}$, that is we do not assume any quadric to begin with.

To bring out the correlation more clearly we consider all the points and planes twice over; we have a space S consisting of points (x) and planes (ξ); to the points (x) correspond planes (ξ') and to the planes (ξ) correspond points (x'), and these form a space S'.

To determine the point (x) in S which corresponds to the plane (ξ') in S' we have to solve the equations (1) for x in terms of ξ'. Let D represent the determinant

$$D \equiv \begin{vmatrix} a_{00} & \ldots & a_{03} \\ \ldots\ldots\ldots\ldots \\ a_{30} & \ldots & a_{33} \end{vmatrix},$$

and let A_{rs} denote the cofactor of a_{rs}, so that

$$\sum a_{rs} A_{rs} = D, \text{ and } \sum a_{rs} A_{ts} = 0 \text{ if } r \neq t.$$

Multiplying the equations (1) by $A_{0s}, A_{1s}, A_{2s}, A_{3s}$ respectively and adding, we get

$$D x_s = \sum A_{rs} \xi_r' \quad (s=0,1,2,3). \qquad \ldots\ldots(2)$$

Now if the point (x) lies on the corresponding plane (ξ'), $\sum \xi_r' x_r = 0$, hence by (1)

$$\sum\sum a_{rs} x_s x_r = 0, \qquad \ldots\ldots(3)$$

and therefore (x) lies on a certain quadric F.

If the plane (ξ') passes through the corresponding point (x), $\sum x_s \xi_s' = 0$, hence by (2)

$$\sum\sum A_{rs} \xi_r' \xi_s' = 0,$$

and therefore (ξ') is tangent to a certain quadric-envelope

$$\Phi \equiv \sum\sum A_{rs} \xi_r \xi_s = 0. \qquad \ldots\ldots(4)$$

These two quadrics F and Φ are not in general the same. If (x) lies on F then (ξ') is tangent to Φ but not in general to F.

To the point (x) in S corresponds the plane (ξ') in S' where $\xi_r' = \Sigma a_{rs} x_s$. If x' are the current coordinates of a point on this plane we have then

$$\Sigma\Sigma a_{rs} x_s x_r' = 0.$$

This equation connects the coordinates of a point (x) in S and a point (x') which lies in the plane corresponding to (x). If now (x') is fixed and (x) variable this equation represents the locus of (x), a plane in S which corresponds to the point (x') in S'. If ξ_s are the coordinates of this plane (coefficients of x_s) we have

$$\xi_s = \sum_r a_{rs} x_r',$$

or, interchanging r and s,

$$\xi_r = \sum_s a_{sr} x_s'.$$

Hence if the point (x) is considered as belonging to the space S the corresponding plane is $\xi_r' = \Sigma a_{rs} x_s$, but if (x) is considered as belonging to the space S' the corresponding plane is

$$\xi_r'' = \Sigma a_{sr} x_s.$$

8·42. In order that the correlation may be a polarity the two planes which correspond to any point must coincide. Hence the expressions $\Sigma a_{rs} x_s$ and $\Sigma a_{sr} x_s$ must be proportional and therefore

$$a_{rs} = \rho a_{sr},$$

where ρ is a constant factor.

Interchanging r and s,

$$a_{sr} = \rho a_{rs}.$$

Hence $$\rho^2 = 1 \text{ and } \rho = \pm 1.$$

If $\rho = +1$, $a_{rs} = a_{sr}$ and we have a *polarity*.

8·43. If $\rho = -1$, $a_{rs} = -a_{sr}$, $a_{rr} = 0$. In this case F and Φ do not exist since their coefficients all vanish. Every point lies on its corresponding plane and *vice versa*. The correlation in this case is called a *null-system*.

8·431. The term "null-system" is derived from statics. Any system of forces in three dimensions can be reduced to three forces X, Y, Z, acting along three arbitrarily chosen rectangular axes, together with three couples L, M, N in the three coordinate-planes. If $P \equiv [x', y', z']$ is any point the moments about axes through P parallel to the coordinate-axes are $L - Zy' + Yz'$, $M - Xz' + Zx'$,

$N - Yx' + Xy'$, and the plane through P whose direction-cosines are proportional to these three moments has the property that the sum of the moments about any line through P in this plane is zero; it is a *null* plane. Using homogeneous coordinates, if $P \equiv [x, y, z, w]$ *and* the plane is $\xi'x + \eta'y + \zeta'z + \omega'w = 0$ we have

$$\rho\xi' = \qquad\quad - Zy + Yz + Lw,$$
$$\rho\eta' = \quad Zx \qquad - Xz + Mw,$$
$$\rho\zeta' = - Yx + Xy \qquad\quad + Nw,$$
$$\rho\omega' = -Lx - My - Nz.$$

These equations determine a null-system.

8·432. In a null-system, if (x) and (y) are any two points on a line l, the plane corresponding to a variable point on l, say $(x + ty)$, is

$$\xi_r' = \Sigma a_{rs}(x_s + ty_s) = \Sigma a_{rs}x_s + t\Sigma a_{rs}y_s = \xi_r + t\eta_r,$$

and this passes through the fixed line l' in which the planes (ξ) and (η) intersect. Thus to a line l corresponds a unique line l', as in a polarity. If the coordinates of l are (p) and those of l' are (p') we have

$$p_{23}' = \xi_0\eta_1 - \xi_1\eta_0$$
$$= (a_{01}x_1 + a_{02}x_2 + a_{03}x_3)(a_{10}y_0 + a_{12}y_2 + a_{13}y_3)$$
$$\qquad - (a_{10}x_0 + a_{12}x_2 + a_{13}x_3)(a_{01}y_1 + a_{02}y_2 + a_{03}y_3)$$
$$= a_{01}(a_{01}p_{01} + a_{02}p_{02} + a_{03}p_{03} + a_{23}p_{23} + a_{31}p_{31} + a_{12}p_{12})$$
$$\qquad - (a_{01}a_{23} + a_{02}a_{31} + a_{03}a_{12})p_{23}, \text{ etc.}$$

The lines l and l' will coincide if

$$\Sigma a_{ij}p_{ij} = 0.$$

This linear equation determines a *linear complex* of the most general form. Hence a linear complex consists of the self-corresponding lines of a null-system. If the polar-plane of P contains Q it contains the line PQ; then reciprocally the polar-plane of Q contains P and therefore also contains the line PQ. The self-corresponding lines are thus the lines through any point and lying in the polar-plane of the point.

We have assumed that

$$a_{01}a_{23} + a_{02}a_{31} + a_{03}a_{12}$$

is not zero. If this vanishes, p_{23}' is proportional to a_{01}. Thus there is a line whose coordinates are (a), and this line corresponds to every line of the null-system except such as make $\Sigma a_{ij}p_{ij}=0$; but these are the lines which intersect (a), and the lines which correspond to these are indeterminate. In this case $\Sigma a_{ij}p_{ij}=0$ represents a *special* linear complex consisting of all the lines which cut a fixed line (the *directrix*).

8·51. Canonical equation of a quadric.

If we choose as the fundamental tetrahedron $A_0A_1A_2A_3$ one which is self-polar with respect to the quadric, i.e. so that the polar of the point [1, 0, 0, 0] is the plane [1, 0, 0, 0], and so on, we have $a_{rs}=0$ where $r\neq s$, and the equation reduces to the form

$$a_0x_0^2+a_1x_1^2+a_2x_2^2+a_3x_3^2=0.$$

If the coefficients are all of the same sign, the quadric can have no real points, and is therefore *virtual*. There are two other cases according to the number of negative signs.

Changing the notation, suppose the equation to be

$$a^2x^2+b^2y^2-c^2z^2-d^2w^2=0.$$

This may be written

$$(ax-cz)(ax+cz)=(dw-by)(dw+by)$$

and is satisfied by either

$$\begin{aligned}(ax-cz)&=\lambda(dw-by)\\ \lambda(ax+cz)&=(dw+by)\end{aligned}\Big\} \ \text{or} \ \begin{aligned}(ax-cz)&=\mu(dw+by)\\ \mu(ax+cz)&=(dw-by)\end{aligned}\Big\},$$

where λ and μ are any parameters. Hence the systems of lines represented by these pairs of equations lie entirely in the surface. The surface in this case has real generating lines.

If the equation is

$$a^2x^2+b^2y^2+c^2z^2-d^2w^2=0$$

the plane $w=0$ meets the surface in no real points and hence there can be no real lines on the surface.

8·52. Specialised and degenerate quadrics.

If one of the coefficients of the canonical equation, say a_0, is zero, the polar-plane of any point x' is

$$a_1x_1'x_1+a_2x_2'x_2+a_3x_3'x_3=0$$

and passes through the fixed point $A_0\equiv(1, 0, 0, 0)$, which lies on the quadric. If P is any point on the quadric, A_0 and P are conjugate and therefore the line $A_0 P$ lies entirely in the quadric. The quadric therefore consists of a *cone* with vertex A_0.

If two of the coefficients vanish, say a_0 and a_1, the equation reduces to

$$a_2 x_2{}^2 + a_3 x_3{}^2 = 0$$

which represents *two planes*.

If three of the coefficients vanish it reduces to *two coincident planes*.

8·53. Projective classification of conics.

Before completing the projective classification of quadrics it will be useful to do the same for conics. We consider the general conic

$$S \equiv ax^2 + by^2 + cz^2 + 2fyz + 2gzx + 2hxy = 0.$$

The conic has one projective invariant, viz. the discriminant

$$D \equiv \begin{vmatrix} a & h & g \\ h & b & f \\ g & f & c \end{vmatrix}.$$

That this is an invariant is proved in the same way as for the quadric (see 8·34), and its vanishing is the condition that the conic should degenerate to two straight lines. We denote as usual the cofactors of each element of this determinant by the corresponding capital letter. We have identically

$$aS \equiv (ax + hy + gz)^2 + (Cy^2 - 2Fyz + Bz^2).$$

The discriminant of $Cy^2 - 2Fyz + Bz^2$ is $BC - F^2 \equiv aD$, and if $D = 0$, S is the sum or difference of two squares. If also $B = 0$, then F and H vanish since $F^2 = BC$ and $H^2 = AB$. If $D = 0$ and A and B both vanish, then all the minors of the determinant vanish. Since also $G^2 = AB$, it follows that A, B, C are all of the same sign (or zero). Hence we deduce the results:

8·531. *When the matrix $[D]$ is of rank 2, $S = 0$ represents two distinct straight lines, which are real if no one of A, B, C is positive imaginary if no one of A, B, C is negative* (one may be zero).

8·532. *When the matrix $[D]$ is of rank 1, $S = 0$ represents two coincident lines.*

We shall assume now that $D \neq 0$. Consider the pencil of lines $z = \lambda y$. The intersections with the conic are given by the equation

$$ax^2 + 2(\lambda g + h)xy + (c\lambda^2 + 2f\lambda + b)y^2 = 0. \quad \ldots\ldots(1)$$

The discriminant of this quadratic in x/y is

$$(\lambda g + h)^2 - a(c\lambda^2 + 2f\lambda + b) = -B\lambda^2 + 2F\lambda - C. \quad \ldots\ldots(2)$$

The roots of equation (1) are real or imaginary according as (2) is positive or negative. If the quadratic (2) has real roots there are two critical values of λ which separate the lines of the pencil into those which cut the conic and those which do not. The condition for this is $BC - F^2 < 0$, i.e. $aD < 0$. In this case the conic is real and the vertex [1, 0, 0] of the pencil is outside.

If $aD > 0$ the roots of (2) are imaginary, and this quadratic is of fixed sign, $+$ or $-$ according as B and C are both $-$ or both $+$. In the former case the lines of the pencil all cut the conic in real points; the conic is real, and the point [1, 0, 0] is inside. In the other case the lines of the pencil all cut the conic in imaginary points, and the conic is virtual.

If $a = 0$ the point [1, 0, 0] lies on the conic, and the conic is real.

Hence we deduce the result

8·533. *When the matrix* [D] *is of rank* 3, $S = 0$ *represents a proper conic, which is virtual if* aD, bD, cD, A, B, C *are all positive; otherwise it is real.*

Ex. When the conic $f(x, y, z) = 0$ is real, prove that the point [x, y, z] is inside or outside according as $D.f(x, y, z)$ is positive or negative.

8·54. Projective classification of quadrics.

The general quadric

$$\Sigma\Sigma a_{rs} x_r x_s = 0$$

has one projective invariant, viz. the discriminant

$$\Delta \equiv |\, a_{rs}\,|,$$

and its vanishing is the condition that the quadric should be a cone. The cone may be real or virtual, and in either case there is one real point, viz. the vertex, whose coordinates are [$A_{i0}, A_{i1}, A_{i2}, A_{i3}$], where i may be 0, 1, 2 or 3 (provided these four minors are not all zero). If $A_{0j} = 0$ it follows (when $\Delta = 0$) that A_{1j}, A_{2j} and A_{3j} also vanish; the vertex of the cone then lies on $x_j = 0$. The condition for a virtual cone is that any plane, not through the vertex, should cut the cone in a virtual conic. Hence by 8·533 if none of the minors vanish all the quantities $a_{ii}A_{jj}$ and $a_{ii}a_{jj}-a_{ij}{}^2$ ($i \neq j = 0$, 1, 2, 3) must be positive.

If all the minors vanish the quadric degenerates to two planes, their intersection being a real line; and if all the minors of the second order vanish, the quadric degenerates to two coincident planes.

Hence we deduce the results:

8·541. *When the matrix* $[\Delta]$ *is of rank 3, the quadric is a cone, which is virtual if no one of the quantities* $a_{ii}A_{jj}$, $a_{ii}a_{jj}-a_{ij}{}^2$ *is negative, otherwise it is real.*

8·542. *When the matrix* $[\Delta]$ *is of rank 2, the quadric degenerates to two distinct planes, which are imaginary if no one of the quantities* $a_{ii}a_{jj}-a_{ij}{}^2$ *is negative, otherwise they are real.*

8·543. *When the matrix* $[\Delta]$ *is of rank 1, the quadric degenerates to two coincident planes.*

We shall now suppose that $\Delta \neq 0$. The discriminant of the canonical equation is $a_0a_1a_2a_3$, and the sign of the discriminant is an invariant for real transformations. If $\Delta > 0$, the signs of the coefficients a_0, a_1, a_2, a_3 are either (1) all the same, in which case the quadric is virtual, or (2) two positive and two negative, in which case the quadric has real generating lines. If $\Delta < 0$, the signs are either one negative and three positive, or one positive and three negative; in either case the quadric has real points but imaginary lines.

Suppose first that $\Delta > 0$. The quadric is then either virtual or with real generating lines according as any one plane section is virtual or real. Taking the sections $x_0 = 0, ..., x_3 = 0$, and applying 8·533, we obtain the results:

8·544. *When* $\Delta > 0$, *the quadric is virtual if all the quantities* $a_{ii}A_{jj}, a_{ii}a_{jj}-a_{ij}{}^2$ *are positive, otherwise it has real generating lines.*

8·545. *When* $\Delta < 0$, *the quadric has real points but imaginary lines.*

Ex. 1. If a quadric has real generating lines prove that the tangent-planes through a given line are real or imaginary according as the line cuts the quadric in real or imaginary points, and *vice versa* for a real quadric with imaginary generating lines.

Ex. 2. For a quadric with real generating lines prove that of two non-intersecting polar-lines either both cut the surface in real points or neither does.

Ex. 3. A real quadric $S \equiv x^2 + y^2 + z^2 - w^2 = 0$ whose generating lines are imaginary divides space into two regions, the interior for $S < 0$ and the exterior for $S > 0$. Every line through an interior point cuts the quadric in real points, and through a line which lies entirely in the exterior region pass two real tangent-planes.

Ex. 4. A quadric whose generating lines are real does not divide space projectively into two regions.

By the real collineation $\rho x' = z, \rho y' = w, \rho z' = x, \rho w' = y$ the expression $S \equiv x^2 + y^2 - z^2 - w^2$ is transformed into $-S$. This transformation is not a perspective transformation or homology in which there is a fixed centre and a plane of fixed points; in this transformation, which is sometimes called a *skew involution*, there are two lines of fixed points $x = z, y = w$ and $x = -z, y = -w$. There is no actual homology which will change S into $-S$.

Ex. 5. Verify that

$$F(x, y, z, w) \equiv \frac{1}{a} (ax + hy + gz + pw)^2 + \frac{1}{aC} \{Cy - Fz + (aq - hp) w\}^2$$
$$+ \frac{1}{CD_1} (D_1 z - R_1 w)^2 + \frac{\Delta}{D_1} w^2,$$

where $C = ab - h^2$, $F = gh - af$, and D_1, R_1 are the cofactors of d, r in the determinant Δ.

8·6. We have now to consider the quadric in its metrical aspect, that is in relation to the plane at infinity.

If C is the pole of the plane at infinity, the polar-planes of all points at infinity pass through C. And since the harmonic conjugate of the point at infinity on the join of two points P, Q is the mid-point of PQ, all chords through C are bisected at C. C is therefore the centre of symmetry, or *centre*, of the quadric, planes through C are *diametral planes*, chords through C are *diameters*.

8·61. Diametral planes.

Since the equation of the polar-plane of $[x', y', z', w']$ is symmetrical in the two sets of coordinates, it can be written also in the form

$$x' \frac{\partial F}{\partial x} + y' \frac{\partial F}{\partial y} + z' \frac{\partial F}{\partial z} + w' \frac{\partial F}{\partial w} = 0.$$

Let $w=0$ represent the plane at infinity, then the polar-plane of the point at infinity $[x', y', z', 0]$ is

$$x'\frac{\partial F}{\partial x} + y'\frac{\partial F}{\partial y} + z'\frac{\partial F}{\partial z} = 0.$$

This is the *diametral plane conjugate to the direction* $[x', y', z']$.

8·62. For all values of x', y', z' this represents a plane through the common point of the three planes

$$\frac{\partial F}{\partial x} = 0, \quad \frac{\partial F}{\partial y} = 0, \quad \frac{\partial F}{\partial z} = 0,$$

or shortly F_x, F_y, F_z. This point is the centre C. The three planes $F_x = 0$, $F_y = 0$, $F_z = 0$ are the diametral planes conjugate to the directions of the coordinate-axes.

8·63. The conic at infinity on a quadric.

The plane at infinity $w = 0$ cuts the quadric in a conic whose equations are

$$f(x, y, z) \equiv ax^2 + by^2 + cz^2 + 2fyz + 2gzx + 2hxy = 0 \atop w = 0 \Big\}.$$

The discriminant of the first equation is

$$D \equiv \begin{vmatrix} a & h & g \\ h & b & f \\ g & f & c \end{vmatrix}.$$

If $D = 0$, the conic breaks up into two straight lines. The plane at infinity is then a tangent-plane. No significance is attachable to the sign of D when it is not zero, since its sign would be changed by changing the signs of all the coefficients. If $D \neq 0$ the conic is either a real proper conic or virtual. For a conic at infinity there is no distinction corresponding to ellipse, parabola or hyperbola, for this refers to the nature of its intersections with the line at infinity in its plane. As the equation by itself represents a cone we see also that cones are distinguished only as real or virtual. The section of a real cone may be any type of conic.

By 8·53 we obtain the following criteria for the nature of the conic at infinity:

If the matrix $[D]$ is of rank 3, the conic at infinity is virtual if aD, bD, cD, A, B, C are all positive, otherwise it is real; if the matrix $[D]$ is of rank 2, the conic becomes two straight lines which are real if no one of A, B, C is positive, imaginary if no one of A, B, C is negative; and if the matrix $[D]$ is of rank 1, the conic reduces to two coincident lines.

8·64. Metrical classification of quadrics.

(A) If the three planes F_x, F_y, F_z have in common a unique finite point C, the surface is called a *central quadric*. This is the general case, no special condition being assigned except that $D \neq 0$.

If $\Delta = 0$ the quadric is specialised as a *cone*.

If $\Delta < 0$ the quadric has real points and imaginary lines, and is either an ellipsoid or a hyperboloid of two sheets.

If $\Delta > 0$ the quadric is either virtual, or a hyperboloid of one sheet.

To distinguish these further we consider whether the section by the plane at infinity is real or virtual. Hence by the last paragraph we obtain the results:

$\Delta > 0$, aD, bD, cD, A, B, C all positive. *Virtual quadric.*

              „   „   not all positive. *Hyperboloid of one sheet.*

$\Delta < 0$,   „   „   all positive. *Ellipsoid.*

              „   „   not all positive. *Hyperboloid of two sheets.*

$\Delta = 0$,   „   „   all positive. *Virtual cone.*

              „   „   not all positive. *Real cone.*

(B) If the three planes F_x, F_y, F_z have in common a unique point at infinity, all diameters pass through this point at infinity and are therefore parallel. Further, since the polar-plane of C passes through C, C is a point on the surface, and the plane at infinity is the tangent-plane at C. The surface is a paraboloid. The analytical condition that C should be a point at infinity is $D = 0$. The conic at infinity then becomes a pair of straight lines, real or imaginary according as A, B, C are negative or positive.

       $D = 0$, A, B, C all positive. *Elliptic paraboloid.*

                „   all negative. *Hyperbolic paraboloid.*

(C) If the planes F_x, F_y, F_z have in common a unique finite line, there is no unique centre, but a line of centres or axis. The surface is a cylinder, elliptic or hyperbolic. The condition is that the matrix

$$\begin{bmatrix} a & h & g & p \\ h & b & f & q \\ g & f & c & r \end{bmatrix}$$

should be of rank 2. It is equivalent to the two conditions $\Delta = 0$, $D = 0$. As in the case of the paraboloid, the conic at infinity again breaks up into two straight lines, real in the case of the hyperbolic cylinder, imaginary in the case of the elliptic cylinder.

$\Delta = 0$, $D = 0$, A, B, C all positive. *Elliptic cylinder.*
  „   all negative. *Hyperbolic cylinder.*

If the matrix $[\Delta]$ is of rank 2, the quadric degenerates to *two planes*, real or imaginary according as A, B, C are all negative or all positive (three conditions).

(D) If the planes F_x, F_y, F_z have in common a unique line at infinity the surface is a parabolic cylinder. The condition is that the matrix

$$\begin{bmatrix} a & h & g & p \\ h & b & f & q \\ g & f & c & r \\ 0 & 0 & 0 & 1 \end{bmatrix}$$

should be of rank 2, and this is equivalent to the condition that the matrix $[D]$ should be of rank 1. The conic at infinity then reduces to two coincident straight lines. This is equivalent to the three conditions $D = 0$, $A = 0$, $B = 0$.

$\Delta = 0$, $D = 0$, $A = 0$, $B = 0$, $C = 0$. *Parabolic cylinder.*

(E) If the planes F_x, F_y, F_z coincide, forming a finite plane, we have a plane of centres. The surface degenerates to *two parallel planes*. The condition is that the matrix

$$\begin{bmatrix} a & h & g & p \\ h & b & f & q \\ g & f & c & r \end{bmatrix}$$

should be of rank 1. This is equivalent to the conditions that the matrix $[\Delta]$ should be of rank 2 and A, B, C be all zero.

(F) If the planes F_x, F_y, F_z all coincide with the plane at infinity, the surface degenerates to the *plane at infinity and another plane*. The condition is that the matrix [D] should be of rank o, i.e. that all the coefficients a, b, c, f, g, h should vanish.

(G) If the matrix
$$\begin{bmatrix} a & h & g & p \\ h & b & f & q \\ g & f & c & r \end{bmatrix}$$

is of rank o the quadric reduces to the *plane at infinity twice*.

8·65. Principal diametral planes.

A diametral plane is said to be a principal plane when it is perpendicular to its conjugate direction, and this direction is called a principal direction.

The direction-cosines of the diametral plane conjugate to the direction $[l, m, n]$ are proportional to
$$\frac{\partial f}{\partial l}, \quad \frac{\partial f}{\partial m}, \quad \frac{\partial f}{\partial n},$$
i.e. $\quad al+hm+gn, \quad hl+bm+fn, \quad gl+fm+cn.$
Hence if $[l, m, n]$ is a principal direction
$$\frac{al+hm+gn}{l}=\frac{hl+bm+fn}{m}=\frac{gl+fm+cn}{n}=\lambda, \text{ say.}$$

These give two homogeneous equations in l, m, n, and therefore there are a finite number of principal directions.

We have
$$(a-\lambda)l+hm \quad +gn \quad =o,$$
$$hl \quad +(b-\lambda)m+fn \quad =o,$$
$$gl \quad +fm \quad +(c-\lambda)n=o.$$

Eliminating l, m, n
$$\begin{vmatrix} a-\lambda & h & g \\ h & b-\lambda & f \\ g & f & c-\lambda \end{vmatrix}=o,$$

i.e. $\quad \lambda^3-(a+b+c)\lambda^2+(A+B+C)\lambda-D=o,$

where
$$A=bc-f^2, \qquad D=\begin{vmatrix} a & h & g \\ h & b & f \\ g & f & c \end{vmatrix}.$$
$$B=ca-g^2,$$
$$C=ab-h^2,$$

This equation, which is called the *Discriminating cubic*, gives three values for λ, and hence we get three sets of values of l, m, n.

8·71. *The roots of the discriminating cubic are all real.*

Assume $a > b > c$, and let

$$\phi(\lambda) \equiv \lambda^3 - (a+b+c)\lambda^2 + (A+B+C)\lambda - D$$
$$\equiv (\lambda-a)\{(\lambda-b)(\lambda-c)-f^2\} - \{(\lambda-b)g^2 + (\lambda-c)h^2 + 2fgh\}.$$

Consider also the function

$$\psi(\lambda) \equiv (\lambda-b)(\lambda-c) - f^2.$$

When
$$\lambda \quad = -\infty, \quad c, \quad b, \quad +\infty,$$
$$\psi(\lambda) = +\infty, \quad -f^2, \quad -f^2, \quad +\infty.$$

Hence $\psi(\lambda) = 0$ has real roots α, β such that

$$-\infty < \beta < c \leqslant b < \alpha < +\infty.$$

Then $\qquad \phi(\alpha) = -\{(\alpha-b)g^2 + (\alpha-c)h^2 + 2fgh\},$

and since $(\alpha-b)(\alpha-c) = f^2$, $\phi(\alpha)$ is a perfect square, viz.

$$(\alpha-b)\phi(\alpha) \equiv -\{(\alpha-b)g + fh\}^2;$$

and similarly for $\phi(\beta)$. Hence $\phi(\alpha) < 0$ and $\phi(\beta) > 0$.

Hence substituting in $\phi(\lambda)$, when

$$\lambda = \quad -\infty, \quad \beta, \quad \alpha, \quad +\infty,$$
$$\phi(\lambda) \text{ is} \quad - \quad + \quad - \quad +.$$

Hence the equation $\phi(\lambda) = 0$ has three real roots, separated by α and β. We have supposed that $\alpha \neq \beta$. When $\alpha = \beta$, we must have $(b-c)^2 + 4f^2 = 0$, hence $b = c$ and $f = 0$. Then

$$\phi(\lambda) = (\lambda-a)(\lambda-b)^2 - (g^2+h^2)(\lambda-b)$$
$$= (\lambda-b)\{(\lambda-a)(\lambda-b) - (g^2+h^2)\}$$
$$= (\lambda-b)\{\lambda^2 - (a+b)\lambda + ab - g^2 - h^2\}$$
$$= (\lambda-b)[\{\lambda - \tfrac{1}{2}(a+b)\}^2 - \tfrac{1}{4}(a-b)^2 - g^2 - h^2].$$

Hence the roots of $\phi(\lambda) = 0$ are all real. One root is $b(=\alpha)$, one is $< b$ and the remaining one is $> b$.

8·72. Multiple roots of the discriminating cubic.

The occurrence of repeated roots of the equation is connected with the rank of the matrix $[D]$. The result is stated most conveniently for the general equation of this form, viz.

$$D(\lambda) \equiv \begin{vmatrix} a_{11}+\lambda & a_{12} & \cdots & a_{1n} \\ a_{12} & a_{22}+\lambda & \cdots & a_{2n} \\ \cdots\cdots\cdots\cdots\cdots\cdots\cdots \\ a_{1n} & a_{2n} & \cdots & a_{nn}+\lambda \end{vmatrix} = 0,$$

which is called the *characteristic equation* of the square matrix $[a_{nn}]$; it is sometimes called the *secular equation* as it arises in astronomy in connection with the secular perturbations of planets. The general theorem, which is due to Weierstrass, is as follows:

If λ is a p-repeated root of the characteristic equation, when this value is substituted the matrix $[D(\lambda)]$ is of rank $n - p$; and conversely and the proof depends upon the lemma:

8·721. *If m is the rank of a symmetrical matrix, the product of two principal determinants of order m is equal to the square of another determinant of order m of the matrix*; and further *the sum of all the principal determinants of order m cannot vanish.*

We shall consider only the determinant of the third order

$$D \equiv \begin{vmatrix} a & h & g \\ h & b & f \\ g & f & c \end{vmatrix}.$$

If the matrix $[D]$ is of rank 3, $D \neq 0$. If it is of rank 2, $D = 0$, but the minors of the second order do not all vanish. In this case we have $BC - F^2 = aD$, etc., i.e. $BC = F^2$, $CA = G^2$, $AB = H^2$, as stated in the lemma. Further, $A + B + C$ cannot vanish, for on squaring we obtain $A^2 + B^2 + C^2 + 2(F^2 + G^2 + H^2)$, which can only vanish if all the second-order minors vanish.

If $[D]$ is of rank 1, all the second-order minors vanish, so that $bc = f^2$, $ca = g^2$, $ab = h^2$. But $a + b + c$ cannot vanish, for on squaring this we obtain $\Sigma a^2 + 2\Sigma f^2$, which can only vanish if all the elements vanish.

Now consider the equation

$$D(\lambda) = \begin{vmatrix} a+\lambda & h & g \\ h & b+\lambda & f \\ g & f & c+\lambda \end{vmatrix} = 0.$$

For the values λ_1, λ_2, λ_3 the determinant vanishes and $[D(\lambda)]$ is of rank 2 at most. The condition for equal roots is that both $D(\lambda) = 0$ and $D'(\lambda) = 0$. But, differentiating with regard to λ, we have

$$D'(\lambda) = \alpha + \beta + \gamma,$$

where α, β, γ are the principal minors, i.e. $\alpha \equiv (b+\lambda)(c+\lambda)-f^2$, etc. Now by the above results if $[D(\lambda)]$ is of rank 2, $\alpha+\beta+\gamma$ cannot vanish, and hence the roots are all unequal.

If $[D(\lambda)]$ is of rank 1, α, β and γ all vanish and $D'(\lambda)=0$. Hence there are equal roots. The condition for three equal roots is further

$$0 = D''(\lambda) \equiv 2\{(a+\lambda)+(b+\lambda)+(c+\lambda)\},$$

but this, being the sum of the principal minors of first order, cannot vanish unless the determinant is of rank 0.

8·722. *Hence* (1) *if* $[D(\lambda)]$ *is of rank 3, λ is not a root,* (2) *if* $[D(\lambda)]$ *is of rank 2, λ is a simple root,* (3) *if* $[D(\lambda)]$ *is of rank 1, λ is a double root, and* (4) *if* $[D(\lambda)]$ *is of rank 0, λ is a triple root.*

8·73. *The three principal directions are in general mutually at right angles.* We have

$$\frac{\frac{\partial f}{\partial l}}{l} = \frac{\frac{\partial f}{\partial m}}{m} = \frac{\frac{\partial f}{\partial n}}{n} = 2\lambda,$$

where λ is a root of the discriminating cubic.

Since f is a homogeneous quadratic in l, m, n

$$l_1\frac{\partial f}{\partial l_2}+m_1\frac{\partial f}{\partial m_2}+n_1\frac{\partial f}{\partial n_2} = l_2\frac{\partial f}{\partial l_1}+m_2\frac{\partial f}{\partial m_1}+n_2\frac{\partial f}{\partial n_1}.$$

Hence $2\lambda_2(l_1 l_2+m_1 m_2+n_1 n_2) = 2\lambda_1(l_2 l_1+m_2 m_1+n_2 n_1).$

Therefore, provided $\lambda_1 \neq \lambda_2$,

$$l_1 l_2+m_1 m_2+n_1 n_2 = 0.$$

In the alternative case $\lambda_1 = \lambda_2$ we shall see (8·9) that the surface, if real, is a surface of revolution, and any diameter perpendicular to the axis of rotation is a principal axis.

8·74. The discriminating cubic is the discriminant of the quadratic equation

$$(ax^2+by^2+cz^2+2fyz+2gzx+2hxy)-\lambda(x^2+y^2+z^2)=0.$$

This equation represents a system of conics at infinity passing through the points of intersection H, H', K, K' of the conic at infinity and the circle at infinity. The discriminant expresses that the conic breaks up into two straight lines. Corresponding to the three roots of the discriminating cubic we have the three pairs of lines HH', KK'; HK, $H'K'$; HK', $H'K$. If these pairs of lines intersect in A, B, C, ABC is a self-polar triangle with regard to both conics; and if O is the centre of the quadric (the pole of the plane at infinity), $OABC$ is a self-polar tetrahedron with regard to the quadric. Also the lines OA, OB, OC are mutually orthogonal and are therefore the principal axes. Referred to this tetrahedron the equation of the quadric will be of the form

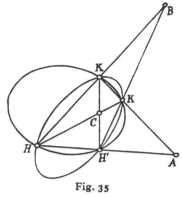

Fig. 35

$$ax^2 + by^2 + cz^2 + dw^2 = 0.$$

8·8. Transformation of rectangular coordinates.

The equation of a quadric is simplified when the axes are suitably chosen. The transformation to new axes can best be made in two stages, first keeping the directions of the axes fixed and changing the origin, and then keeping the origin fixed and rotating the axes.

8·81. Reduction of the general equation to axes through the centre.

Let the coordinates of the centre be $[X, Y, Z]$. The equations of transformation are then

$$\left. \begin{aligned} x &= x' + X \\ y &= y' + Y \\ z &= z' + Z \end{aligned} \right\}.$$

Then $F(x, y, z) = F(x' + X, y' + Y, z' + Z)$

$$= f(x', y', z') + \left(x' \frac{\partial F}{\partial X} + y' \frac{\partial F}{\partial Y} + z' \frac{\partial F}{\partial Z} \right)$$

$$+ F(X, Y, Z).$$

By this transformation the coefficients of the terms of the second degree are unaltered, i.e. a, b, c, f, g, h are invariants.

The coordinates of the centre satisfy the equations $\frac{\partial F}{\partial x} = 0$, $\frac{\partial F}{\partial y} = 0$, $\frac{\partial F}{\partial z} = 0$, hence the coefficients of x', y', z' all vanish.

The remaining term can be simplified. Using homogeneous coordinates $[x, y, z, w]$, we have

$$X\frac{\partial F}{\partial X} + Y\frac{\partial F}{\partial Y} + Z\frac{\partial F}{\partial Z} + W\frac{\partial F}{\partial W} = 2F_1(X, Y, Z, W).$$

Hence $\qquad F(X, Y, Z) = F_1(X, Y, Z, W) = \frac{1}{2}\frac{\partial F_1}{\partial W}$,

where W is put $= 1$ after differentiating.

The last equation is the best form for calculating, but a concise expression for the new constant d' can be obtained as follows. We have

$$\frac{1}{2}\frac{\partial F_1}{\partial W} = pX + qY + rZ + d = d',$$

$$\frac{1}{2}\frac{\partial F_1}{\partial X} = aX + hY + gZ + p = 0,$$

$$\frac{1}{2}\frac{\partial F_1}{\partial Y} = hX + bY + fZ + q = 0,$$

$$\frac{1}{2}\frac{\partial F_1}{\partial Z} = gX + fY + cZ + r = 0.$$

Eliminating X, Y, Z between these four equations we obtain

$$\begin{vmatrix} a & h & g & p \\ h & b & f & q \\ g & f & c & r \\ p & q & r & d-d' \end{vmatrix} = 0 = \Delta - Dd',$$

hence $d' = \Delta/D$.

The reduced equation is therefore

$$f(x', y', z') + \Delta/D = 0,$$

provided $D \neq 0$. If $D = 0$ the centre is at infinity and this transformation cannot be applied.

8·82. Rotation of axes. Invariants.

Omitting the dashes we shall now write x, y, z for coordinates referred to the centre, and x', y', z' for coordinates referred to new rectangular axes whose direction-cosines referred to the old are $[l_1, m_1, n_1]$, $[l_2, m_2, n_2]$, $[l_3, m_3, n_3]$. The scheme of transformation is

	x'	y'	z'
x	l_1	l_2	l_3
y	m_1	m_2	m_3
z	n_1	n_2	n_3

The equations of transformation being homogeneous, the constant term is unaltered, and $f(x, y, z)$ is transformed into a homogeneous quadratic expression in x', y', z'. We have then

$$ax^2 + by^2 + cz^2 + 2fyz + 2gzx + 2hxy \equiv a'x'^2 + b'y'^2 + c'z'^2$$
$$+ 2f'y'z' + 2g'z'x' + 2h'x'y'.$$

In the orthogonal transformation with origin fixed the expression $x^2 + y^2 + z^2$, which represents OP^2, is transformed into

$$x'^2 + y'^2 + z'^2,$$

i.e. $$x^2 + y^2 + z^2 \equiv x'^2 + y'^2 + z'^2.$$

Hence

$$(a-\lambda)x^2 + (b-\lambda)y^2 + (c-\lambda)z^2 + 2fyz + 2gzx + 2hxy$$
$$\equiv (a'-\lambda)x'^2 + (b'-\lambda)y'^2 + (c'-\lambda)z'^2$$
$$+ 2f'y'z' + 2g'z'x' + 2h'x'y'$$

for any value of λ. If the left-hand side of this identity breaks up into factors, so also will the right-hand side for the same value of λ. Hence the two equations

$$\begin{vmatrix} a-\lambda & h & g \\ h & b-\lambda & f \\ g & f & c-\lambda \end{vmatrix} = 0 \quad \text{and} \quad \begin{vmatrix} a'-\lambda & h' & g' \\ h' & b'-\lambda & f' \\ g' & f' & c'-\lambda \end{vmatrix} = 0$$

must be identical, i.e.

$$\lambda^3 - (a+b+c)\lambda^2 + (A+B+C)\lambda - D$$
$$\equiv \lambda^3 - (a'+b'+c')\lambda^2 + (A'+B'+C')\lambda - D'.$$

In the orthogonal transformation we have therefore the three absolute invariants

$$a + b + c \equiv I,$$
$$A + B + C \equiv J,$$

and $$D.$$

Also since the orthogonal transformation is a particular case of the general projective transformation, Δ is at least a relative invariant. The modulus of the transformation is

$$\begin{vmatrix} l_1 & m_1 & n_1 \\ l_2 & m_2 & n_2 \\ l_3 & m_3 & n_3 \end{vmatrix}.$$

The square of this is

$$\begin{vmatrix} \Sigma l_1^2 & \Sigma l_1 l_2 & \Sigma l_1 l_3 \\ \Sigma l_1 l_2 & \Sigma l_2^2 & \Sigma l_2 l_3 \\ \Sigma l_1 l_3 & \Sigma l_2 l_3 & \Sigma l_3^2 \end{vmatrix} = 1, \quad \begin{array}{l} \text{since} \quad \Sigma l_1^2 = \Sigma l_2^2 = \Sigma l_3^2 = 1, \\ \text{while} \quad \Sigma l_1 l_2 = \text{etc.} = 0. \end{array}$$

Hence Δ is an absolute invariant for the orthogonal transformation.

The equation $\qquad \lambda^3 - I\lambda^2 + J\lambda - D = 0$

is the discriminating cubic; since its coefficients are all invariants the three roots λ_1, λ_2, λ_3 are all invariants.

8·83. Reduction of the equation to the principal axes.

Let the canonical equation of the quadric be

$$a'x'^2 + b'y'^2 + c'z'^2 + \Delta/D = 0.$$

We have the three invariants

$$a' + b' + c' = I,$$
$$b'c' + c'a' + a'b' = J,$$
$$a'b'c' = D.$$

Hence a', b', c' are the roots of the discriminating cubic, and the reduced equation is

$$\lambda_1 x'^2 + \lambda_2 y'^2 + \lambda_3 z'^2 + \Delta/D = 0.$$

8·84. Reduction of the paraboloid.

In the case of the paraboloid the centre is a point at infinity. $D = 0$, and one root of the discriminating cubic is zero. The direction $[l, m, n]$ of the corresponding axis, *the axis* of the paraboloid, is determined by any two of the equations

$$al + hm + gn = 0,$$
$$hl + bm + fn = 0,$$
$$gl + fm + cn = 0.$$

The axis itself cuts the surface in one finite point, the vertex A, such that the tangent-plane at A is perpendicular to the axis. Its coordinates $[X, Y, Z]$ may be found from the equations

$$\frac{\partial F}{\partial X}\Big/l = \frac{\partial F}{\partial Y}\Big/m = \frac{\partial F}{\partial Z}\Big/n,$$

$$F(X, Y, Z) = 0.$$

Let the reduced equation be

$$a'x'^2 + b'y'^2 + 2r'z' = 0.$$

Then $a' + b' = I$, $a'b' = J$, $-a'b'r'^2 = \Delta$.

Hence a', b' are the two finite roots of the discriminating cubic, and the reduced equation is

$$\lambda_1 x'^2 + \lambda_2 y'^2 \pm 2\sqrt{(-\Delta/\lambda_1\lambda_2)}\,z' = 0.$$

8·85. Elliptic and hyperbolic cylinders.

In the case of an elliptic or hyperbolic cylinder, $\Delta = 0$ and $D = 0$. The three planes $F_x = 0$, $F_y = 0$, $F_z = 0$ have a line in common, the axis or line of centres. Transformed to any point $[X, Y, Z]$ on this line as origin, the equation reduces, as in 8·81, to the form

$$f(x, y, z) + d' = 0,$$

where $d' = F(X, Y, Z) = pX + qY + rZ + d.$

Then using along with this any two of the equations

$$aX + hY + gZ + p = 0,$$
$$hX + bY + fZ + q = 0,$$
$$gX + fY + cZ + r = 0$$

and eliminating X, Y, Z (as is possible since $D = 0$), we find

$$d' = A_1/A = \dots = F_1/F = \dots,$$

where $A_1, \dots$ and $A, \dots$ are the cofactors of $a, \dots$ in the determinants Δ and D respectively.

The discriminating cubic has one root zero, and if λ_1, λ_2 are the two finite roots the equation reduces finally to the form

$$\lambda_1 x^2 + \lambda_2 y^2 + d' = 0.$$

If $d' = 0$ the surface degenerates to two intersecting planes, and $F(x, y, z)$ can be resolved into factors.

8·86. Parabolic cylinder.

In the case of a parabolic cylinder the three planes $F_x = 0$, $F_y = 0$, $F_z = 0$ have a line at infinity in common. The matrix $[D]$ is of rank 1, and $f(x, y, z)$ is a perfect square

$$= (x\sqrt{a} + y\sqrt{b} + z\sqrt{c})^2.$$

The equation is then of the form $Y^2 = 4kX$, where $X = 0$ and $Y = 0$ represent planes, but these two planes are probably not at right angles. Introducing an unknown coefficient λ we may write the equation

$$(\Sigma x\sqrt{a} + \lambda)^2 + 2\Sigma(p - \lambda\sqrt{a})x + (d - \lambda^2) = 0,$$

and then determine λ by the condition for orthogonality

$$\Sigma(p - \lambda\sqrt{a})\sqrt{a} = 0,$$

so that $\qquad \lambda = \Sigma p\sqrt{a}/\Sigma a.$

Then if we write $\qquad \Sigma x\sqrt{a} + \lambda = y'\sqrt{\Sigma a}$

and $\quad \Sigma(p - \lambda\sqrt{a})x + \tfrac{1}{2}(d - \lambda^2) = -x'\sqrt{\{\Sigma(q\sqrt{c} - r\sqrt{b})^2/\Sigma a\}}$

the equation reduces to the form

$$y'^2 = 4kx',$$

where $\qquad k = \tfrac{1}{2}\{\Sigma(q\sqrt{c} - r\sqrt{b})^2/(\Sigma a)^3\}^{\frac{1}{2}}.$

If the value of λ makes the coefficients of x, y, z in the expression for x' all vanish, the surface reduces to two parallel planes; and if $k = 0$ it reduces to two coincident planes.

8·9. Quadrics of revolution.

In a surface of revolution every plane perpendicular to the axis cuts the surface in a circle whose centre is on the axis. Let A (Fig. 36) represent the point at infinity on the axis OA, and UV the polar of A with respect to the circle at infinity. Then every plane section through UV is a circle, and its centre, which lies on OA, is the pole of UV with respect to the curve of section. Hence the planes OAU and OAV are tangent-planes to the quadric at U and V. *The condition for a quadric of revolution is therefore that its conic at infinity should have double contact with the circle at infinity.*

The conic at infinity is

$$f(x, y, z) \equiv ax^2 + by^2 + cz^2 + 2fyz + 2gzx + 2hxy = 0,$$

and the circle at infinity is

$$x^2 + y^2 + z^2 = 0.$$

If these have double contact, then, for some value of λ,

$$f(x, y, z) + \lambda(x^2 + y^2 + z^2) = 0$$

represents two coincident lines. The conditions for this are that the matrix

$$\begin{bmatrix} a+\lambda & h & g \\ h & b+\lambda & f \\ g & f & c+\lambda \end{bmatrix}$$

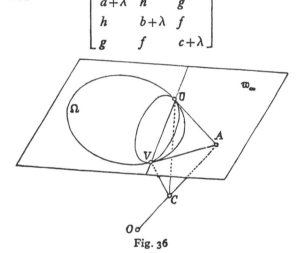

Fig. 36

should, for some value of λ, be of rank 1. This gives the following equations in λ:

$$\lambda^2 + (b+c)\lambda + (bc - f^2) = 0, \quad f\lambda - (gh - af) = 0,$$
$$\lambda^2 + (c+a)\lambda + (ca - g^2) = 0, \quad g\lambda - (hf - bg) = 0,$$
$$\lambda^2 + (a+b)\lambda + (ab - h^2) = 0, \quad h\lambda - (fg - ch) = 0.$$

The case where $\lambda = 0$ should be rejected for this makes

$$f(x, y, z) = 0$$

represent two coincident straight lines and the quadric is a parabolic cylinder. If f, g, h do not vanish we get the conditions

$$\frac{f}{F} = \frac{g}{G} = \frac{h}{H}.$$

If $f=0$ then either $g=0$ or $h=0$. If $f=0=g$ while $h\neq0$ we find $\lambda=-c$, and then $h^2=(a-c)(b-c)$. If $f=g=h=0$, then two of a, b, c must be equal.

The condition for a sphere is that $f(x, y, z)$ should coincide with the circle at infinity, and this is expressed analytically by the condition that the matrix should, for some value of λ, be of rank 0, which leads at once to $f=g=h=0$, $a=b=c$.

In the case of a surface of revolution two roots of the discriminating cubic are equal. The converse, however, is not true unless it is understood that the quadric is real, for two conditions are required in order that a quadric should be a surface of revolution, viz. that the conic at infinity should have double contact with the circle at infinity. The condition for equal roots merely requires single contact; single contact with the virtual circle is, however, impossible in the case of a real conic.

Similarly, in the case of a sphere the roots of the cubic are all equal. But this implies only two conditions, whereas five are required for a sphere. The equality of the three roots merely implies three-point contact, but for a sphere the conic at infinity must coincide with the circle at infinity.

The condition for equal roots, or the discriminant, of the cubic

$$\lambda^3 - I\lambda^2 + J\lambda - D = 0$$

is

$$I^2J^2 - 4I^3D - 4J^3 - 27D^2 + 18IJD = 0,$$

and the conditions for three equal roots are

$$\frac{3}{I} = \frac{I}{J} = \frac{J}{3D}.$$

Now $\quad I^2 - 3J = (\Sigma a)^2 - 3(\Sigma bc - \Sigma f^2) = (\Sigma a^2 - \Sigma bc) + 3\Sigma f^2.$

But $\Sigma a^2 - \Sigma bc$ is a positive definite form. Hence if $I^2-3J=0$, and the coefficients are real, we must have $f=0$, $g=0$, $h=0$ and $0 = \Sigma a^2 - \Sigma bc = \{a - \frac{1}{2}(b+c)\}^2 + \frac{3}{4}(b-c)^2$, therefore $a=b=c$.

8·95. EXAMPLES.

1. Find the pole of the plane $2x - 8y - 3z = 2$ with regard to the surface $x^2 - 2y^2 + z^2 - 2yz + 6x - 4z + 5 = 0$.

Ans. $[-1, 3, 2]$.

2. Reduce to their principal axes:

(i) $x^2 + y^2 + z^2 - 4yz - 4zx - 4xy = 3$.

(ii) $y^2 + z^2 + yz + zx - xy - 2x + 2y - 2z + 1 = 0$.

(iii) $4x^2 + y^2 + z^2 + 2yz + 4zx + 4xy - 24x + 32 = 0$.

(iv) $x^2 + 3y^2 + 3z^2 - 2yz - 2x - 2y + 6z + 3 = 0$.

(v) $2y^2 - 2yz + 2zx - 2xy - x - 2y + 3z - 2 = 0$.

(vi) $2x^2 - 7y^2 + 2z^2 - 10yz - 8zx - 10xy + 6x + 12y - 6z + 5 = 0$.

(vii) $2y^2 + 4zx + 2x - 4y + 6z + 5 = 0$.

(viii) $9x^2 + 4y^2 + 4z^2 + 8yz + 12zx + 12xy + 4x + y + 10z + 1 = 0$.

(ix) $x^2 + 2y^2 + z^2 - 4yz - 6zx - 2x + 8y - 2z + 9 = 0$.

(x) $x^2 + 3y^2 + z^2 + 4zx + 2x + 12y - 2z + 9 = 0$.

(xi) $x^2 + 4y^2 + 9z^2 - 12yz + 6zx - 4xy + 4x - 8y + 12z + 4 = 0$.

(xii) $2x^2 + 5y^2 + 2z^2 - 2yz + 4zx - 2xy + 14x - 16y + 14z + 26 = 0$.

(xiii) $16x^2 + 9y^2 + 4z^2 + 7x + 2y - 12z - 12yz - 16zx + 24xy = 0$.

(xiv) $x^2 + 4y^2 + 9z^2 - 12yz + 6zx - 4xy + 6x + 4y + 10z - 23 = 0$.

(xv) $2x^2 + 2y^2 - 4z^2 - 2yz - 2zx - 5xy - 2x - 2y + z = 0$.

(xvi) $16x^2 + 4y^2 + 4z^2 + 4yz - 8zx + 8xy + 4x + 4y - 16z - 24 = 0$.

Ans. (i) $X^2 + Y^2 - Z^2 = 1$.

(ii) $-X^2 + 2Y^2 + 3Z^2 = 4$, centre $[1, -1, 1]$.

(iii) Parabolic cylinder. $\sqrt{3}\, Y^2 = 4X$ or
$$(2x + y + z - 4)^2 = 8(x - y - z - 2).$$

(iv) $X^2 + 2Y^2 + 4Z^2 = 1$, centre $[1, 0, -1]$.

(v) Hyperbolic cylinder. $3X^2 - Y^2 = \tfrac{1}{2}$.

(vi) Cone. $X^2 + 2Y^2 - 4Z^2 = 0$, vertex $[\tfrac{1}{3}, -\tfrac{1}{3}, \tfrac{4}{3}]$.

(vii) Cone. $X^2 + Y^2 - Z^2 = 0$, vertex $[-\tfrac{3}{2}, 1, -\tfrac{1}{2}]$.

(viii) Parabolic cylinder. $17 Y^2 = 7X$ or
$$(3x + 2y + 2z + 1)^2 = 2x + 3y - 6z.$$

(ix) Cone. $4 \cdot 80 X^2 + 1 \cdot 68 Y^2 - 2 \cdot 48 Z^2 = 0$, vertex $[1, -2, 0]$.

(x) $3X^2 + 3Y^2 - Z^2 = 1$, centre $[1, -2, -1]$.

(xi) Two coincident planes. $(x - 2y + 3z + 2)^2 = 0$.

(xii) Elliptic cylinder. $X^2 + 2Y^2 = 1$.

(xiii) Parabolic cylinder. $29Y^2 = 9X$ or
$$(4x + 3y - 2z + 1)^2 = x + 4y + 8z + 1.$$

(xiv) Parabolic cylinder. $7Y^2 = 2\sqrt{6}X$ or
$$(x - 2y + 3z + 1)^2 + 4(x + 2y + z - 6) = 0.$$

(xv) Hyperbolic paraboloid. $3(X^2 - Y^2) = 2Z$, vertex $[0, 0, 0]$.

(xvi) Elliptic paraboloid. $3X^2 + Y^2 = 2Z$, vertex $[-1, 2, -1]$.

3. Show that the equation
$$2x^2 - y^2 - 22z^2 + 20yz + 10zx - 4xy + 2x - 20y + 14z + 14 = 0$$
represents a cone of revolution; find the coordinates of the vertex, the vertical semi-angle, and the direction of the axis.

Ans. $[-1, 2, 1]$, $\tan^{-1}3$, $[1 : 2 : -5]$.

4. Prove that the equation $\sqrt{x} + \sqrt{y} + \sqrt{z} = 0$ represents a circular cone whose axis is $x = y = z$ and vertical semi-angle $\cot^{-1}\sqrt{2}$.

5. Find the conditions that $F(x, y, z) = 0$ should represent a paraboloid of revolution.

Ans.
$$agh + f(g^2 + h^2) = 0,$$
$$bhf + g(h^2 + f^2) = 0,$$
$$cfg + h(f^2 + g^2) = 0.$$

6. Find the conditions that $F(x, y, z) = 0$ should represent a circular cylinder.

Ans. $agh + f(g^2 + h^2) = 0$, etc., and $p/f + q/g + r/h = 0$.

7. Show that the equation
$$fyz + gzx + hxy + pxw + qyw + rzw = 0$$
represents a quadric passing through the vertices of the tetrahedron of reference and find the conditions that the lines of intersection of the tangent-planes at the vertices with the opposite faces should lie in one plane.

Ans. $fp = gq = hr$.

8. Show that
$$x^2 + y^2 + z^2 + w^2 - yz - zx - xy - xw - yw - zw = 0$$
represents a quadric inscribed in the tetrahedron of reference, and that the lines joining each vertex to the point of contact with the opposite face are concurrent.

9. Show that by suitable choice of unit-point the equation of a quadric which touches the six edges of the tetrahedron of reference can be written

$$x^2 + y^2 + z^2 + w^2 \pm 2yz \pm 2zx \pm 2xy \pm 2xw \pm 2yw \pm 2zw = 0.$$

If the signs are all $+$, the quadric degenerates to two coincident planes; if the signs are all $-$, it is a proper quadric (not a cone); and these cases each give rise to seven others by changing the sign of one or more of the coordinates; in the remaining forty-eight cases the quadric is a cone.

10. If a quadric which is not a cone touches the edges of a skew quadrilateral show that the four points of contact are coplanar.

11. If a quadric which is not a cone touches the six edges of a tetrahedron show that the lines joining the pairs of points of contact of opposite edges are concurrent.

12. Three mutually perpendicular lines through a point O of a given quadric cut the surface in P, Q, R. Show that the plane PQR cuts the normal at O in a fixed point (Frégier point).

13. Three lines through a fixed point O of a given quadric, parallel to three conjugate diameters of another quadric, cut the first quadric in P, Q, R. Show that the plane PQR passes through a fixed point.

14. Find the locus of the Frégier points when O is varied.

Ans. A concentric quadric. If the given quadric is

$$ax^2 + by^2 + cz^2 = 1$$

the equation of the locus is

$$(a+b+c)^2 \Sigma\{ax^2/(-a+b+c)^2\} = 1.$$

15. If the quadrics U and V both have ring-contact with the quadric S, show that U and V intersect one another in two conics.

16. Show that any two tangent-cones of the same quadric intersect in two conics.

17. Show that two quadrics which have the same tangent-cone intersect in two conics.

CHAPTER IX

GENERATING LINES AND
PARAMETRIC REPRESENTATION

9·1. We have seen that in determining the intersection of a line with a quadric we obtain an equation of the second degree. This equation either has two definite roots or becomes an identity. Hence if three points of the line lie on the quadric, *all* its points must lie on the quadric. This implies three conditions. But a line in space has four degrees of freedom, hence there is still one degree of freedom, and thus an infinity of lines lie on the quadric.

If we apply the same reasoning to the case of a cubic surface we see that a line either cuts the surface in three points or lies entirely on the surface. If, then, four points of the line lie on the surface it lies entirely on the surface. But this implies four conditions, and hence a finite number of lines are determined as lying on a cubic surface. It will be proved in Chap. XVII that this number is 27, though they are not necessarily all real.

Similarly, in general, no lines at all lie on a surface of the fourth or higher order. Of course there may be cubic surfaces which contain an infinity of lines (ruled cubics), and quartic surfaces which contain a finite number, or even an infinity of lines, etc., but these are special cases.

9·11. We have seen also that through any point on a quadric there pass two generating lines, real, coincident, or imaginary. Let A be any point on the surface, a and a' the two generating lines through A. These form the intersection of the quadric with the tangent-plane at A. If B is another point on the surface it does not lie in this plane. Let b and b' be the two generators through B. b meets the plane (aa') in a point lying on the surface and therefore lying on either a or a', say a', i.e. b cuts a' and cannot cut a. Similarly b' cuts a, but not a'. If C is a third point on the surface, and c, c' the generators through C, one of them, c' say, cuts a in a point P. The generators through P are a and c', of which a cuts b' and therefore c' cuts b. Hence all the generators a, a'; b, b'; c, c' form two distinct systems. Every generator of the one system, e.g. a, cuts all the generators b', c', ... of the other system, but no two generators of the same system intersect.

9·2. The equation of a hyperbolic paraboloid was obtained in the form $yz = cx$, or homogeneously $yz = cxw$. The equation of a hyperboloid of one sheet can be written in the form

$$x^2/a^2 - z^2/c^2 = 1 - y^2/b^2,$$

i.e. $$(x/a - z/c)(x/a + z/c) = (1 - y/b)(1 + y/b).$$

Both of these equations are of the form

$$\alpha\beta = \gamma\delta,$$

where α, β, γ, δ represent planes. This is the form of *the equation of a quadric when two pairs of generators are taken as four of the edges of the tetrahedron of reference.* Take $A \equiv [1, 0, 0, 0]$ and $B \equiv [0, 1, 0, 0]$ two points of the surface, and let a, a' be the generators through A, and b, b' those through B. Let a cut b' in $C \equiv [0, 0, 1, 0]$ and a' cut b in $D \equiv [0, 0, 0, 1]$. Then $x = 0$ cuts the surface in a and a' where also $zw = 0$. Hence the equation of the surface is of the form $xy = kzw$.

9·21. Equations of the generating lines of the quadric $xy = zw$.

Let $P \equiv [x', y', z', w']$ be any point on the surface, so that $x'y' = z'w'$, and let l and l' be the two generators through P. Then since l cuts a' and b' it is the intersection of the planes (Pa') and (Pb'), i.e. $yz' = y'z$ and $xw' = x'w$.

Write $y' = \lambda z'$, then $w' = \lambda x'$. The equations of l are then

$$y = \lambda z, \quad \lambda x = w.$$

Similarly the equations of l' are

$$y = \lambda'w, \quad \lambda'x = z,$$

where $y' = \lambda'w'$.

Conversely, these two pairs of equations represent, for varying values of the parameters λ and λ', two systems of straight lines lying on the surface.

From this we can deduce independently that every generator

of the one system meets every generator of the other system, but no generator of the same system. For the equation

$$(y - \lambda z) + k(\lambda x - w) = 0$$

represents a plane containing the generator λ of the first system; and the equation

$$(y - \lambda' w) + k'(\lambda' x - z) = 0$$

represents a plane containing the generator λ' of the second system. But if $k = \lambda'$ and $k' = \lambda$ these two planes coincide, i.e. the plane

$$\lambda\lambda' x + y - \lambda z - \lambda' w = 0$$

contains the generators λ and λ'.

But taking two generators of the same system we have four equations

$$y = \lambda z, \ \lambda x = w; \ y = \mu z, \ \mu x = w$$

which have no solution in common.

9·22. The hyperbolic paraboloid.

Taking the equation of the hyperbolic paraboloid in the form

$$yz = cxw,$$

where $w = 0$ is the plane at infinity, the two systems of generators are

$$\left. \begin{array}{c} y = \lambda x \\ \lambda z = cw \end{array} \right\}, \quad \left. \begin{array}{c} z = \mu x \\ \mu y = cw \end{array} \right\}.$$

But for all values of λ the equation $\lambda z = cw$ represents parallel planes, hence all the generators of the one system are parallel to the plane $z = 0$. Similarly all the generators of the other system are parallel to the plane $y = 0$. This is clear also from the figure in 6·33.

The surface, in fact, cuts the plane at infinity in the two lines $y = 0$, $w = 0$ and $z = 0$, $w = 0$, and all the generators cut one or other of these two lines at infinity.

9·3. The two equations $y = \lambda z$, $\lambda x = w$ represent two projectively related sheafs of planes. Thus a regulus is generated by the intersection of two projective pencils of planes. Denote the two lines $x = 0 = w$ and $y = 0 = z$ by a' and b' respectively, the planes $\lambda x = w$, $y = \lambda z$ by α and β, and let l be the line of intersection of α and β. Then there is a (1, 1) correspondence between the planes α and β. Now $\alpha \equiv (a'l)$ and $\beta \equiv (b'l)$. a' and b' therefore cut l, say in P and Q. We have then on a' and b' the points P and Q in $(1, 1)$ correspondence, and $(PQ) \equiv l$. The regulus is therefore generated also by the lines joining points on two projective ranges. It follows that if a' and b' are cut by four generators in P_1, P_2, P_3, P_4 and Q_1, Q_2, Q_3, Q_4 these two ranges have equal cross-ratios. Also since (la') is the tangent-plane at P, the tangent-planes at the four points P form a pencil of planes with the same cross-ratio as that of the tangent-planes at the four points Q.

9·31. *A quadric surface is generated by a straight line which meets three given skew lines.*

Let a, b, c be three given lines, all mutually skew. Take any plane α through a and let it cut c in R, then a plane $B \equiv (Rb)$ is determined which cuts α in a line l. As l lies in α it cuts a, and as it lies also in β it cuts b, also it cuts c in R. Hence l cuts a, b and c. But l is the intersection of the planes α and β which are in $(1, 1)$ correspondence, hence by 9·3 l generates a regulus.

Without assuming this, we may show that the surface generated by l is of the second order. Let p be any line. R being a variable point on c, the planes (Ra) and (Rb) cut p in points P and Q which are in $(1, 1)$ correspondence. If U and V are the self-corresponding points in this homography, the planes (Ua) and (Ub) cut c in the same point, therefore the line of intersection of (Ua) and (Ub) cuts a, b and c and is therefore a generator of the surface. U therefore lies on the surface, and similarly also V. Hence U and V are the two points in which p cuts the surface.

9·311. *Equation of the surface generated by the transversals of three given lines.*

Let l, m, n be the three given lines. Take any two points A, B on l. The planes (An) and (Bn) cut the line m in points C and D. AC and BD then cut n in points E and F. Take $ABCD$ as tetrahedron of reference. Let $E \equiv [1, 0, p, 0]$ and $F \equiv [0, q, 0, 1]$. Let $L \equiv [1, \lambda, 0, 0]$ be any point on l. The planes (Lm) and (Ln) intersect in a line which cuts m in $M \equiv [0, 0, 1, \mu]$, say, and n in $N \equiv [1, q\nu, p, \nu] \equiv [1, \lambda, u, \mu u]$. Then $q\nu = \lambda$, $p = u$, $\nu = \mu u$, therefore $\nu = \mu p$, $\lambda = pq\mu$.

Let $P \equiv [x, y, z, w]$ be any point on LM. Then $P \equiv [1, \lambda, t, \mu t]$.
Hence
$$\rho x = 1,$$
$$\rho y = \lambda = pq\mu,$$
$$\rho z = t,$$
$$\rho w = \mu t.$$

Eliminating ρ, μ and t we have
$$\rho^2 yz = pq\mu t = pq\rho w . \rho x,$$
therefore
$$yz = pqxw.$$

9·32. Quadric surface with two given generators of different systems.

A surface passing through the line $u = 0 = v$ is represented by $\alpha u + \beta v = 0$, where α, β are expressions of the first degree. But this passes also through $\alpha = 0 = \beta$. If this second generator is given, the surface is still not determined and its equation may be more generally $(\lambda \alpha + \beta) u + (\mu \alpha + \beta) v = 0$. Hence the general equation of a quadric with generators $u = 0 = v$ and $u' = 0 = v'$ is
$$auu' + bvv' + cuv' + du'v = 0.$$

The quadric has been made to satisfy six conditions and there are still three disposable constants, the ratios of a, b, c, d.

Ex. Show that the equation of a hyperboloid of one sheet referred to a set of axes through the centre parallel to three generating lines is
$$fyz + gzx + hxy = 1.$$

9·4.
The condition that a variable line should cut a given line imposes one degree of restraint. The assemblage of all lines which cut one fixed line is, as we have seen (8·432), a *special linear complex*. The assemblage of all lines which cut two fixed lines is a *linear congruence*. That of all lines which cut three fixed lines is a *regulus*. If a line is to cut four given lines a, b, c, d it is deprived of all freedom, and only a finite number of lines exist satisfying these conditions. The number of lines can be determined by taking the special case in which the given lines intersect in pairs. Suppose a, b intersect in A and determine a plane α, while c, d intersect in B and determine a plane β. Then if the line l meets both a and b, either it passes through A or lies in the plane α; similarly if it meets both c and d, either it passes

through B or lies in the plane β. If A lies in the plane β, or B in the plane α, there is an infinity of lines cutting the four. Excluding this case, if the line l meets all four lines it either passes through A and B or lies in the two planes α and β. Hence *there are in general two lines which meet four given lines in space.* We may determine these lines in the general case as follows. The three lines a, b, c determine a quadric, which is cut by d in two points P and Q. Through each of these there passes a generator of the other system, which therefore cuts a, b, c and also d. The two common transversals may be real, coincident, or imaginary.

This may be shown also by using Plücker coordinates. Let (p), (q), (r), (s) be four straight lines, and (l) a line meeting them. We have then four equations of the form

$$p_{01}l_{23} + p_{23}l_{01} + \ldots = 0,$$

linear and homogeneous in (l), and also the equation

$$l_{01}l_{23} + l_{02}l_{31} + l_{03}l_{12} = 0.$$

These five equations determine two sets of values of the ratios of the six l's.

9·41. A problem which is important in non-euclidean geometry is to determine the common transversals of two pairs of lines which are polars with regard to a given quadric. When this quadric is the "absolute" the common transversals are the common perpendiculars of the lines.

Let the equation of the given quadric be

$$ax^2 + by^2 + cz^2 + dw^2 = 0,$$

and let one pair of lines be $x = 0 = w$ and $y = 0 = z$. Let the other pair be given by their Plücker coordinates p, p'; let the first be the join of the points $[x_1, y_1, z_1, w_1]$ and $[x_2, y_2, z_2, w_2]$. Then the second is the intersection of the lines

$$ax_1 x + by_1 y + cz_1 z + dw_1 w = 0,$$

$$ax_2 x + by_2 y + cz_2 z + dw_2 w = 0.$$

Hence

$$p_{01}' = bcp_{23}, \quad p_{23}' = adp_{01},$$

$$p_{02}' = cap_{31}, \quad p_{31}' = bdp_{02},$$

$$p_{03}' = abp_{12}, \quad p_{12}' = cdp_{03}.$$

If (q) is the common transversal we have

$$q_{01}=0, \quad q_{23}=0,$$

$$p_{31}q_{02}+p_{12}q_{03}+p_{02}q_{31}+p_{03}q_{12}=0, \quad \ldots\ldots(1)$$

and
$$bdp_{02}q_{02}+cdp_{03}q_{03}+cap_{31}q_{31}+abp_{12}q_{12}=0. \quad \ldots\ldots(2)$$

Also
$$q_{02}q_{31}+q_{03}q_{12}=0. \quad \ldots\ldots(3)$$

Express q_{12} and q_{31} from (1) and (2) in terms of q_{02} and q_{03} and substitute in (3) and we obtain

$$(ap_{31}p_{12}-dp_{02}p_{03})(bq_{02}{}^2-cq_{03}{}^2)$$
$$+\{a(bp_{12}{}^2-cp_{31}{}^2)+d(bp_{02}{}^2-cp_{03}{}^2)\}q_{02}q_{03}=0,$$

a quadratic in q_{02}/q_{03}. The condition for real roots is

$$\{a(bp_{12}{}^2-cp_{31}{}^2)+d(bp_{02}{}^2-cp_{03}{}^2)\}^2+4bc(ap_{31}p_{12}-dp_{02}p_{03})^2>0.$$

This is equivalent also to

$$\{b(ap_{12}{}^2-dp_{02}{}^2)+c(ap_{31}{}^2-dp_{03}{}^2)\}^2+4ad(bp_{02}p_{12}-cp_{03}p_{31})^2>0,$$

and to

$$\{a(bp_{12}{}^2+cp_{31}{}^2)+d(bp_{02}{}^2+cp_{03}{}^2)\}^2-4abcd\,p_{01}{}^2p_{23}{}^2>0.$$

The roots are real if $abcd<0$ or if $bc>0$ or $ad>0$. Hence the two common transversals are always real if the quadric has imaginary generators. If the quadric has real generators and the signs of a, b, c, d are $+--+$ the transversals are again real; in this case the two lines $x=0=w$ and $y=0=z$ cut the quadric in imaginary points. If, however, the signs are either $++--$ or $+-+-$, so that these two lines cut the quadric in real points, a further condition is required.

The condition that the line joining (x_1) and (x_2) should cut the quadric in real points is that the roots of

$$(ax_1{}^2+\ldots)+2(ax_1x_2+\ldots)\lambda+(ax_2{}^2+\ldots)\lambda^2=0$$

should be real. This gives

$$(ax_1x_2+\ldots)^2-(ax_1{}^2+\ldots)(ax_2{}^2+\ldots)>0,$$

i.e.
$$P\equiv abp_{12}{}^2+acp_{13}{}^2+adp_{10}{}^2+bcp_{23}{}^2+bdp_{02}{}^2+cdp_{03}{}^2<0.$$

Now the condition for real transversals is

$$\{P-(adp_{01}{}^2+bcp_{23}{}^2)\}^2-4abcd\,p_{01}{}^2p_{23}{}^2>0,$$

i.e.
$$P^2-2(adp_{01}{}^2+bcp_{23}{}^2)P+(adp_{01}{}^2-bcp_{23}{}^2)^2>0.$$

Hence if ad and bc are both negative the roots are real if $P>0$.

9·51. Freedom-equations of the hyperboloid of one sheet.

Writing the canonical equation in the form

$$x^2/a^2 - y^2/b^2 = w^2 - z^2/c^2, \qquad \ldots\ldots(1)$$

each side factorises.

Write

$$x/a+y/b = \lambda\,(w-z/c) \quad \text{and} \quad x/a+y/b = \mu\,(w+z/c)$$
then
$$x/a-y/b = \lambda^{-1}(w+z/c) \quad \text{and} \quad x/a-y/b = \mu^{-1}(w-z/c) \Bigg\} \ldots(2)$$

These four equations are equivalent to three, and we may solve any three of them for the ratios of x, y, z, w. We obtain then

$$\left.\begin{aligned}
\rho x/a &= \lambda\mu + 1 \\
\rho y/b &= \lambda\mu - 1 \\
\rho z/c &= \quad \lambda - \mu \\
\rho w &= \quad \lambda + \mu
\end{aligned}\right\} \qquad \ldots\ldots(3)$$

These are freedom-equations in terms of the two parameters λ and μ, and the two systems of generators of the surface are represented by $\lambda =$ const. and $\mu =$ const.

9·52. Similarly the paraboloid

$$x^2/a^2 - y^2/b^2 = 2zw/c$$

is represented by the freedom-equations

$$\left.\begin{aligned}
\rho x/a &= \lambda + \mu \\
\rho y/b &= \lambda - \mu \\
\rho z/c &= 1 \\
\rho w &= 2\lambda\mu
\end{aligned}\right\}.$$

9·6. Parametric equations of a curve.

If the homogeneous coordinates x, y, z, w are expressed as functions of a single parameter λ they have just one degree of freedom and represent points on a curve. If the equations are algebraic the curve is called an *algebraic curve*, and if, further, they are rational we have a *rational algebraic curve*. To every value of λ there corresponds then one set of values of the co-ordinates and therefore one point. We shall assume for the present (see 14·13) that, conversely, to every point, with certain possible exceptions, corresponds one value of the parameter, so that between the points of the curve and the values of the parameter there is an algebraic $(1, 1)$ correspondence. The exceptional points are double-points on the curve.

The intersections of the curve with an arbitrary plane are found by substituting the coordinates in terms of λ. If r is the highest degree in λ of the four parametric equations, this will lead to an equation of degree r in λ. Hence the curve is cut by an arbitrary plane in r points, and is said to be of *order r*.

9·61. Thus for example the general freedom-equations of the second degree
$$\rho x = a_0 \lambda^2 + 2a_1 \lambda + a_2, \text{ etc.},$$

represent a conic. Eliminating ρ, λ^2, λ linearly between the four equations we obtain

$$\begin{vmatrix} x & a_0 & a_1 & a_2 \\ y & b_0 & b_1 & b_2 \\ z & c_0 & c_1 & c_2 \\ w & d_0 & d_1 & d_2 \end{vmatrix} = 0,$$

which represents the plane of the conic. Again, eliminating ρ and λ between the first three equations we get a homogeneous equation of the second degree in x, y, z which represents a cone with vertex $[0, 0, 0, 1]$. If $w = 0$ is the plane at infinity the conic will be a hyperbola, parabola, or ellipse according as $d_1^2 - d_0 d_2 >$, $=$, or < 0.

9·62. Again, freedom-equations of the third degree represent a cubic curve, which, however, does not in general lie in one plane:
$$\rho x = a_0 \lambda^3 + a_1 \lambda^2 + a_2 \lambda + a_3, \text{ etc.}$$

Eliminating λ^3 between the first equation and each of the others in turn we get three equations

$$\rho(b_0 x - a_0 y) = (b_0 a_1 - a_0 b_1)\lambda^2 + (b_0 a_2 - a_0 b_2)\lambda \\ + (b_0 a_3 - a_0 b_3), \text{ etc.},$$

and eliminating ρ and λ between these three equations we obtain a homogeneous equation of the second degree in $(b_0 x - a_0 y)$, etc., which represents a quadric cone with vertex $[a_0, b_0, c_0, d_0]$. Similarly eliminating the constant term we get three equations

$$\rho(b_3 x - a_3 y) = (b_3 a_0 - a_3 b_0)\lambda^3 + (b_3 a_1 - a_3 b_1)\lambda^2$$
$$+ (b_3 a_2 - a_3 b_2)\lambda, \text{ etc.}$$

Eliminating ρ/λ and λ we obtain a homogeneous equation of the second degree in $(b_3 x - a_3 y)$, etc., which represents a quadric cone with vertex $[a_3, b_3, c_3, d_3]$. Each of these cones passes through the vertex of the other as we see by putting $\lambda = 0$ in the freedom-equations of the former and letting $\lambda \to \infty$ in those of the latter. Hence these two cones have a common generating line and their remaining intersection is a cubic curve.

9·63. Conversely any conic can be represented by rational freedom-equations; for let l be any line not in the plane of the conic and meeting it in a point O. Then any plane α through l cuts the conic in one other point P, hence there is a $(1, 1)$ correspondence between the points P and the planes of the pencil with axis l, and these can be uniquely related to a parameter λ.

Ex. Find freedom-equations for the conic in which the plane $x/a - 2y/b + w = 0$ cuts the ellipsoid $x^2/a^2 + y^2/b^2 + z^2/c^2 = w^2$.

By inspection one point on the conic is $[3a, 4b, 0, 5]$. Take the line $4bx = 3ay$, $z = 0$ as axis of a pencil of planes $4x/a - 3y/b - \lambda z/c = 0$. We find that this plane cuts the conic besides in the point whose coordinates are given by

$$x/a = 3 - \lambda^2,$$
$$y/b = 4,$$
$$z/c = 4\lambda,$$
$$w = 5 + \lambda^2.$$

9·64. Further, any non-plane cubic is rational, for if we take two fixed points on the curve a plane through these will cut the curve in one other point. Hence there is a $(1, 1)$ correspondence between the points of the curve and the planes of the pencil. A *plane* cubic curve, however, is not in general rational, but it is rational if it possesses a double-point, for any line through the double-point and lying in the plane of the cubic will cut the curve in just one other point.

We shall reserve the further discussion of cubic curves in space till a later chapter.

9·7. Parametric equations of a surface.

If the coordinates of a point are functions of two parameters, the point has two degrees of freedom and its locus is a surface. If the functions are algebraic and rational the surface is called a rational algebraic surface. The *order* of a surface is equal to the number of points in which it is met by an arbitrary line. Hence if the parametric equations are of degree r the order of the surface appears to be in general equal to r^2. It may, however, be less than this (see 9·73).

9·71. It is convenient to consider the parametric equations from another standpoint. If we consider the parameters λ, μ as cartesian coordinates in a plane we have a representation of the surface on the plane of λ, μ. To every pair of values of λ, μ corresponds a unique set of values of the ratios $x : y : z : w$, hence to every point in the (λ, μ)-plane corresponds uniquely a point on the surface. The converse, however, is not so obvious, and to fix the ideas more clearly we shall confine our attention again to the hyperboloid (see 9·51). The equations (3) express (x, y, z, w) rationally in terms of (λ, μ), and two of the equations (2) conversely express (λ, μ) rationally in terms of (x, y, z, w) when these coordinates satisfy the equation of the surface (1). Thus in this case there is a birational relation between (λ, μ) and the ratios of (x, y, z, w), and therefore a $(1, 1)$ correspondence between the points of the quadric and the points of the plane.

To a plane section $lx + my + nz + pw = 0$ of the quadric corresponds a conic

$$la(\lambda\mu + 1) + mb(\lambda\mu - 1) + nc(\lambda - \mu) + p(\lambda + \mu) = 0.$$

Writing λ/ν instead of λ, and μ/ν instead of μ, i.e. replacing λ, μ by homogeneous coordinates λ, μ, ν, all the conics which correspond to plane sections of the quadric pass through the points common to the four conics

$$\lambda\mu + \nu^2 = 0, \ \lambda\mu - \nu^2 = 0, \ \nu(\lambda - \mu) = 0, \ \nu(\lambda + \mu) = 0.$$

More than two conics do not in general have any point in common, but these four conics have in common the two points $L \equiv [1, 0, 0]$, and $M \equiv [0, 1, 0]$. These two points are exceptional points in the representation, since the corresponding values of

x, y, z, w are all zero and therefore there is no unique point on the quadric which corresponds to either of them. Since the generators of the quadric are represented by equations of the form $\lambda = h\nu$ and $\mu = k\nu$, the two points L and M are in fact the points through which pass the lines in the $(\lambda\mu)$-plane which represent generators of the quadric. The line LM or $\nu = 0$ is also exceptional, for to this corresponds the single point $x/a = \lambda\mu$, $y/b = \lambda\mu$, $z = 0$, $w = 0$, i.e. the point $O \equiv [a, b, 0, 0]$. The tangent-plane at this point is $x/a - y/b = 0$. The generators through this point are the lines of intersection of this plane with the two planes $z = \mp cw$; to these planes correspond $\lambda = 0$ and $\mu = 0$, and to the two generators correspond the points $[0, 1, 0]$ and $[1, 0, 0]$. Thus the exceptional points L and M represent the two generating lines through a particular point O of the quadric.

9·72. Stereographic projection.

This representation can be viewed as an actual projection of the quadric upon a plane, the centre of projection being a point O on the surface. Let the generators through O cut the plane of projection in L and M, then every plane section of the quadric which does not pass through O is a conic which is projected into a conic passing through L and M. The generating lines which meet OL are projected into straight lines passing through L, and those which meet OM into straight lines passing through M. The generators OL and OM are represented only by the points L and M, to the point O corresponds the whole line LM, and a plane section through O is represented by a straight line together with the line LM.

Such a projection is called *stereographic*. The term is often restricted to the projection of a sphere in which the centre of projection is a point O on the sphere and the plane of projection α is parallel to the tangent-plane at O. In this special case, since the section by the tangent-plane at O is a point-circle, the generating lines through O are two imaginary lines passing through the circular points I, J in this plane. These are also the circular points in the parallel plane α. Hence every plane section of the sphere (not through O) is represented by a conic passing through I and J, i.e. a circle.

This representation of a quadric surface on a plane is not confined to the case in which the generating lines are real. By stereographic projection any quadric can be represented rationally on a plane, all plane sections being represented by conics passing through two fixed points which are real or imaginary according as the quadric has or has not real generating lines. Thus if the ellipsoid $x^2/a^2 + y^2/b^2 + z^2/c^2 = w^2$ is projected from the point $[0, 0, c]$ on to the plane $z = 0$ the point $P \equiv [x, y, z, w]$ on the ellipsoid is represented by the point $P' \equiv [x', y']$ on the plane $z = 0$, and we have

$$\frac{x}{x'} = \frac{y}{y'} = \frac{cw - z}{c} = w - \frac{z}{c}.$$

Hence

$$\left(\frac{x'^2}{a^2} + \frac{y'^2}{b^2}\right)\left(w - \frac{z}{c}\right)^2 = w^2 - \frac{z^2}{c^2},$$

i.e.

$$\left(\frac{x'^2}{a^2} + \frac{y'^2}{b^2} - 1\right)w = \left(\frac{x'^2}{a^2} + \frac{y'^2}{b^2} + 1\right)\frac{z}{c}.$$

Writing

$$\rho w = \frac{x'^2}{a^2} + \frac{y'^2}{b^2} + 1,$$

we have

$$\rho z/c = x'^2/a^2 + y'^2/b^2 - 1,$$

and $\rho(w - z/c) = 2$, therefore

$$\rho x = 2x', \quad \rho y = 2y';$$

or putting

$$x' = \lambda a, \quad y' = \mu b,$$

we have

$$\rho x/a = 2\lambda,$$

$$\rho y/b = 2\mu,$$

$$\rho z/c = \lambda^2 + \mu^2 - 1,$$

$$\rho w = \lambda^2 + \mu^2 + 1.$$

9·73. Let us now return to the general parametric equations of degree r, and consider the points in which the surface is cut by an arbitrary line. If $u = 0 = v$ are the equations of the line, $u + kv = 0$ represents any plane through the line. When the values of the coordinates are substituted in terms of λ, μ, the equation $u + kv = 0$ represents a pencil of curves of order r in the $(\lambda\mu)$-plane, and this pencil has r^2 base-points, the points of intersection of the curves which correspond to the plane sections

$u=$ o and $v=$ o. Thus to the points of intersection of the line $u=$ o $=v$ with the surface correspond in general r^2 points in the $(\lambda\mu)$-plane, hence in general the surface is of order r^2. If, however, all the curves corresponding to $x=$ o, $y=$ o, $z=$ o, $w=$ o pass through s fixed points, these points do not correspond to any definite points on the surface, and the order of the surface is then r^2-s.

9·731. Thus the general parametric equations of the second degree in three homogeneous parameters represent a quartic surface*. This is not the most general quartic surface, for a quartic surface is not in general rational. If we take a linear homogeneous equation in the parameters, $a\lambda+b\mu+c\nu=$ o and express ν in terms of λ and μ, we obtain parametric equations of the second degree in two homogeneous parameters, and these represent a conic. Hence this quartic surface contains a double infinity of conics.

If the four conics in the $(\lambda\mu\nu)$-plane which correspond to $x=$ o, $y=$ o, $z=$ o, $w=$ o have *one* point in common, the parametric equations represent a cubic surface, also possessing an infinity of conics, and therefore also an infinity of straight lines, since a plane which cuts the surface in a conic will have a straight line for the remainder of its intersection. The surface is therefore a *ruled cubic*.

Ex. 1.
$$\rho x=\lambda,$$
$$\rho y=\mu,$$
$$\rho z=\lambda^2+\lambda\mu,$$
$$\rho w=\lambda^2+\mu^2+\mu.$$

The equation of the surface is found, by eliminating ρ, λ, μ, to be
$$yz\,(x-y)=x\,(x+y)\,(y+z-w).$$

If $\mu=k\lambda$, we get
$$\rho x=\mathrm{i},$$
$$\rho y=k,$$
$$\rho z=(k+\mathrm{i})\,\lambda,$$
$$\rho w=(k^2+\mathrm{i})\,\lambda+\mathrm{i},$$

---

* This is Steiner's Surface. See 17·93.

parametric equations of the first degree in λ, and therefore representing a straight line. This line is the intersection of the two planes

$$\left. \begin{array}{c} y = kx \\ (k^2 + 1)\,z = (k+1)\,(w-x) \end{array} \right\}.$$

In fact $x = 0 = y$ is a double-line on the surface, since $x = 0$ cuts the surface where $y^2 z = 0$, and therefore every plane through this line cuts the surface in this double line and one other line.

Ex. 2. Show that the tangential equation of the cubic surface in Ex. 1 is

$$\omega\xi^2 - \zeta\xi\,(\eta + \omega) + (\zeta + \omega)\,(\eta + \omega)^2 = 0.$$

If the four conics have *two* points in common the parametric equations represent a general quadric.

If they have *three* points in common, say $[1, 0, 0]$, $[0, 1, 0]$ and $[0, 0, 1]$, the parametric equations will all be of the type

$$\rho x = a\mu\nu + b\nu\lambda + c\lambda\mu,$$

and writing λ', μ', ν' for $\mu\nu$, $\nu\lambda$, $\lambda\mu$, they become linear and therefore represent a plane.

The general cubic is a rational surface and can be represented by parametric equations of the third degree, such that the four cubic curves in the $(\lambda\mu\nu)$-plane which correspond to $x = 0$, $y = 0$, $z = 0$, $w = 0$ all pass through six fixed points.

To prove this and at the same time explain a method by which the parametric equations can be found, let two lines a and b be chosen on the surface, mutually skew, and take any plane π, not containing either of the lines. Let P' be any point in π. Then through P' there passes one line which cuts both a and b, and this line cuts the surface in one other point P. Hence there is a $(1, 1)$ correspondence between the points P of the surface and the points P' of the plane.

9·9. EXAMPLES.

1. Show that the generators of the surface $x^2 + y^2 - z^2 = 1$ which intersect on the plane of xy are at right angles.

2. Prove that the hyperboloid $x^2/a^2 + y^2/b^2 - z^2/c^2 = 1$ has generators which intersect at right angles unless c is greater than both a and b.

3. Show that the coordinates of any point on the surface $x^2+y^2=z^2+w^2$ can be expressed as

$$x=\sin(\alpha-\beta),\ y=\cos(\alpha-\beta),\ z=\cos(\alpha+\beta),\ w=\sin(\alpha+\beta).$$

4. Show that the angle between the generating lines of the quadric $x^2/a+y^2/b+z^2/c=1$ through the point $[x',\ y',\ z']$ is $\cos^{-1}(\lambda_1+\lambda_2)/(\lambda_1-\lambda_2)$, where $\lambda_1,\ \lambda_2$ are the roots of the quadratic

$$\frac{x'^2}{a(a+\lambda)}+\frac{y'^2}{b(b+\lambda)}+\frac{z'^2}{c(c+\lambda)}=0.$$

5. If AA', BB', CC' are three concurrent lines, not coplanar, prove that there is a quadric for which BC', CA', AB' are generators of the one system and $B'C$, $C'A$, $A'B$ are generators of the other system.

6. Prove that the normals to a quadric at all points of a generating line generate a hyperbolic paraboloid.

7. Show that the four quadrics, each of which contains three of four given skew lines, have two common generating lines.

8. If a parallelepiped has three edges coinciding with generating lines of the same system of the hyperboloid

$$x^2/a^2+y^2/b^2-z^2/c^2=1,$$

show that the two remaining vertices lie on the hyperboloid

$$x^2/a^2+y^2/b^2-z^2/c^2+3=0.$$

9. If a parallelepiped has three edges coinciding with generating lines of the same system of the hyperboloid

$$x^2/a^2+y^2/b^2-z^2/c^2=1,$$

show that it has three other edges coinciding with generators of the other system.

10. Show that all parallelepipeds which have six of their edges coinciding with generators of the same hyperboloid have the same volume.

11. A straight line moves so that four points marked upon it move in four fixed planes; show that the straight line has one degree of freedom and that every point on it describes a conic.

12. Show that the lines joining the vertices of a tetrahedron to the corresponding vertices of the polar tetrahedron with respect to a given quadric all belong to the same regulus.

13. Show that the lines of intersection of corresponding planes of a tetrahedron and its polar with respect to a given quadric all belong to the same regulus.

14. Show that the four altitudes of a tetrahedron belong to the same regulus of a hyperboloid of one sheet; and that the four perpendiculars to the faces at their orthocentres are generators of the other system.

15. Show that the centroid of a tetrahedron is at the mid-point of the join of the circumcentre and the centre of the hyperboloid on the altitudes. What does this theorem become when the tetrahedron is orthocentric?

16. If a quadric is circumscribed about a tetrahedron show that the four lines of intersection of the tangent-planes at a vertex with the opposite face belong to the same regulus.

17. If a quadric is inscribed in a tetrahedron show that the four lines joining a vertex to the point of contact of the opposite face belong to the same regulus.

18. Two generators of the same system of a quadric being given it is required to find a generator meeting them in points at which the tangent-planes are perpendicular. Show that the problem admits of two solutions or of an infinite number, but that in the latter case the quadric is not of general type.
                                        (Math. Trip. II, 1914.)

19. If the generators of the same system of a hyperboloid at four points A, B, C, D meet the opposite faces of the tetrahedron respectively in A', B', C', D', prove that a quadric exists touching these faces at these points.   (Math. Trip. II, 1915.)

20. With the notation of 2·511, if (a), (b), (c) are given points and (x) a variable point on the plane, $(abcx) = 0$, where $(abcx)$ denotes the determinant whose rows are a_0, a_1, a_2, a_3; b_0, ... ; c_0, ... ; x_0, ...

21. If (a), (a'), (b), (b') are given points and (x) a variable point, show that the coordinates of the point of intersection of the plane $aa'x$ with the line bb' are

$$b_i(aa'xb') - b_i'(aa'xb) \quad (i = 0, 1, 2, 3).$$

22. Prove that the equation of the quadric which has as three generators the lines joining the pairs of points (a), (a'); (b), (b'); (c), (c') is

$$(aa'bx)(b'cc'x) = (aa'b'x)(bcc'x).$$

23. Given six points (a), (b'), (c), (a'), (b), (c'), forming a skew hexagon, show that the three quadrics U, V, W with generators (i) aa', bb', cc', (ii) ab', bc', ca', (iii) ac', ba', cb' are connected by a linear relation $\lambda U + \mu V + \nu W = 0$.

CHAPTER X

PLANE SECTIONS OF A QUADRIC

10·1. A quadric has the property that it is cut by any straight line in two points, real, coincident, or imaginary. The surface is therefore said to be of the second order, the order being equal to the degree of the equation.

Consider the section by any plane α. Every line in α cuts the surface, and therefore the section, in two points. Hence the section is a curve of the second order, i.e. a conic.

The plane at infinity cuts the surface in a conic at infinity C. Let the section α cut this conic in H, K. Then the section is an ellipse, a parabola, or a hyperbola, according as H, K are imaginary, coincident, or real and distinct.

If the conic C is virtual, H and K are always imaginary. Every section is therefore either an ellipse or virtual. The surface is either a real ellipsoid, or a virtual surface.

If C is real, the surface is a hyperboloid. If O is its centre, the cone with vertex O and base C is the asymptotic cone. Sections by planes which touch this cone are parabolas, sections by planes through O cutting the cone in real lines are hyperbolas, and sections by planes through O cutting the cone in imaginary lines are real ellipses in the case of the hyperboloid of one sheet, and virtual ellipses in the case of the hyperboloid of two sheets.

If C breaks up into two straight lines, their point of intersection is a double-point on the curve of intersection, i.e. every line lying in the plane at infinity and passing through this point meets the surface in two coincident points and is therefore a tangent. The plane at infinity is therefore a tangent-plane. The surface is a paraboloid (hyperbolic or elliptic according as the two lines at infinity are real or imaginary).

All parallel sections cut the plane at infinity in the same two points, and have therefore their corresponding asymptotes parallel. They are therefore similar conics, similarly placed, or homothetic.

10·2. The centre of a plane section.

The centre C of any section is the pole, with regard to the curve of intersection, of the line at infinity in the plane. This line at infinity is therefore part of the polar of C with regard to the surface; hence the given plane and the polar of C both intersect the plane at infinity in the same line, i.e. the polar-plane of C is parallel to the given plane.

If the surface is a central one, let its equation be

$$ax^2 + by^2 + cz^2 = 1,$$

and let the plane of section be

$$lx + my + nz = p.$$

If $[X, Y, Z]$ is the centre of the section, its polar is

$$aXx + bYy + cZz = 1.$$

These two planes cut the plane at infinity in the lines

$$lx + my + nz = 0, \quad w = 0,$$

and $\qquad aXx + bYy + cZz = 0, \quad w = 0.$

In order that these may coincide

$$\frac{aX}{l} = \frac{bY}{m} = \frac{cZ}{n}, \text{ hence} = \frac{p}{l^2/a + m^2/b + n^2/c}.$$

These equations determine X, Y, Z.

10·21. If the surface is a paraboloid

$$ax^2 + by^2 = 2cz,$$

the polar of $[X, Y, Z]$ is

$$aXx + bYy = c(z + Z),$$

and we have to identify the equations

$$lx + my + nz = 0,$$

and $\qquad aXx + bYy - cz = 0.$

Hence $\qquad \dfrac{aX}{l} = \dfrac{bY}{m} = -\dfrac{c}{n} = \dfrac{p - nZ}{l^2/a + m^2/b}.$

10·31. Axes of a central plane section.

Let C be the centre of the section. Equal diameters of the section are equally inclined to either of its axes. The magnitudes and positions of the axes can be investigated by considering the limiting case of a pair of equal diameters when they come to coincide.

Consider first for simplicity a central section. The centre of the section then coincides with the centre of the surface. The extremities of all semi-diameters of length r lie on a sphere with centre C. The lines joining C to the points in which this sphere cuts the surface form a cone. The generators of the cone are the common diameters of the sphere and the surface. Since every section through C has C for centre and cuts the sphere in a concentric circle, and since a conic and a concentric circle have just two common diameters, every plane through C cuts the cone in two generators and the cone is of the second order.

Let the equation of the surface be

$$ax^2 + by^2 + cz^2 = 1,$$

and the plane $\qquad lx + my + nz = 0.$

Take the concentric sphere

$$x^2 + y^2 + z^2 = r^2.$$

The cone formed by the common diameters is then

$$(ar^2 - 1)x^2 + (br^2 - 1)y^2 + (cr^2 - 1)z^2 = 0.$$

Now if the given plane touches the cone the two equal diameters coincide with one of the axes of the section. The condition for a tangent-plane is

$$l^2/(ar^2 - 1) + m^2/(br^2 - 1) + n^2/(cr^2 - 1) = 0,$$

i.e. $\qquad \Sigma l^2 (br^2 - 1)(cr^2 - 1) = 0,$

which is a quadratic in r^2.

The two roots $r_1{}^2$ and $r_2{}^2$ are the squares of the semi-axes of the section.

To find their direction-cosines λ, μ, ν, find the equation of the tangent-plane at $[\lambda, \mu, \nu]$ to the cone, viz.

$$(ar^2 - 1)\lambda x + (br^2 - 1)\mu y + (cr^2 - 1)\nu z = 0,$$

and identify with the equation of the given plane. Then

$$\lambda : \mu : \nu = l/(ar^2 - 1) : m/(br^2 - 1) : n/(cr^2 - 1).$$

10·32. Axes of non-central section.

In the general case when C is not the centre of the surface, an arbitrary plane through C cuts the sphere in a circle with centre C and the surface in a conic not having C for centre. These two curves cut in four points, and their joins with C form four distinct generators of the cone. The cone is therefore of the fourth order. The section which has C for centre, however, cuts the cone in two pairs of coincident generators, and we obtain the axes of the section by choosing r so that these two lines coincide.

Let the equation of the surface be

$$ax^2 + by^2 + cz^2 = 1,$$

and that of the plane

$$lx + my + nz = p.$$

If $[X, Y, Z]$ is the centre of the section,

$$\frac{aX}{l} = \frac{bY}{m} = \frac{cZ}{n} = \frac{p}{l^2/a + m^2/b + n^2/c} = \frac{aX^2 + bY^2 + cZ^2}{p}.$$

Now transform to $[X, Y, Z]$ as origin. The equation of the quadric becomes

$$ax'^2 + by'^2 + cz'^2 + 2(aXx' + bYy' + cZz') \\ + (aX^2 + bY^2 + cZ^2 - 1) = 0,$$

i.e.

$$ax'^2 + by'^2 + cz'^2 + 2(lx' + my' + nz')\,p/(l^2/a + m^2/b + n^2/c) - K = 0,$$

where $$K \equiv 1 - p^2/(l^2/a + m^2/b + n^2/c),$$

and the equation of the plane becomes

$$lx' + my' + nz' = 0.$$

Hence the axes of the section are the same as those of the conic

$$ax^2 + by^2 + cz^2 = K \atop lx + my + nz = 0 \Big\}.$$

Hence the quadratic equation for the squares of the semi-axes is $$\Sigma l^2(br^2 - K)(cr^2 - K) = 0,$$

and the direction-cosines of the axes are

$$\lambda : \mu : \nu = l/(ar^2 - K) : m/(br^2 - K) : n/(cr^2 - K).$$

10·33. The directions of the axes of a plane section may also be investigated as follows. Let the equations of the surface and the plane in homogeneous coordinates be

$$ax^2 + by^2 + cz^2 = w^2,$$
$$lx + my + nz = pw.$$

The points at infinity on the section are determined by the equations

$$\left.\begin{array}{l} w=0 \\ ax^2 + by^2 + cz^2 = 0 \\ lx + my + nz = 0 \end{array}\right\}.$$

The second equation represents the cone joining the origin to the conic at infinity on the surface, and the plane, which passes through its vertex, cuts it in two straight lines, which are the lines joining the origin to the points at infinity on the section. Hence the *asymptotes* of the section are parallel to the two lines

$$\left.\begin{array}{l} ax^2 + by^2 + cz^2 = 0 \\ lx + my + nz = 0 \end{array}\right\}.$$

Eliminating z and expressing the condition for real roots, we find that the section is

$$\left.\begin{array}{l} \text{an ellipse} \\ \text{a hyperbola} \\ \text{a parabola} \end{array}\right\} \text{according as } bcl^2 + cam^2 + abn^2 \begin{array}{l} > \\ < \\ = \end{array} 0.$$

The axes are harmonic conjugates with regard to the asymptotes, and are also at right angles, i.e. harmonic conjugates with regard to the absolute lines in their plane, viz. the two lines

$$\left.\begin{array}{l} x^2 + y^2 + z^2 = 0 \\ lx + my + nz = 0 \end{array}\right\}.$$

If therefore the axes are the two lines

$$\left.\begin{array}{l} a'x^2 + b'y^2 + c'z^2 = 0 \\ lx + my + nz = 0 \end{array}\right\},$$

we have the two conditions (cf. 5·6)

$$\Sigma\,(bc' + b'c)\,l^2 = 0,$$
$$\Sigma\,(b' + c')\,l^2 = 0,$$

hence $\quad a' : b' : c' = l^2\{-(b-c)\,l^2 + (c-a)\,m^2 + (a-b)\,n^2\}$

$$: m^2\{(b-c)\,l^2 - (c-a)\,m^2 + (a-b)\,n^2\}$$

$$: n^2\{(b-c)\,l^2 + (c-a)\,m^2 - (a-b)\,n^2\}.$$

Ex. Show that the section will be a rectangular hyperbola if $\Sigma\,(b+c)\,l^2 = 0$.

10·4. Circular sections.

The most interesting of the plane sections of a quadric are the circular sections. To show that circular sections actually exist consider the conic at infinity C on the quadric, and the circle at infinity Ω. These two conics, in the plane at infinity, intersect in four points, which are conjugate imaginary in pairs, H, H' and K, K', and they determine three pairs of common chords, HH' and KK', HK and $H'K'$, HK' and $H'K$, of which the first pair are real and the others conjugate imaginaries. Any plane through one of these

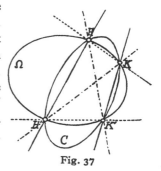

Fig. 37

chords, say HH', has the two points H and H' as the circular points, and as the section of the surface passes through these points it is a circle. There are therefore three pairs of sets of parallel circular sections, one pair real, the others imaginary.

10·41. Consider the central quadric

$$ax^2 + by^2 + cz^2 = 1.$$

The conic at infinity is

$$w = 0,\ ax^2 + by^2 + cz^2 = 0,$$

and the equations

$$ax^2 + by^2 + cz^2 - \lambda(x^2 + y^2 + z^2) = 0,\ w = 0$$

represent a conic through the four points H, H', K, K'. Choosing λ so that these equations may represent two straight lines we have

$$(\lambda - a)(\lambda - b)(\lambda - c) = 0.$$

Hence $\lambda = a$, b, or c. For each of these values the quadratic breaks up into factors and represents two planes through the centre, the central planes of circular section. Thus for

$$\lambda = a,\ (b-a)y^2 + (c-a)z^2 = 0, \qquad \ldots\ldots\text{(i)}$$
$$\lambda = b,\ (a-b)x^2 + (c-b)z^2 = 0, \qquad \ldots\ldots\text{(ii)}$$
$$\lambda = c,\ (a-c)x^2 + (b-c)y^2 = 0. \qquad \ldots\ldots\text{(iii)}$$

Hence the central planes of circular section all pass through one of the principal axes, and in pairs are equally inclined to a principal plane. If $a < b < c$ the planes corresponding to $\lambda = b$ are the real planes, whether a, b, c are positive or negative.

For the ellipsoid the real central planes of circular section are those which contain the mean axis.

For the hyperboloid of one sheet, say $a < 0 < b < c$, the real central planes are those which contain the major axis of the principal elliptic section.

For the hyperboloid of two sheets, say $a < b < 0 < c$, the real central planes do not cut the surface in real sections, but parallel sections sufficiently remote from the centre will cut the surface in real circles. A plane parallel to the plane of xy cuts the surface in an ellipse $ax^2 + by^2 = k$ and a real plane of circular section then contains the minor axis of this ellipse.

10·42. Circular sections of the paraboloids.

The case of the paraboloids requires some modification. A paraboloid $ax^2 + by^2 = 2cz$ cuts the plane at infinity in two straight lines $w = 0$, $ax^2 + by^2 = 0$. These are real, say HH' and KK', in the case of the hyperbolic paraboloid, imaginary, say HK and $H'K'$, for the elliptic paraboloid. The planes of circular section are found by choosing λ so that

$$ax^2 + by^2 - \lambda(x^2 + y^2 + z^2)$$

factorises. The values of λ are 0, a, b.

In the case of the hyperbolic paraboloid the real planes correspond to $\lambda = 0$, but this gives planes which cut the surface in a line at infinity and another line. We have seen that this pair forms a degenerate case of a circle with centre at infinity. The hyperbolic paraboloid possesses no proper circular sections other than its rectilinear generators. The rectilinear generators of a hyperboloid do not of course correspond to circular sections, since they consist of pairs of finite lines, not a finite line and a line at infinity as in the case of the paraboloid.

In the case of the elliptic paraboloid there are real proper circular sections corresponding to $\lambda = b$ if a is numerically greater than b.

10·5. Models of these surfaces can be constructed of cardboard by fixing together two series of circular sections. If the planes are hinged at their lines of intersection, these models are deformable, being capable of being squeezed or expanded into different shapes. As an example take the sphere $x^2+y^2+z^2=b^2$, and form two series of parallel sections perpendicular to the plane of xz and inclined to the plane of xy at angles $\pm\alpha$. Then if P is any point on the surface, and PL and PM are the traces on the plane of zx of the two planes through P, and if $OL=p$, $OM=q$, we have

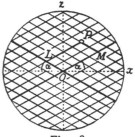

$$x=(-p+q)\cos\alpha,$$

$$z=(p+q)\sin\alpha,$$

Fig. 38

while $\qquad y^2=b^2-(q-p)^2\cos^2\alpha-(q+p)^2\sin^2\alpha.$

Now change the inclination of the planes to θ. For the same material point P, y is unchanged, but

$$x=(q-p)\cos\theta,\quad z=(q+p)\sin\theta.$$

Then eliminating p and q we get

$$y^2=b^2-x^2\cos^2\alpha\sec^2\theta-z^2\sin^2\alpha\operatorname{cosec}^2\theta,$$

which represents an ellipsoid with semi-axes $b\sec\alpha\cos\theta$, b, $b\operatorname{cosec}\alpha\sin\theta$. In the extreme cases when $\theta=0$ or $90°$ the planes

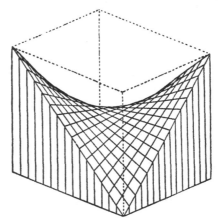

Fig. 39. Hyperbolic paraboloid

flatten out and we obtain two ellipses, one in the xy-plane with semi-axes $b\sec\alpha$ and b, the other in the yz-plane with semi-axes b and $b\operatorname{cosec}\alpha$.

10·6. *Any two circles which belong one each to the two sets of real circular sections lie on a sphere.*

Taking the two sections

$$u \equiv \sqrt{(b-a)}.x + \sqrt{(c-b)}.z - p = 0$$

and
$$v \equiv \sqrt{(b-a)}.x - \sqrt{(c-b)}.z - q = 0,$$

the equation
$$k(ax^2 + by^2 + cz^2 - 1) + uv = 0$$

represents a quadric containing both circles. The equation contains no terms in yz, zx or xy, and the coefficients of x^2, y^2 and z^2 will all be equal if $k = 1$.

Otherwise, let α, β be two circular sections. Their planes intersect in a line which cuts the quadric in two points U, V. α passes through H, H', and β through K, K'. Since a conic is determined by five points the two sections can be determined, apart from the surface, by taking a fifth point, A and B respectively, on each. Thus α is determined by the five points U, V, H, H', A, and β by the five points U, V, K, K', B. Any quadric through the eight points U, V, H, H', K, K', A, B will then contain the two sections, since it contains five points of each, and it can be made to pass through a ninth point L which we may choose on the circle at infinity. It then contains five points H, H', K, K', L on the circle at infinity and therefore contains it and is a sphere.

The condition that two circles should lie on the same sphere is simply that they should have two common points, and this condition is evidently satisfied by two circles of different systems on a quadric.

10·7. Umbilics. A circle which lies on a quadric, in the limiting case when its radius becomes zero, is a point-circle. These point-circles are called *umbilics*. There are four real umbilics, lying in pairs at the ends of two diameters. In addition there are four pairs of imaginary umbilics, i.e. twelve in all.

If $[X, Y, Z]$ are the coordinates of an umbilic of the quadric $ax^2 + by^2 + cz^2 = 1$ the conjugate diametral plane is a circular section. Hence identifying

$$aXx + bYy + cZz = 0$$

with
$$\sqrt{(b-a)}.x \pm \sqrt{(c-b)}.z = 0,$$

we have
$$Y = 0 \text{ and } aX = \lambda\sqrt{(b-a)}, \quad cZ = \pm\lambda\sqrt{(c-b)}.$$

But
$$aX^2 + bY^2 + cZ^2 = 1,$$

hence
$$\{c(b-a) + a(c-b)\}\lambda^2 = ac,$$

and therefore
$$\lambda = \pm\sqrt{\{ac/b(c-a)\}}.$$

We have therefore

$$X = \pm\sqrt{\{c(b-a)/ab(c-a)\}}, \quad Y = 0, \quad Z = \pm\sqrt{\{a(c-b)/bc(c-a)\}}.$$

10·71. The twelve umbilics, together with the four absolute points or points of intersection of the conic at infinity on the quadric with the circle at infinity, form a remarkable configuration of sixteen points on the quadric. Denote the absolute points by P_1, P_2, P_3, P_4; these lie in the plane $w = 0$. Four umbilics U_1, U_2, U_3, U_4 lie in the plane $x = 0$; four, V_1, etc., in the plane $y = 0$; and four, W_1, etc., in the plane $z = 0$. The tangent-plane at an umbilic U_1 cuts the surface in two generating lines which together form also a point-circle and therefore pass through two of the points P. Also through each of the points P there pass two generating lines, i.e. eight lines in all, and all the umbilics lie on these eight lines. On each line there is an absolute point P_1, say, and three umbilics, viz. the points in which one generator through P_1 cuts the generators of the opposite system through the other absolute points P_2, P_3 and P_4. Thus the sixteen points lie in sets of four on eight lines.

Again, the tangent-plane at U_1 contains the two generators through U_1 and therefore contains seven of the sixteen points, viz. five umbilics and two absolute points. Similarly the tangent-plane at P_1 contains the two generators through P_1 and therefore also contains seven points, viz. six umbilics and one absolute point.

A *configuration* is a figure consisting of points, lines, and planes, such that on every line there are the same number of points, on every plane the same number of points and the same number of lines, through every point the same number of lines and the same number of planes, and so on. If N_0, N_1, N_2 denote the total numbers of points, lines and planes; N_{01}, N_{02} the numbers of points in a line and in a plane; N_{10}, N_{20} the numbers of lines and planes through a point; N_{12} the number of lines in a plane, and N_{21} the number of planes through a line, the con-figuration may be denoted by

$$
\begin{array}{ccc}
N_0 & N_{01} & N_{02} \\
N_{10} & N_1 & N_{12} \\
N_{20} & N_{21} & N_2
\end{array}
$$

This configuration is therefore represented by

$$
\begin{array}{ccc}
16 & 4 & 7 \\
2 & 8 & 2 \\
7 & 4 & 16
\end{array}
$$

10·9. EXAMPLES.

1. Find the coordinates of the centre of the section of the surface $3x^2 - 2y^2 + z^2 = 6$ by the plane $6x - 8y + 3z + 11 = 0$.

Ans. $[2, 4, 3]$.

2. Find the coordinates of the centre of the section of the surface $3x^2 + 2y^2 + 4z^2 = 24$ made by the plane $3x - y + 2z = 9$. From the coordinates thus obtained show that the plane cuts the surface in a real curve.

Ans. $[2, -1, 1]$.

3. Find the equations of the real circular sections of the quadrics:

    (i) $4x^2 + 2y^2 + z^2 + 3yz + zx - 1 = 0$.

    (ii) $2x^2 + 5y^2 - 3z^2 + 4xy - 1 = 0$.

    (iii) $2x^2 + 5y^2 + 2z^2 - yz - 4zx - xy + 4 = 0$.

Ans. (i) $x + y - z = 0$, $x - y + 2z = 0$.

    (ii) $x + 2y + 2z = 0$, $x + 2y - 2z = 0$.

    (iii) $x + y + z = 0$, $2x - y + 2z = 0$.

4. Find the real circular sections of the paraboloids:

    (i) $10x^2 + 2y^2 = z$.

    (ii) $3x^2 + 7y^2 + 8z^2 + 12yz + 4zx + 8xy = 2x - 2y + z$.

Ans. (i) $2x \pm z = \lambda$.

    (ii) $4x + 2y + 5z = \lambda$, $2y + z = \mu$.

5. Find the real central circular sections of the ellipsoid

$$13x^2 + 3y^2 + 5z^2 = 4.$$

Prove that the sphere $5(x^2 + y^2 + z^2) + 4x - 6y = 0$ cuts this ellipsoid in a pair of circles, and find the equations of their planes.

Ans. $y = \pm 2x$; $2x - y + 1 = 0$, $2x + y - 2 = 0$.

6. Find the directions and lengths of the axes of the section of the ellipsoid $14x^2 + 6y^2 + 9z^2 = 3$ by the plane $x + y + z = 0$.

Ans. $r = \frac{1}{2}$, $[-3, 1, 2]$; $r^2 = 9/22$, $[1, -5, 4]$.

7. Show that the plane $x + 2y + 3z = 1$ cuts the hyperboloid $2x^2 + y^2 - 2z^2 = 1$ in a parabola, and find the direction-ratios of its axis.

Ans. $[1, 4, -3]$.

8. Show that the plane $x + 2y + 3z = 0$ cuts the hyperboloid $-6x^2 + 7y^2 - 14z^2 = 7$ in a hyperbola, and find the direction-ratios of the axes.

Ans. $[7, 1, -3]$, $[9, -24, 13]$.

9. Find the equation of the cone with vertex at the origin and passing through the curve of intersection of the quadric

$$ax^2 + by^2 + cz^2 = 1$$

with the concentric sphere of radius r. Prove that every tangent-plane of this cone cuts the quadric in a conic having one axis $= r$.

10. If the plane $lx + my + nz = 0$ cuts the surface $F(x, y, z) = 0$ in a rectangular hyperbola show that

$$a(m^2 + n^2) + b(n^2 + l^2) + c(l^2 + m^2) - 2fmn - 2gnl - 2hlm = 0.$$

11. Find the area of a given central section of an ellipsoid.
Ans. $\pi abc (\Sigma l^2 / \Sigma a^2 l^2)^{\frac{1}{2}}$.

12. Show that the envelope of plane central sections of constant area of an ellipsoid is a quadric cone.

13. A sphere of constant radius cuts an ellipsoid in plane sections; find the surface generated by the line of intersection of the planes.

Ans. Three cylinders.

14. Show that the planes, whose sections with a given quadric have their centres on a given straight line, are parallel to another fixed line and envelop a parabolic cylinder.

15. Show that there is a doubly infinite system of spheres which cut a given central quadric in pairs of circles, and that the locus of point-spheres of the system consists of three conics (the focal conics).

16. Prove that the plane section of the ellipsoid

$$x^2/a^2 + y^2/b^2 + z^2/c^2 = 1$$

whose centre is at the point $[\frac{1}{3}a, \frac{1}{3}b, \frac{1}{3}c]$ passes through three of the extremities of the principal axes of the ellipsoid.

(Math. Trip. I, 1914.)

CHAPTER XI

TANGENTIAL EQUATIONS

11·1. A set of four numbers $[x_0, x_1, x_2, x_3]$, homogeneous point-coordinates, represents a point; and a set of four numbers $[\xi_0, \xi_1, \xi_2, \xi_3]$, homogeneous plane-coordinates, represents a plane. The point and the plane are incident when

$$\xi_0 x_0 + \xi_1 x_1 + \xi_2 x_2 + \xi_3 x_3 = 0.$$

If (ξ) is fixed and (x) variable this is the equation of the plane (ξ); if (x) is fixed and (ξ) variable, it is the equation of the point (x).

11·21. Tangent-plane of a surface.

A single homogeneous equation in $x_0, \ldots, x_3$ represents a two-dimensional assemblage of points, two-way locus, or (in general) a surface,

$$F(x_0, x_1, x_2, x_3) = 0.$$

If (x') is any point on the surface and $(x' + \delta x')$ is a neighbouring point on the surface, so that

$$F(x_0', \ldots) = 0 \text{ and } F(x_0' + \delta x_0', x_1' + \delta x_1', x_2' + \delta x_2', x_3' + \delta x_3') = 0,$$

expanding by Taylor's theorem, we have

$$F(x_0' + \delta x_0', \ldots) = F(x_0', \ldots) + \Sigma \frac{\partial F}{\partial x_i'} \delta x_i' + \ldots.$$

Hence, neglecting squares of the increments,

$$\Sigma \frac{\partial F}{\partial x_i'} \delta x_i' = 0.$$

The matrix

$$\begin{bmatrix} x_0' & x_1' & x_2' & x_3' \\ dx_0' & dx_1' & dx_2' & dx_3' \end{bmatrix}$$

represents a line-element of the surface through (x') when $F(x_0', \ldots) = 0$ and $\Sigma \frac{\partial F}{\partial x_i'} dx_i' = 0$, and determines the direction of a tangent-line at (x'). If (x) is any point on this tangent, this line will be that determined by the points (x') and (x) provided (x) satisfies the same equation as (dx'). Hence the coordinates of any point on any tangent-line at (x') satisfy the equation

$$\Sigma \frac{\partial F}{\partial x_i'} x_i = 0.$$

This is the equation of the tangent-plane at (x').

11·22. Point of an envelope.

Similarly a single homogeneous equation in (ξ) represents a two-dimensional assemblage of planes, two-way envelope, or (in general) a surface

$$\Phi(\xi_0, \xi_1, \xi_2, \xi_3) = 0.$$

The matrix
$$\begin{bmatrix} \xi_0' & \xi_1' & \xi_2' & \xi_3' \\ d\xi_0' & d\xi_1' & d\xi_2' & d\xi_3' \end{bmatrix}$$

where $\Phi(\xi_0', \ldots) = 0$ and $\Sigma \frac{\partial \Phi}{\partial \xi_i'} d\xi_i' = 0$, represents a line-element in the plane (ξ'), and the coordinates of any plane which passes through any line-element in this plane satisfy the equation

$$\Sigma \frac{\partial \Phi}{\partial \xi_i'} \xi_i = 0.$$

This is the equation of the point of contact of the tangent-plane (ξ').

11·3. Tangential equation derived from point-equation, and *vice-versa*.

A surface, which can be considered either as a two-way locus or as a two-way envelope, has a point-equation and a plane or tangential equation; and if the one is given the other can be deduced.

Let $F(x_0, x_1, x_2, x_3) = 0$ be the point-equation of the surface. The condition that (ξ) should be a tangent-plane is found by identifying this plane with the tangent-plane at (x'). Hence

$$\lambda \xi_i = \frac{\partial F}{\partial x_i'} \ (i = 0, 1, 2, 3),$$

and further
$$\Sigma \xi_i x_i' = 0.$$

Between these five equations we can eliminate x_i' and λ, and thus obtain the tangential equation in (ξ).

By an exactly similar process the point-equation can be obtained from the tangential equation.

11·31. Tangential equation of the quadric

$$\Sigma\Sigma\, a_{rs} x_r x_s = 0.$$

The equation of the tangent-plane at (x') is

$$\Sigma\Sigma\, a_{rs} x_r x_s' = 0.$$

Hence
$$\lambda\xi_r = \sum_s a_{rs} x_s' \quad (r = 0,\, 1,\, 2,\, 3),$$

and since (x') lies on the tangent-plane

$$\Sigma\, \xi_r x_r' = 0.$$

Eliminating x_r' and λ we obtain the equation

$$\begin{vmatrix} a_{00} & a_{01} & a_{02} & a_{03} & \xi_0 \\ a_{10} & a_{11} & a_{12} & a_{13} & \xi_1 \\ a_{20} & a_{21} & a_{22} & a_{23} & \xi_2 \\ a_{30} & a_{31} & a_{32} & a_{33} & \xi_3 \\ \xi_0 & \xi_1 & \xi_2 & \xi_3 & 0 \end{vmatrix} = 0,$$

which is homogeneous and of the second degree in ξ_r. If capital letters denote the cofactors of the corresponding small letters in the determinant
$$\Delta \equiv |\, a_{rs}\, |$$

the equation can be written

$$\Sigma\Sigma\, A_{rs} \xi_r \xi_s = 0.$$

As the point-equation is obtained from the tangential equation by exactly the same process we verify a known theorem in determinants, that the cofactors of the capital letters in the determinant $\Delta' \equiv |\, A_{rs}\, |$ are proportional to the corresponding small letters in the determinant $\Delta \equiv |\, a_{rs}\, |$. It can be proved in fact that $\Delta' = \Delta^3$, and if a_{rs}' denotes the cofactor of A_{rs}, then $a_{rs}' = \Delta^2 a_{rs}$, or $a_{rs}'/\Delta' = a_{rs}/\Delta$.

11·4.
The general equation of the second degree in plane-coordinates thus represents a quadric-envelope. Some special forms of the equation may be noted.

If $\Phi = 0$ and $\Psi = 0$ represent two quadric-envelopes, the equation
$$\Phi + \lambda\Psi = 0$$
represents a quadric-envelope which touches all the planes common to the two given quadric-envelopes. In particular, if Φ breaks up into linear factors $\alpha\beta$, $\Phi = 0$ represents two bundles of planes with vertices $\alpha = 0$ and $\beta = 0$, and the equation
$$\alpha\beta + \lambda\Psi = 0$$
represents a quadric-envelope inscribed in each of the tangent-cones to Ψ with vertices α and β.

Further, the equation
$$U \equiv \alpha^2 + \lambda\Psi = 0$$
represents a quadric-envelope inscribed in the tangent-cone to Ψ with vertex α, and having *ring-contact* with it and with Ψ around the conic of contact of the tangent-cone with Ψ. For the point of contact of a tangent-plane (ξ') of Ψ is $\Sigma\xi \dfrac{\partial\Psi}{\partial\xi'} = 0$, and its point of contact with U is $2\alpha\alpha' + \lambda\Sigma\xi \dfrac{\partial\Psi}{\partial\xi'} = 0$. But if (ξ') passes through α, then $\alpha' = 0$, and (ξ') touches Ψ and U at the same point.

11·5. Order and class of a surface.

In the dual correlation, to a point corresponds a plane, and *vice versa*, and to a line corresponds a line. To a surface considered as a two-way locus corresponds in general a surface considered as a two-way envelope. An arbitrary line cuts a surface in a finite number of points equal to the degree of the equation in point-coordinates, the *order* of the surface. Dually, through an arbitrary line there are a finite number of tangent-planes to the surface equal to the degree of the equation in plane-coordinates, the *class* of the surface.

The points which a plane has in common with the surface form a *plane curve* whose order is equal to that of the surface. Dually, the planes which a bundle (planes through a point) has in common with a surface (two-way envelope) form a *cone*. Thus cone is dual to plane curve. A plane curve is a one-way locus of points, and a cone is a one-way envelope of planes. At any point on a curve there is just one tangent-line, and every plane through this line is a tangent-plane. In any tangent-plane to a cone there is just one proper tangent-line (the generating line), and every point on this line is a point of the cone. Thus as a cone is a two-way locus of points, a plane curve is a two-way envelope of planes.

11·51. In plane geometry the order of a plane curve is defined as the number of points in which it is cut by an arbitrary line in its plane. In space a line does not in general cut a given curve, and the order is defined more generally as the number of points in which it is cut by an arbitrary plane. Dually, for a cone there are in general through an arbitrary line no tangent-planes, but through a line which contains the vertex there are a finite number of tangent-planes; this is equal also to the number of tangent-planes through an arbitrary point, and is called the class of the cone. The order of the cone is of course equal to the number of points in which it is cut by an arbitrary line, and the class of a plane curve is equal to the number of tangent-planes through an arbitrary line, which is the same as the number of tangent-lines through an arbitrary point in its plane; dually, for the cone the order is equal to the number of generating lines in an arbitrary plane through its vertex. Thus to plane curve corresponds cone, and to tangent corresponds generating line.

A plane curve requires two equations in point-coordinates, one of which, linear, will represent its plane. Dually, a cone requires two equations in plane-coordinates, one of which, linear, will represent its vertex. A cone is represented by a single equation in point-coordinates, and a plane curve by a single equation in plane-coordinates.

A single tangential equation therefore may represent either a surface or a plane curve. We shall see afterwards that it may represent any curve, not necessarily plane.

11·52. Tangential equation of a plane curve.

A plane curve, as we have seen, is dual to a cone, its plane corresponding to the vertex of the cone. The algebraic method of finding the tangential equation of a plane curve is therefore the same as that of finding the point-equation of a cone.

If $t=0$ is the point-equation of the plane of the curve, and $u=0$ represents any tangent-plane, the two equations together represent a tangent-line to the curve in its plane, and any plane through this line, say $u+\lambda t=0$, is also a tangent-plane. The point-coordinates being $[x, y, z, w]$ and the plane-coordinates $[\xi, \eta, \zeta, \omega]$, if the plane of the curve is $w=0$ or, in plane-coordinates, $[0, 0, 0, 1]$, then if $[\xi, \eta, \zeta, \omega]$ or

$$\xi x+\eta y+\zeta z+\omega w=0$$

is a tangent-plane so also is

$$\xi x+\eta y+\zeta z+\omega' w=0$$

or $[\xi, \eta, \zeta, \omega']$ for all values of ω'. The tangential equation therefore does not contain ω, but is homogeneous in ξ, η, ζ. And conversely *a homogeneous equation in ξ, η, ζ represents a curve in the plane $w=0$*.

As an example let us find the tangential equation of the circle at infinity whose point-equations are $x^2+y^2+z^2=0$, $w=0$. The plane $\xi x+\eta y+\zeta z+\omega w=0$ cuts the plane at infinity in the line

$$\xi x+\eta y+\zeta z=0, \quad w=0,$$

and the condition that this should be a tangent to the conic

$$x^2+y^2+z^2=0, \quad w=0$$

is

$$\xi^2+\eta^2+\zeta^2=0.$$

Similarly, more generally, the tangential equation of the conic at infinity on the quadric

$$ax^2+by^2+cz^2+2fyz+2gzx+2hxy+2px+2qy+2rz+d=0$$

is

$$A\xi^2+B\eta^2+C\zeta^2+2F\eta\zeta+2G\zeta\xi+2H\xi\eta=0,$$

where capital letters denote the cofactors of the corresponding small letters in the determinant

$$D \equiv \begin{vmatrix} a & h & g \\ h & b & f \\ g & f & c \end{vmatrix}.$$

11·6. Tangential equations of a cone.

When $\Delta=0$, which is the condition for a cone, the tangential equation, derived from the point-equation as in 11·31, degenerates. If $A_{00}\neq0$ it can be written

$$\Sigma\Sigma A_{00}A_{rs}\xi_r\xi_s=0.$$

But by a theorem in determinants

$$\begin{vmatrix} A_{00} & A_{01} \\ A_{10} & A_{11} \end{vmatrix} = \Delta \begin{vmatrix} a_{22} & a_{23} \\ a_{32} & a_{33} \end{vmatrix}, \text{ etc.,}$$

hence when $\Delta = 0$ $A_{00}A_{rs} = A_{0r}A_{0s}$,

and the equation becomes

$$\Sigma\Sigma A_{0r}A_{0s}\xi_r\xi_s = 0,$$

i.e. $(\Sigma A_{0r}\xi_r)^2 = 0.$

This equation thus represents the vertex of the cone, taken twice; and it may similarly be represented by any one of the four equations $\sum_r A_{ir}\xi_r = 0$ $(i = 0, 1, 2, 3).$ ......(1)

While a plane through the vertex has the property of a tangent-plane in cutting the quadric in a pair of lines, the actual tangent-planes of the cone are further specified as cutting the cone in coincident lines. Another equation in ξ_r is therefore required along with the former one to represent this one-dimensional system of tangent-planes. For this we may take the tangential equation of any plane section of the cone not passing through the vertex, or of any quadric which is inscribed in the cone. For example, the section by the plane $x_0 = 0$ is

$$\sum_1^3\sum_1^3 a_{rs}x_r x_s = 0,$$

and its tangential equation is

$$\sum_{123}\{(a_{22}a_{33} - a_{23}^2)\xi_1^2 + 2(a_{31}a_{12} - a_{11}a_{23})\xi_2\xi_3\} = 0. \quad ......(2)$$

This equation, together with any one of the equations (1), then form tangential equations of the cone.

11·7. Equations in line-coordinates.

A line in space has four degrees of freedom, and a single equation connecting the coordinates of a line represents a three-dimensional assemblage of lines. This is in general a *complex*. Two equations represent a two-dimensional assemblage, in general a *congruence*; and three equations represent a one-dimensional assemblage or *line-series*, such as the lines on a quadric surface.

A single equation may also, however, represent a surface or a curve, for there are ∞^3 lines tangent to a given surface, or cutting a given curve. Such an equation will be called the line-equation of the surface or curve.

11·71. Line-equation of a conic.

As an example of the line-equation of a conic let us take the conic at infinity on the general quadric

$$\sum_{1}^{4}\sum_{1}^{4} a_{rs} x_r x_s = 0,$$

$x_0 = 0$ being the plane at infinity.

Let $p_{ij} \equiv a_i b_j - a_j b_i$ be the line-coordinates of any line. The point-coordinates of any point on the line are $a_i - \lambda b_i$. Hence the line cuts the plane $x_0 = 0$ where $\lambda = a_0/b_0$, and the coordinates of the point of intersection are

$$x_i = (a_i b_0 - a_0 b_i)/b_0 = p_{i0}/b_0.$$

As this point lies also on the quadric we have

$$\sum_{1}^{3}\sum_{1}^{3} a_{rs} p_{r0} p_{s0} = 0.$$

In particular the line-equation of the circle at infinity is

$$p_{01}{}^2 + p_{02}{}^2 + p_{03}{}^2 = 0.$$

Ex. Prove that the line-equation of any curve in the plane $x_0 = 0$ is homogeneous in p_{01}, p_{02}, p_{03}; and conversely that any homogeneous equation in p_{01}, p_{02}, p_{03} represents a curve in the plane $x_0 = 0$.

11·72. Line-equation of a quadric.

This is the condition that the line (p) should touch the quadric. Let the line be the intersection of the two planes

$$\alpha \equiv \Sigma \xi_i x_i = 0,$$

$$\beta \equiv \Sigma \eta_i x_i = 0.$$

Then for some value of the ratio λ/μ the equation $\lambda\alpha + \mu\beta = 0$ represents a tangent-plane to the quadric. If (y) is the point of contact we have

$$a_{i0} y_0 + a_{i1} y_1 + a_{i2} y_2 + a_{i3} y_3 = \lambda \xi_i + \mu \eta_i \quad (i = 0, 1, 2, 3),$$

also $\Sigma \xi_i y_i = 0$ and $\Sigma \eta_i y_i = 0.$

Eliminating y_0, y_1, y_2, y_3, λ and μ between these six equations we get

$$\begin{vmatrix} a_{00} & a_{01} & a_{02} & a_{03} & \xi_0 & \eta_0 \\ a_{10} & a_{11} & a_{12} & a_{13} & \xi_1 & \eta_1 \\ a_{20} & a_{21} & a_{22} & a_{23} & \xi_2 & \eta_2 \\ a_{30} & a_{31} & a_{32} & a_{33} & \xi_3 & \eta_3 \\ \xi_0 & \xi_1 & \xi_2 & \xi_3 & 0 & 0 \\ \eta_0 & \eta_1 & \eta_2 & \eta_3 & 0 & 0 \end{vmatrix} = 0.$$

When this determinant is expanded it can be written as a *homogeneous equation of the second degree* in the line-coordinates $\varpi_{ij} \equiv \xi_i \eta_j - \xi_j \eta_i$.

11·73. Polar of a line (p) with respect to a given quadric.
The polar plane of (x') with respect to the quadric

$$\Sigma\Sigma a_{rs} x_r x_s = 0$$

is $\qquad\qquad \Sigma\Sigma a_{rs} x_r x_s' = 0.$

Hence the polar line of the line joining (x') and (x'') is

$$\left.\begin{array}{l} \Sigma\Sigma a_{rs} x_r x_s' = 0 \\ \Sigma\Sigma a_{rs} x_r x_s'' = 0 \end{array}\right\} .$$

Now $\qquad\qquad p_{ij} = x_i' x_j'' - x_j' x_i'',$

and if ϖ_{ij}' are the line-coordinates of the polar line

$$\begin{aligned} \varpi_{ij}' &= \Sigma a_{is} x_s' . \Sigma a_{js} x_s'' - \Sigma a_{js} x_s' . \Sigma a_{is} x_s' \\ &= \underset{r\ s}{\Sigma\Sigma} (a_{ir} a_{js} - a_{jr} a_{is}) p_{rs}. \end{aligned}$$

11·74. From this we can deduce in another way the line equation of the quadric. The line (p) is a tangent to the quadric when it meets its polar. The condition for this is

$$\Sigma\Sigma p_{ij} \varpi_{ij}' = 0.$$

Hence substituting the values of ϖ_{ij}' we obtain the line-equation of the quadric in the form

$$\underset{i\ j\ r\ s}{\Sigma\Sigma\Sigma\Sigma} (a_{ir} a_{js} - a_{jr} a_{is}) p_{ij} p_{rs} = 0,$$

where i, j and r, s take the successive pairs of values 0, 1; 0, 2; 0, 3; 2, 3; 3, 1; 1, 2 independently.

In terms of the coefficients of the tangential equation the line-equation is similarly

$$\Sigma\Sigma\Sigma\Sigma(A_{ir}A_{js}-A_{jr}A_{is})\varpi_{ij}\varpi_{rs}=0.$$
$$\scriptstyle i\ j\ r\ s$$

If the quadric is a cone with vertex $[0, 0, 0, 1]$ so that $a_{0i}=0$ ($i=0, 1, 2, 3$), the line-equation contains only p_{23}, p_{31}, p_{12}, or, if it is expressed in terms of ϖ_{ij}, it is homogeneous in ϖ_{01}, ϖ_{02}, ϖ_{03}.

Ex. 1. Show that the line-equation of any cone with vertex $\xi_0=0$ is homogeneous in ϖ_{01}, ϖ_{02}, ϖ_{03}; and conversely.

Ex. 2. If the point-equation of the quadric is

$$\Sigma\, a_r x_r{}^2=0,$$

show that its line-equation is

$$\Sigma\Sigma\, a_r a_s p_{rs}{}^2=0.$$

11·8. We can now amplify the classification of quadrics in the cases of degeneracy.

When $[\Delta]$ is of rank 3, the quadric as a locus is a cone. If the point-equation is

$$a_1 x_1{}^2+a_2 x_2{}^2+a_3 x_3{}^2=0$$

the plane-equation degenerates to $\xi_0{}^2=0$, which represents just the vertex of the cone twice. The determinant $[\nabla]$ is of rank 1. The cone, however, is represented completely in plane-coordinates by the two equations

$$\xi_0=0 \text{ and } \xi_1{}^2/a_1+\xi_2{}^2/a_2+\xi_3{}^2/a_3=0.$$

The line-equation is

$$a_2 a_3 p_{23}{}^2+a_3 a_1 p_{31}{}^2+a_1 a_2 p_{12}{}^2=0.$$

When $[\Delta]$ is of rank 2, the quadric as a locus degenerates to two planes. If the point-equation is

$$a_0 x_0{}^2+a_1 x_1{}^2=0,$$

the plane-equation completely degenerates, $[\nabla]$ being of rank 0. Consider the quadric

$$a_0 x_0{}^2+a_1 x_1{}^2+\epsilon\Sigma\Sigma a_{rs} x_r x_s=0.$$

Forming the tangential equation and arranging in powers of ϵ we find

$$\epsilon^3\Sigma\Sigma A_{rs}\xi_r\xi_s+\epsilon^2 K-\epsilon a_0 a_1(a_{33}\xi_2{}^2+a_{22}\xi_3{}^2-2a_{23}\xi_2\xi_3)=0.$$

Divide by ϵ and let $\epsilon \to 0$ and we get

$$a_{33}\xi_2^2 + a_{22}\xi_3^2 - 2a_{23}\xi_2\xi_3 = 0,$$

which represents two points on the line joining $\xi_2 = 0$ and $\xi_3 = 0$, i.e. on the line $x_0 = 0 = x_1$.

In this case the point-equation and the plane-equation must be separately given, and the matrices of the coefficients are in general each of rank 2. We have the following cases:

$[\Delta]$ of rank 2, $[\nabla]$ of rank 2. The quadric consists as a locus of two distinct planes through a line l, and as an envelope of two distinct points on the line l.

$[\Delta]$ of rank 2, $[\nabla]$ of rank 1. Two distinct planes through a line l, and two coincident points on l.

$[\Delta]$ of rank 1, $[\nabla]$ of rank 2. Two coincident planes; two distinct points in this plane.

$[\Delta]$ of rank 1, $[\nabla]$ of rank 1. Two coincident planes; two coincident points on this plane.

11·9. EXAMPLES.

1. Show that the envelope of planes which cut a given quadric in sections whose centres lie on a given plane is a paraboloid.

2. Pairs of orthogonal tangent-planes to a given quadric pass through a fixed point; show that their lines of intersection generate a quadric cone.

3. Show that

$$fqrx^2 + grpy^2 + hpqz^2 + fghw^2 + (fp - gq - hr)(pyz + fxw)$$
$$+ (gq - hr - fp)(qzx + gyw) + (hr - fp - gq)(rxy + hzw) = 0$$

represents a quadric which touches the four faces of the tetrahedron of reference.

Show also that the conditions that the lines joining each point of contact to the opposite vertex should be concurrent are

$$fp = gq = hr.$$

Show that by suitable choice of unit-point the equation can be written

$$x^2 + y^2 + z^2 + w^2 + 2l(yz + xw) + 2m(zx + yw) + 2n(xy + zw) = 0,$$

where

$$2lmn - l^2 - m^2 - n^2 + 1 = 0.$$

4. If $F(x, y, z, w) = 0$ represents a cone with vertex $[A, B, C, D]$ show that its tangential equations are

$$A\xi + B\eta + C\zeta + D\omega = 0$$

and

$$(bc - f^2)\xi^2 + (ca - g^2)\eta^2 + (ab - h^2)\zeta^2 + 2(gh - af)\eta\zeta$$
$$+ 2(hf - bg)\zeta\xi + 2(fg - ch)\xi\eta = 0.$$

5. A quadric cone has vertex A and a circular base of which P is any point. Show that the envelope of the plane through P perpendicular to AP is another quadric cone, and that it cuts the plane of the circle in a conic the foci of which are the orthogonal projections of the vertices of the two cones.

(Math. Trip. II, 1913.)

CHAPTER XII

FOCI AND FOCAL PROPERTIES

12·1. A focus of a conic is a point such that every pair of lines through it which are conjugate with regard to the conic are also at right angles. In general through any point there is just one pair of conjugate lines which are at right angles. Pairs of conjugate lines through a point P form an involution whose double-lines are the tangents from P to the conic; pairs of orthogonal lines through P also form an involution and its double-lines are the lines joining P to the circular points I, J. For an ordinary point P these two involutions are distinct, and since one of them at least, the involution of orthogonal lines, is elliptic, they have always a real pair of lines in common (3·93). If, however, P is a focus F, the two involutions coincide; the lines FI and FJ are tangents to the conic. As these two lines also form a point-circle, we may say also that the foci of a conic are point-circles having double contact with the conic; the chord of contact is the corresponding directrix.

12·11. Focal axes.

For a quadric surface these ideas may be extended in two different ways. First, we may consider pairs of planes through a line, orthogonal and conjugate with regard to the quadric. Pairs of conjugate planes form an involution whose double-planes are the tangent-planes to the quadric, and pairs of orthogonal planes form an elliptic involution whose double-planes are tangents to the circle at infinity (i.e. cut the plane at infinity in lines tangent to the circle at infinity). In general these two involutions have one pair of common planes, which are always real. But for certain lines the two involutions coincide, and every pair of orthogonal planes are also conjugate. Such lines are called *focal axes*. The pair of tangent-planes through a focal axis cut the plane at infinity in two lines which touch the circle at infinity, hence they form a circular cylinder whose radius is zero. *A focal axis is therefore the axis of a circular cylinder of zero radius which has double contact with the quadric.*

12·12. Foci.

Second, we may consider triads of planes through a point, mutually orthogonal and conjugate with regard to the quadric. If the point P does not lie on the quadric any triad of mutually conjugate planes through P are also mutually conjugate with regard to the tangent-cone from P. But of these sets there is just one which is orthogonal, viz. the three principal planes of the cone. If the point P lies on the quadric the tangent-cone degenerates to a plane; the three planes then consist of the tangent-plane and the two planes which contain the normal and one of the bisectors of the angles between the two generators through P. Thus in general there is just one such triad of planes through a given point. But for certain points the triad becomes indeterminate, i.e. an infinity of triads are possible. Lying on the quadric the only such points are the umbilics; and for a point not on the quadric the triad becomes indeterminate only when the tangent-cone is circular. Such points, through which there is an infinity of mutually orthogonal and conjugate planes, are called *foci*. A focus has thus the property that *the tangent-cone from a focus to the quadric is circular*.

The tangent-cone cuts the plane at infinity in a conic C' which has double contact with the circle at infinity Ω, hence it follows that the cone with vertex F and containing Ω touches the tangent-cone along two generators and has therefore double contact with the quadric. But this cone is a point-sphere. Hence *a focus is the centre of a sphere of zero radius having double contact with the quadric*. The chord of contact is called the corresponding *directrix*.

Let U, V be the points of contact of C' and Ω, and let the tangents at U and V intersect in T. Then FT is the axis of rotation of the tangent-cone. FU and FV touch the quadric in P and Q, say. Then PQ is the directrix for the focus F. Also the two planes FTU and FTV form a circular cylinder of zero radius having double contact with the quadric at P and Q, and therefore FT is a focal axis. Hence *through any focus there is one focal axis, which is the axis of rotation of the tangent-cone from F*. Since the planes FTU and FTV are tangent-planes to the quadric at P and Q, *the focal axis FT is the polar of the directrix PQ*.

Again, since the tangent-cone has ring-contact with the quadric on the conic S, their sections C' and C by the plane at infinity have double contact on the line of intersection l of the plane at infinity with the plane of S. Let L be the common pole of l with respect to C and C'; and let O be the pole of the plane

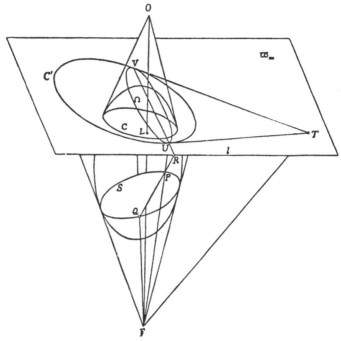

Fig. 40

at infinity with respect to the quadric, i.e. the centre. Then OF is the polar of l and therefore passes through L. The point at infinity R on the directrix PQ is the intersection of PQ with UV and lies on l. The polar-plane of R with respect to the quadric is TFO, hence its polar with respect to the conic C is LT. But since it lies on l its polar with respect to C' is also LT, and, since it lies on UV, LT is also its polar with respect to Ω. Hence since R has the same polar with respect to both C and Ω, it is the point at infinity on one of the principal axes of the quadric; its polar-plane OFT is then the corresponding principal plane.

Hence *a focus F and its focal axis FT lie in a principal plane of the quadric, while the corresponding directrix is parallel to the corresponding principal axis.* Since T is the pole of UV with respect to Ω, the line FT is perpendicular to the plane FUV, i.e. *the plane through a focus F and the corresponding directrix is perpendicular to the focal axis through F.*

A focal axis which does not lie in a principal plane does not pass through any focus. A focal axis may be obtained from *any* chord UV of Ω. Through each of the tangents UT and VT can be drawn two tangent-planes to the quadric, and the four lines of intersection of these form four focal axes through T, i.e. four parallel focal axes, two real and two imaginary. The corresponding chords of contact PQ with the quadric are not in general coplanar with UV.

12·21. Focal axes.

We shall treat the problem now analytically, confining our attention for the present to central quadrics. Let the equation of the quadric be

$$\frac{x^2}{A}+\frac{y^2}{B}+\frac{z^2}{C}=1,$$

and suppose a focal axis to be determined as the intersection of the two planes

$$u \equiv lx+my+nz+p=0,$$

$$u' \equiv l'x+m'y+n'z+p'=0.$$

A plane through the intersection of these is represented by

$$(l+\lambda l')x+(m+\lambda m')y+(n+\lambda n')z+(p+\lambda p')=0.$$

Consider this plane and a similar plane with parameter μ instead of λ. The condition that these two planes should be orthogonal is

$$\lambda\mu(l'^2+m'^2+n'^2)+(\lambda+\mu)(ll'+mm'+nn')+(l^2+m^2+n^2)=0,$$

and the condition that they should be conjugate with regard to the quadric is

$$\lambda\mu(Al'^2+Bm'^2+Cn'^2-p'^2)+(\lambda+\mu)(All'+Bmm'+Cnn'-pp')$$
$$+(Al^2+Bm^2+Cn^2-p^2)=0.$$

If their line of intersection is a focal axis these two conditions must be identical for all values of λ and μ. Hence

$$\frac{Al'^2 + Bm'^2 + Cn'^2 - p'^2}{l'^2 + m'^2 + n'^2} = \frac{All' + Bmm' + Cnn' - pp'}{ll' + mm' + nn'}$$

$$= \frac{Al^2 + Bm^2 + Cn^2 - p^2}{l^2 + m^2 + n^2} \cdot \dots(1)$$

If one plane is kept fixed, say l', m', n', p' are given, we have two equations, one linear and one quadratic, to determine the ratios of l, m, n, p. If (l) is any one plane satisfying these equations, then $(l+\lambda l')$ will also satisfy them. Thus these equations determine two pencils of planes having their axes on the plane (l'). Hence in general *there are two focal axes in any plane*.

Equating each of the ratios (1) to t, we get the three equations

$$(A-t)l^2 + (B-t)m^2 + (C-t)n^2 - p^2 = 0,$$
$$(A-t)l'^2 + (B-t)m'^2 + (C-t)n'^2 - p'^2 = 0,$$
$$(A-t)ll' + (B-t)mm' + (C-t)nn' - pp' = 0.$$

Solving for $A-t$, $B-t$, $C-t$, and writing $mn' - m'n = p_{01}$, $pl' - p'l = p_{23}$, etc. (the line-coordinates of the focal axis), we have

$$(A-t)p_{02}p_{03} = p_{31}p_{12},$$
$$(B-t)p_{03}p_{01} = p_{12}p_{23},$$
$$(C-t)p_{01}p_{02} = p_{23}p_{31}.$$

Then eliminating t we obtain the two equations

$$p_{01}(Ap_{02}p_{03} - p_{31}p_{12}) = p_{02}(Bp_{03}p_{01} - p_{12}p_{23}) = p_{03}(Cp_{01}p_{02} - p_{23}p_{31}),$$

which represent the whole two-dimensional assemblage of focal axes, or the *focal congruence*.

Ex. Show that the equations of the focal congruence can also be written

$$\frac{(p_{02}p_{12} + p_{03}p_{13})p_{23}}{B-C} = \frac{(p_{03}p_{23} + p_{01}p_{21})p_{31}}{C-A} = \frac{(p_{01}p_{31} + p_{02}p_{32})p_{12}}{A-B} = p_{01}p_{02}p_{03}.$$

12·22. Focal conics.

The principal planes of the quadric are exceptional. Taking (l') as the plane $x=0$, so that $l'=1$, $m'=0=n'=p'$, then the equations (1) of 12·21 reduce to

$$A = A = \frac{Al^2 + Bm^2 + Cn^2 - p^2}{l^2 + m^2 + n^2},$$

which give only the one equation

$$(A-B)m^2 + (A-C)n^2 + p^2 = 0.$$

The plane $x=0$ therefore contains an infinity of focal axes, whose envelope is the conic

$$\frac{y^2}{B-A} + \frac{z^2}{C-A} = 1, \quad x=0.$$

Similarly we have conics in the planes $y=0$ and $z=0$.

These three conics are called the *focal conics* of the quadric. If $A > B > C$ the conic in the plane of yz is virtual, that in the plane of zx is a hyperbola, and that in the plane of xy is an ellipse. This holds whether A, B, C are positive or negative.

12·23. Foci.

Consider now any point $P \equiv [X, Y, Z]$, and a plane through P,

$$l(x-X) + m(y-Y) + n(z-Z) = 0.$$

Let the pole of this plane with regard to the quadric

$$x^2/A + y^2/B + z^2/C = 1$$

be $Q \equiv [X', Y', Z']$. Then, identifying the equation

$$xX'/A + yY'/B + zZ'/C = 1$$

with that of the given plane, we have

$$\frac{X'}{Al} = \frac{Y'}{Bm} = \frac{Z'}{Cn} = \frac{1}{lX + mY + nZ}.$$

Also if PQ is perpendicular to the plane,

$$\frac{X'-X}{l} = \frac{Y'-Y}{m} = \frac{Z'-Z}{n} = \frac{\lambda}{lX + mY + nZ}, \text{ say.}$$

Hence $\quad Al - X(lX + mY + nZ) = \lambda l$, etc.,

i.e.
$$(X^2 - A + \lambda)l + \qquad XYm + \qquad XZn = 0,$$
$$XYl + (Y^2 - B + \lambda)m + \qquad YZn = 0,$$
$$XZl + \qquad YZm + (Z^2 - C + \lambda)n = 0.$$

Eliminating l, m, n,

$$\begin{vmatrix} X^2 - A + \lambda & YX & ZX \\ XY & Y^2 - B + \lambda & ZY \\ XZ & YZ & Z^2 - C + \lambda \end{vmatrix} = 0.$$

In general this gives three real values for λ, and these determine one set of three planes through P, mutually orthogonal and conjugate with regard to the quadric. If, however, every element of the determinant vanishes, i.e. if the determinant is of rank o, the three roots will be equal, and every set of three mutually orthogonal planes through P are also mutually conjugate. Such a point is called a *principal focus*. The conditions for this are

$$YZ = 0, \quad ZX = 0, \quad XY = 0, \quad A - X^2 = B - Y^2 = C - Z^2 = \lambda.$$

Hence two of X, Y, Z must vanish, say $Y = 0 = Z$, and then $B = C = \lambda$ and $X^2 = A - B$. The quadric is then a quadric of revolution, and there are two principal foci (real or imaginary), which are the foci of the meridian sections which lie on the axis of rotation.

If the determinant is of rank 1, two roots of the equation in λ will be equal. P is then an ordinary focus. The conditions for this are

$$(\lambda - B)(\lambda - C) + Y^2(\lambda - C) + Z^2(\lambda - B) = 0,$$
$$(\lambda - C)(\lambda - A) + Z^2(\lambda - A) + X^2(\lambda - C) = 0,$$
$$(\lambda - A)(\lambda - B) + X^2(\lambda - B) + Y^2(\lambda - A) = 0,$$

and $\quad YZ(\lambda - A) = 0, \quad ZX(\lambda - B) = 0, \quad XY(\lambda - C) = 0.$

If $A = B = C$ these equations are all satisfied when $\lambda = A$, i.e. for a sphere every point is a focus, the centre being the only principal focus.

Excluding this case, one at least of X, Y, Z must vanish, say $X = 0$. The equations then become

$$(\lambda - B)(\lambda - C) + Y^2(\lambda - C) + Z^2(\lambda - B) = 0,$$
$$(\lambda - A)(\lambda - C + Z^2) = 0,$$
$$(\lambda - A)(\lambda - B + Y^2) = 0,$$
$$(\lambda - A) YZ = 0,$$

which are satisfied when $\lambda = A$ and

$$Y^2(A - C) + Z^2(A - B) + (A - B)(A - C) = 0.$$

Hence we obtain a locus of foci on the plane $x = 0$ forming the focal conic

$$\frac{y^2}{B - A} + \frac{z^2}{C - A} = 1.$$

The foci of a quadric are thus the points of the three focal conics.

For a quadric of revolution with the axis of x as axis of revolution, $B = C$, and we find either $Y = 0$ and $Z = 0$ or $X = 0$ and $Y^2 + Z^2 = B - A$. In this case therefore the focal conic in the plane $x = 0$ becomes a circle, while the other two degenerate to the axis of rotation which contains also the two principal foci.

12·24. We may investigate the foci also as the centres of spheres of zero radius having double contact with the quadric.

Let $S \equiv (x - X)^2 + (y - Y)^2 + (z - Z)^2 = 0$

represent a point-sphere at $[X, Y, Z]$, and denote the quadric by $F = 0$. Then the equation

$$S - \lambda F = 0$$

represents a quadric passing through the curve of intersection of S and F.

Now if S touches F at two points the curve of intersection has a double-point at each of these points. Draw a plane through these two points and any other point on the curve of intersection. This plane has then five points in common with the curve; but the curve is only of the fourth order, hence it has an infinity of points in common with the curve. The curve of intersection therefore breaks up into two plane curves, each a conic. We can therefore choose λ so that the equation $S - \lambda F = 0$ represents these two planes, say $S - \lambda F \equiv \alpha\beta$. Hence the quadric will have a focus $F \equiv [X, Y, Z]$ if its equation can be written in the form

$$(x - X)^2 + (y - Y)^2 + (z - Z)^2 = \alpha\beta,$$

where α and β are expressions of the first degree in x, y, z.

α and β represent the two planes, real or imaginary, which contain the curve of intersection of the quadric with the point-sphere. They are therefore planes of circular section. Their line of intersection d is the line joining the two points of contact P and Q, and is the *directrix* corresponding to the focus F; being the intersection of two circular planes it is parallel to one of the principal axes, viz. that one which is perpendicular to the principal plane in which F lies. The tangent-planes at P and Q intersect in a line f which is the polar of the directrix. Any pair of planes through f which are harmonic conjugates with regard to the two tangent-planes are conjugate with regard to both the quadric and the point-sphere, and are therefore orthogonal. f is therefore a focal axis and by 12·22 is a tangent to the focal conic. As the tangent-planes pass through F, f also passes through F, which is therefore the point of contact of f with the focal conic. If the directrix cuts the principal plane, which contains the focal conic, in D, f is the polar of D with regard to the principal section. One set of three mutually orthogonal and conjugate planes through F consists of the plane of the focal conic, the plane through f perpendicular to the principal plane, and the plane through F perpendicular to f. But the last plane contains d, since it is the polar of f. Hence DF is perpendicular to f and therefore normal to the focal conic. The principal section being

$$y^2/B + z^2/C = 1,$$

and the focal conic

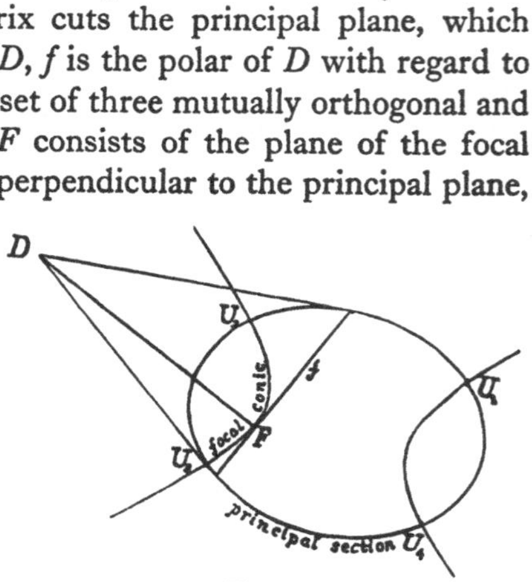

Fig. 41

$$y^2/(B - A) + z^2/(C - A) = 1,$$

these are confocal; they cut orthogonally at the umbilics. These relations are shown in Fig. 41.

12·25. As already remarked, the only foci which lie on the quadric itself are the umbilics. These are therefore the points of intersection of the focal conics with the quadric, or with its principal sections. The circular planes through a directrix d are parallel to the tangent-planes at the umbilics which lie in the principal plane perpendicular to d.

The foci of the focal conic

$$y^2/(B-A)+z^2/(C-A)=1$$

are
$$\left.\begin{array}{l}z=0\\y=\pm\sqrt{(B-C)}\end{array}\right\} \quad \text{and} \quad \left.\begin{array}{l}y=0\\z=\pm\sqrt{(C-B)}\end{array}\right\}.$$

The former are vertices of the focal conic

$$x^2/(A-C)+y^2/(B-C)=1,$$

and the latter are vertices of the remaining focal conic

$$z^2/(C-B)+x^2/(A-B)=1.$$

Thus each of the three focal conics passes through one pair of foci of each of the other two. Even the virtual focal conic thus possesses a pair of real foci (the ends of the minor axis of the focal ellipse), though the corresponding eccentricity is of course imaginary; the eccentricity corresponding to its imaginary foci is real.

Ex. Show that for points on the focal ellipse the circular planes are imaginary in the case of the ellipsoid and real in the case of the hyperboloid of two sheets, and *vice versa* for the focal hyperbola.

12·31. The equation

$$(x-X)^2+(y-Y)^2+(z-Z)^2=\alpha\beta$$

expresses a metrical property of the foci, viz. *the ratio of the square of the distance of any point on the quadric from a focus to the product of its distances from the two circular planes is constant.*

Ex. Show that

$$A\,(x^2/A+y^2/B+z^2/C-1)\equiv\{x^2+(y-Y)^2+(z-Z)^2\}$$
$$-\frac{B-A}{B}\left(y-\frac{BY}{B-A}\right)^2-\frac{C-A}{C}\left(z-\frac{CZ}{C-A}\right)^2,$$

where
$$\frac{Y^2}{B-A}+\frac{Z^2}{C-A}=1.$$

12·32. Taking a focus as origin, the equation of the quadric can be written
$$x^2 + y^2 + z^2 = \alpha\beta,$$
$\alpha\beta = 0$ being the equation of the two circular planes whose intersection is the corresponding directrix. If we choose further the plane $z = 0$ as the plane containing the focus and its directrix, this plane cuts α and β in the same line, say $z = 0$ and
$$u \equiv lx + my + n = 0.$$
Hence the section of the quadric by this plane is
$$z = 0, \quad x^2 + y^2 = u^2.$$
But this represents a conic with focus at the origin and directrix u. Hence *the plane containing a focus and the corresponding directrix cuts the quadric in a conic having these as focus and directrix.*

Also since this plane is normal to the focal conic on which the given focus lies, *every plane normal to a focal conic has for a focus the point where it meets the focal conic normally.*

12·33. We may prove these results also as follows. The focal axis f through the focus F is the axis of rotation of the tangent-cone from F. This cone has *ring-contact* with the quadric. Two quadrics have ring-contact when their curve of intersection reduces to two coincident conics; in this case any plane cuts them in two conics which have double contact at the points where the plane cuts the double conic.

Let α be the plane through F perpendicular to f. Then α cuts the tangent-cone from F in a point-circle, i.e. in two straight lines passing through the circular points I, J in α. It cuts the quadric in a conic having double contact with this line-pair, i.e. FI and FJ are tangents to the conic, and therefore F is a focus of the conic. The plane α also contains the corresponding directrix d of the quadric, and this is the polar of F with respect to the curve of intersection, and is therefore the directrix also for this conic.

12·34. Dandelin's Theorem.

The well-known construction for the foci of a plane section of a circular cone follows from the property of quadrics having ring-contact. The cone being circular, there are two spheres inscribed in the cone and touching the plane. Each sphere has ring-contact with the cone, and therefore the curves of inter-section of the plane with the cone and one of the spheres have double contact. But the intersection of the plane with the sphere is a point-circle, hence this point is a focus of the conic section.

Ex. Prove that the tangent-plane at an umbilic of a quadric cuts any tangent-cone in a conic having a focus at the umbilic.

12·4. Confocal quadrics.

Two quadrics are said to be confocal when they have the same focal conics. Confocal quadrics have therefore the same principal planes. Confining our attention for the present to central quadrics, if

$$x^2/A + y^2/B + z^2/C = 1$$

and

$$x^2/A' + y^2/B' + z^2/C' = 1$$

are confocal, $B - C = B' - C'$, $C - A = C' - A'$, $A - B = A' - B'$ (the third equation following from the first two). Hence if $A' = A - \lambda$, then $B' = B - \lambda$ and $C' = C - \lambda$. The equation

$$\frac{x^2}{A-\lambda} + \frac{y^2}{B-\lambda} + \frac{z^2}{C-\lambda} = 1$$

therefore represents all the quadrics which are confocal with $x^2/A + y^2/B + z^2/C = 1$. These form a *system of confocal quadrics*.

If A, B, C are all unequal we may assume that $A > B > C$. Then if $\lambda < 0$ the quadric is an ellipsoid, for $B > \lambda > C$ a hyperboloid of one sheet, for $A > \lambda > B$ a hyperboloid of two sheets, and for $\lambda > A$ it is virtual.

The critical values, $\lambda = A, B, C$, make the quadric degenerate, e.g. $\lambda = A$ requires $x = 0$ and $y^2/(B-A) + z^2/(C-A) = 1$, which is one of the focal conics. The focal conics are therefore degenerate quadrics of the confocal system.

If two of the quantities A, B, C are equal, say $B = C$, all the quadrics of the systems are of revolution. The focal conic in the plane of yz, which is perpendicular to the axis of rotation, is a circle of radius $\sqrt{(B-A)}$, real or virtual according as $A <$ or $> B$. The principal planes perpendicular to this are inde-terminate, but in each of them the focal conic degenerates to the same pair of points $[\pm\sqrt{(A-B)}, 0, 0]$; these are the real foci of the meridian section if $A > B$, and are principal foci; the virtual focal circle then passes through the two imaginary foci of the meridian section. When $A < B$, the focal circle, which is real, passes through the real foci of the meridian section.

If $A = B = C$, the quadric is a sphere and all the principal foci are collected at the centre.

12·41. Confocals through a given point.

Through a given point $[X, Y, Z]$ there pass in general three quadrics of a given confocal system. The equation of the system being

$$\frac{x^2}{A-\lambda} + \frac{y^2}{B-\lambda} + \frac{z^2}{C-\lambda} = 1,$$

if this equation is satisfied by $[X, Y, Z]$ we obtain a cubic equation in λ:

$$\phi(\lambda) \equiv (\lambda - A)(\lambda - B)(\lambda - C) + \Sigma(\lambda - B)(\lambda - C)X^2 = 0.$$

Assuming that no one of X, Y, Z is zero, and that $A > B > C$, if we substitute

$$\lambda = -\infty, \quad C, \quad B, \quad A, \quad +\infty,$$

the signs of $\phi(\lambda)$ are $\quad - \quad + \quad - \quad + \quad +$.

Hence the roots are all real: the first root, $< C$, gives an ellipsoid; the second, between C and B, gives a hyperboloid of one sheet; and the third, between B and A, gives a hyperboloid of two sheets.

Exceptional cases occur when the point $[X, Y, Z]$ lies (1) in one of the principal planes, then one of the three quadrics reduces to that plane; (2) in one of the principal axes, then two of the quadrics reduce to the principal planes through this axis; (3) on one of the focal conics, then two of the quadrics reduce to the plane of this conic; (4) at the centre, then the three quadrics reduce to the three principal planes; (5) at a vertex of one of the focal conics, then two of the quadrics reduce to the plane of this conic while the third reduces to the other principal plane through the point.

As regards the nature of the surviving confocals through a point, the reader may verify the results indicated in Fig. 42, where E, H_1, and H_2 stand for ellipsoid, hyperboloid of one sheet, and hyperboloid of two sheets respectively.

12·42. *The three confocals through a point are mutually orthogonal.* Let the roots of the cubic equation in λ,

$$\frac{X^2}{A-\lambda} + \frac{Y^2}{B-\lambda} + \frac{Z^2}{C-\lambda} = 1,$$

be denoted by λ_1, λ_2, λ_3. The tangent-planes at $[X, Y, Z]$ are

$$\frac{Xx}{A-\lambda_i}+\frac{Yy}{B-\lambda_i}+\frac{Zz}{C-\lambda_i}=1 \quad (i=1, 2, 3).$$

Now

$$\Sigma\frac{X^2}{(A-\lambda_1)(A-\lambda_2)}\equiv\frac{1}{\lambda_1-\lambda_2}\Sigma\left(\frac{X^2}{A-\lambda_1}-\frac{X^2}{A-\lambda_2}\right)=0.$$

The confocals therefore form a *triple orthogonal system.*

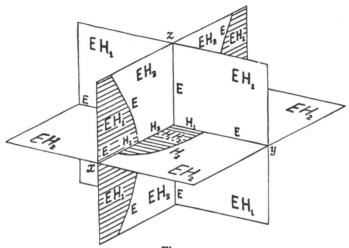

Fig. 42

12·43. Confocal quadrics in tangential coordinates.

The tangential equation of the confocal system is

$$(A-\lambda)l^2+(B-\lambda)m^2+(C-\lambda)n^2=p^2,$$

or $\quad\quad \lambda(l^2+m^2+n^2)-(Al^2+Bm^2+Cn^2-p^2)=0. \quad ......(1)$

This equation is linear in λ, and if l, m, n, p are given, a unique value of λ is in general determined. Hence *one and only one quadric of a confocal system touches an arbitrary plane.*

The equation is satisfied by any set of values of l, m, n, p which satisfy the two equations

$$Al^2+Bm^2+Cn^2-p^2=0 \quad \text{and} \quad l^2+m^2+n^2=0.$$

These equations represent two quadrics of the system. The latter, however, is a special quadric whose only real tangent-plane is $[0, 0, 0, 1]$, i.e. the plane at infinity. The equation expresses the condition that the line of intersection of the plane $lx + my + nz + pw = 0$ with the plane at infinity $w = 0$ should touch the conic $x^2 + y^2 + z^2 = 0$, $w = 0$, i.e. the circle at infinity. *The equation $l^2 + m^2 + n^2 = 0$ is therefore the tangential equation of the circle at infinity.* Thus in addition to the three focal conics, which are degenerate quadrics of the system lying in the principal planes $x = 0$, $y = 0$, $z = 0$, there is a fourth degenerate quadric of the system, the circle at infinity, in the plane $w = 0$.

The equation (1) represents a *linear tangential system of quadrics* determined by a given quadric $Al^2 + Bm^2 + Cn^2 - p^2 = 0$ and the circle at infinity. Any plane which touches these two quadrics, or indeed any two quadrics of the system, will touch them all. Such planes are in general only imaginary.

The tangential equation of the point of contact of the plane $[l', m', n', p']$ is

$$(A - \lambda)l'l + (B - \lambda)m'm + (C - \lambda)n'n - p'p = 0,$$

i.e. the coordinates of the point of contact are

$$[(A - \lambda)l', (B - \lambda)m', (C - \lambda)n', -p'],$$

and by giving all values to λ we get a straight line joining the point $[Al', Bm', Cn', -p']$ to the point at infinity $[l', m', n', 0]$. Hence *when a plane touches two quadrics of the system, and therefore all of them, the points of contact lie on a line.*

12·431. The planes which touch all the quadrics of the system, i.e. the assemblage of common tangent-planes to two quadrics of the system, form a one-dimensional assemblage and generate a *developable*. This is called the *focal developable* and is represented by the two simultaneous equations

$$l^2 + m^2 + n^2 = 0, \quad Al^2 + Bm^2 + Cn^2 - p^2 = 0.$$

We shall return to this in another chapter.

12·44. Confocals touching a given line.

Let l be a given line. The pairs of tangent-planes through l to quadrics of the system form an involution, for if α is any plane through l there is one quadric which touches α, hence a second tangent-plane α' to this quadric is uniquely determined; and conversely to α' corresponds α. The two tangent-planes coincide when the line is a tangent to the quadric, and form the double-elements of the involution. Hence *there are two quadrics of the system which touch a given line.* As the involution contains as one pair the planes through *l* which touch the circle at infinity, and as the double-elements are harmonic conjugates with regard to them, it follows that *the tangent-planes to the two quadrics at their points of contact with l are at right angles.*

12·5. The paraboloids.

As a confocal system of quadrics is the linear tangential system determined by one quadric and the circle at infinity, and since the plane at infinity touches the absolute circle, it follows that if one quadric of the system is a paraboloid, the plane at infinity, which then touches two quadrics of the system, will touch them all, and therefore all the quadrics will be paraboloids. As the focal conics are degenerate quadrics of the system they must be parabolas.

Let the equation of the paraboloid be

$$x^2/A + y^2/B = 2z.$$

Let $F \equiv [X, Y, Z]$ be a focus, and consider a plane through F

$$l(x - X) + m(y - Y) + n(z - Z) = 0.$$

The pole $Q \equiv [X', Y', Z']$ of this plane is given by

$$\frac{X'}{Al} = \frac{Y'}{Bm} = -\frac{1}{n} = \frac{Z'}{lX + mY + nZ},$$

and, if FQ is perpendicular to the plane,

$$\frac{X' - X}{l} = \frac{Y' - Y}{m} = \frac{Z' - Z}{n} = \frac{\lambda}{n}, \text{ say;}$$

then

$$-Al - nX = \lambda l,$$

$$-Bm - nY = \lambda m,$$

$$-(lX + mY + nZ) - nZ = \lambda n.$$

Eliminating $l, m, n,$

$$\begin{vmatrix} \lambda + A & 0 & X \\ 0 & \lambda + B & Y \\ X & Y & \lambda + 2Z \end{vmatrix} = 0.$$

This gives as before a cubic equation in λ, having in general three different roots. If all the elements of the determinant vanish the roots are all equal and we have

$$\lambda = -A = -B = -2Z, \quad X = 0 = Y.$$

The paraboloid is then a paraboloid of rotation and we have a single *principal focus* at $[0, 0, \frac{1}{2}A]$; this is the focus of the meridian section.

If the minors of the determinant all vanish two roots are equal and the tangent-cone with vertex F is circular, its principal axis corresponding to the double-root. The conditions for this are

$$(\lambda + A)(\lambda + 2Z) = X^2,$$
$$(\lambda + B)(\lambda + 2Z) = Y^2,$$
$$(\lambda + A)(\lambda + B) = 0,$$
$$Y(\lambda + A) = 0,$$
$$X(\lambda + B) = 0,$$
$$XY = 0.$$

These are satisfied by

(i) $X = 0, \quad \lambda = -A, \quad Y^2 = (A - B)(A - 2Z),$

(ii) $Y = 0, \quad \lambda = -B, \quad X^2 = (B - A)(B - 2Z).$

Hence we have two equal focal parabolas, with axes along the axis of z and vertices in opposite directions; each passes through the focus of the other. In the case of the elliptic paraboloid one of these cuts the surface orthogonally in the two real umbilics; the other parabola, and both the confocal parabolas in the case of the hyperbolic paraboloid, meet the surface in imaginary points.

12·51. Confocal paraboloids.

The tangential equation of the paraboloid

$$x^2/A + y^2/B = 2z/C$$

is

$$Al^2 + Bm^2 = 2Cnp,$$

and therefore the tangential equation of the system of quadrics confocal with the paraboloid is

$$Al^2 + Bm^2 - 2Cnp - \lambda(l^2 + m^2 + n^2) = 0.$$

The point-equation corresponding to this equation is then found
to be

$$\frac{x^2}{A-\lambda} + \frac{y^2}{B-\lambda} = \frac{2z}{C} - \frac{\lambda}{C^2}.$$

Assuming $A > B$, when $\lambda > A$ the quadric is an elliptic paraboloid with vertex downwards, when $A > \lambda > B$ a hyperbolic
paraboloid, and when $\lambda < B$ again an elliptic paraboloid, but with
vertex upwards. The focal parabolas are limiting cases separating
these series. The third focal conic, which should exist in the
general case, coincides in the case of the paraboloids with the
circle at infinity.

Through any point $[X, Y, Z]$ there are again three quadrics
of the confocal system, corresponding to the roots of the cubic
equation in λ:

$$\phi(\lambda) \equiv \lambda(\lambda - A)(\lambda - B) - C^2 X^2(\lambda - B) - C^2 Y^2(\lambda - A)$$
$$- 2ZC(\lambda - A)(\lambda - B) = 0.$$

When $\lambda = -\infty, \quad B, \quad A, \quad +\infty,$

        $\phi(\lambda)$ is $-$ $+$ $-$ $+$.

Hence one root, $< B$, gives an elliptic paraboloid; the second,
between B and A, gives a hyperbolic paraboloid; and the third,
$> A$, gives another elliptic paraboloid.

Exceptional cases occur here also for special positions of the
point $[X, Y, Z]$, when one or more of the confocals degenerate.

12·6. Foci of a cone or cylinder.

The polar of any point P with respect to a cone is a plane
passing through the vertex O, and this is also the polar-plane of
any point on the line OP. The vertex itself is the pole of any
plane, and the pole of a plane which does not pass through the
vertex is the vertex.

The ideas of foci and focal axes in the case of a cone or cylinder
require modification, for when a cone is considered as the limiting case of a hyperboloid of one sheet, say, the tangent-planes
become planes through the vertex. But only those are considered
as tangent-planes which meet the surface in coincident lines, and
the surface possesses only a single infinity of tangent-planes.
Through an arbitrary line there are no tangent-planes, and

through a point there is no tangent-cone but only a pair of tangent-planes.

In the case of the general quadric the tangent-planes through a focus envelop a circular cone, and this meets the plane at infinity in a conic having double-contact with the circle at infinity. In the case of a cone or a cylinder there are just two tangent-planes through a point, and if this point is a focus these planes meet the plane at infinity in a pair of lines both touching the circle at infinity.

12·61. In the case of a cone there is a proper conic at infinity C, and the tangent-planes are planes through the vertex and touching this conic. C and the circle at infinity Ω have four common tangents which intersect in pairs in six points, of which two are real and two pairs of conjugate imaginaries. If F is one of these points any point on OF is a focus. The foci therefore lie on three pairs of lines through O, and these are degenerate focal conics. The pair of tangent-planes through a focus form a circular cylinder of zero radius and their line of intersection, OF, is therefore also a focal axis. There are thus only six focal axes, and not an infinite number as in the case of the general quadric.

Let the equation of the cone be

$$x^2/A + y^2/B + z^2/C = 0,$$

then if the line $[l, m, n]$ is a focal axis the point at infinity $[l, m, n, 0]$ is the point of intersection of two common tangents to the conic at infinity

$$w = 0, \quad x^2/A + y^2/B + z^2/C = 0$$

and the circle at infinity.

The tangential equations of these two conics are

$$A\xi^2 + B\eta^2 + C\zeta^2 = 0,$$

$$\xi^2 + \eta^2 + \zeta^2 = 0.$$

A pair of intersections of common tangents is a degenerate conic-envelope of the system

$$A\xi^2 + B\eta^2 + C\zeta^2 - \lambda(\xi^2 + \eta^2 + \zeta^2) = 0.$$

This degenerates when $\lambda = A$, B or C. Taking $\lambda = A$ we get

$$(B-A)\eta^2 + (C-A)\zeta^2 = 0,$$

which represents the two points

$$[0, \sqrt{(B-A)}, \pm\sqrt{(A-C)}, 0].$$

Hence we have a pair of focal axes

$$x = 0, \quad \frac{y^2}{B-A} + \frac{z^2}{C-A} = 0,$$

which are also loci of foci or a degenerate focal conic.

Two other pairs lie in the other coordinate-planes, viz.

$$y = 0, \quad \frac{x^2}{A-B} + \frac{z^2}{C-B} = 0$$

and

$$z = 0, \quad \frac{x^2}{A-C} + \frac{y^2}{B-C} = 0.$$

Of these, one pair are real and the others imaginary.

12·62. In the case of a cylinder the intersection with the plane at infinity is a line-pair through a point C. Tangent-planes to the cylinder all pass through C. From C there are two tangents i, j to Ω, and through each of these there are two tangent-planes to the cylinder. These tangent-planes intersect in four lines, besides i and j, two real and two imaginary, passing through C. All points on these lines are foci and the lines in pairs form degenerate focal conics. They are also focal axes since the tangent-planes through any one of them cut the plane at infinity in i and j. If I and J are the points of contact of i and j with Ω, any section of the cylinder by a plane through IJ will be a conic whose foci are the points in which the plane cuts the four focal axes. These planes are principal planes of the cylinder, perpendicular to the direction of the axis C.

12·63. **Confocal cones and confocal cylinders.**

Confocal cones—cones having the same focal axes—have a common vertex. Taking this as origin we find, as in 12·4, the equation of a confocal system

$$\frac{x^2}{A+\lambda} + \frac{y^2}{B+\lambda} + \frac{z^2}{C+\lambda} = 0.$$

The conics in which these cut the plane at infinity have for their tangential equation

$$(A+\lambda)\xi^2+(B+\lambda)\eta^2+(C+\lambda)\zeta^2=0,$$

and therefore form a linear tangential system touching the four common tangents of the circle at infinity

$$\xi^2+\eta^2+\zeta^2=0,$$

and the conic-envelope

$$A\xi^2+B\eta^2+C\zeta^2=0.$$

Hence all the cones of a confocal system have four fixed tangent-planes, i.e. the focal-developable in this case reduces to four planes, their intersections in pairs being the six focal axes.

12·64. The equation

$$\frac{x^2}{A+\lambda}+\frac{y^2}{B+\lambda}=1$$

represents a system of confocal cylinders. The four focal axes are the lines parallel to the axis of the cylinder and passing through the real and imaginary foci of the transverse section, i.e.

$$x=0,\ y=\pm\sqrt{(B-A)}\quad\text{and}\quad y=0,\ x=\pm\sqrt{(A-B)}.$$

The two tangents i and j to Ω from the point at infinity C on the axis of the cylinder are the vestiges of two other focal axes.

12·7. Conjugate focal conics.

Let there be given any real proper conic S, say

$$z=0,\ x^2/A+y^2/B=1,\ \text{with}\ A>B.$$

This can be considered as a degenerate quadric-envelope, its tangential equation being

$$Al^2+Bm^2=p^2.$$

A confocal system of quadrics

$$Al^2+Bm^2-p^2+\lambda(l^2+m^2+n^2)=0$$

is then determined, whose point-equation is

$$\frac{x^2}{A+\lambda}+\frac{y^2}{B+\lambda}+\frac{z^2}{\lambda}=1.$$

One focal conic of this system (for $\lambda=0$) is the given conic S. The other real focal conic (for $\lambda=-B$) is

$$y=0,\ x^2/(A-B)-z^2/B=1,$$

which is a hyperbola or an ellipse according as B is $>$ or <0, i.e. according as S is an ellipse or a hyperbola. Hence when S is given, a second real conic S' is uniquely determined; if one is an ellipse the other is a hyperbola. If S is a parabola, the confocal system consists of paraboloids, and S' is also a parabola, equal to the former. Such pairs of conics are called *conjugate focal conics*. They lie in orthogonal planes, the real foci of the one coinciding with vertices of the other.

Since the focal conics form the locus of vertices of circular

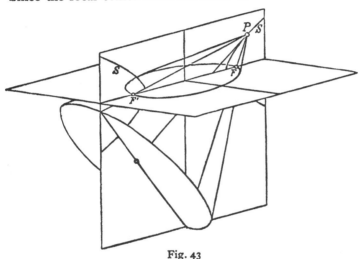

Fig. 43

tangent-cones, each of the two conics is the locus of vertices of circular cones which project the other. The axis of the circular cone which has its vertex P on the conic S is the tangent to S at P. If F, F' are the foci of S, and therefore vertices of S', the angle FPF' is the vertical angle of the cone. If S is either a hyperbola or a parabola, as P moves from the vertex to infinity along the curve the vertical angle varies from 180° to 0, and therefore takes all possible values. If S is an ellipse, however, the least value of the angle, which occurs when P is at the end of the minor axis, is

$$2 \sin^{-1}\sqrt{(B/A)} = 2 \tan^{-1}\sqrt{\{B/(A-B)\}},$$

and is therefore equal to the angle between the asymptotes of the hyperbola. Thus an ellipse or a parabola of given dimensions can be cut out of any circular cone, but in the case of a hyperbola the asymptotes cannot be inclined at an angle greater than the vertical angle of the cone.

Ex. 1. If α is the semi-vertical angle of a circular cone and θ the angle which a given plane makes with the axis, prove that the eccentricity of the section is $\cos \theta / \cos \alpha$.

Ex. 2. Prove that conjugate focal conics have reciprocal eccentricities.

Ex. 3. If the square of one of the eccentricities of one of the focal conics of a quadric is represented by the cross-ratio $(ABCD)$, show that the squares of one of the eccentricities of the other focal conics are $(BCAD)$ and $(CABD)$, and that the squares of the other eccentricities are respectively $(CBAD)$, $(ACBD)$ and $(BACD)$.

12·8. *The foci of a given quadric are foci also of any confocal quadric*, since the focal conics belong to the whole confocal system. The tangent-cones, with any focus F as vertex, to all the quadrics are circular and have a common axis, the tangent to the focal conic at F.

It is true also that *the focal axes of a given quadric are focal axes of any confocal quadric*. If the two planes $[l, m, n, p]$ and $[l', m', n', p']$ are conjugate with regard to each of the quadric-envelopes

$$Al^2 + Bm^2 + Cn^2 - p^2 = 0 \quad \text{and} \quad l^2 + m^2 + n^2 = 0,$$

we have $\qquad All' + Bmm' + Cnn' - pp' = 0$

and $\qquad ll' + mm' + nn' = 0.$

Therefore

$$(A + \lambda) ll' + (B + \lambda) mm' + (C + \lambda) nn' - pp' = 0,$$

i.e. the two planes are conjugate with respect to the confocal quadric

$$Al^2 + Bm^2 + Cn^2 - p^2 + \lambda(l^2 + m^2 + n^2) = 0.$$

Hence if their line of intersection is a focal axis for one quadric it is a focal axis also for any quadric of the confocal system.

12·81. We can now prove that *the focal axes of a quadric are the generating lines of the quadrics of the confocal system*.

Let l be any generating line of a quadric S of the confocal system. Then every plane through l is a tangent-plane to S, and since l is self-conjugate with respect to S, any two planes through l are conjugate with respect to S. l is therefore a focal axis.

Conversely, if l is not a generating line of any quadric of the confocal system there are two quadrics S, S' of the system which touch l, and the tangent-planes α, α' at the points of contact A, A' are orthogonal. Pairs of planes through l conjugate with respect to S consist of the tangent-plane α and any other plane through l; only one such pair, viz. α and α', are orthogonal. Hence l is not a focal axis; and therefore every focal axis must be a generating line of some quadric of the system.

On an arbitrary plane there are two focal axes; these are the lines in which the plane is met by that quadric of the confocal system which touches the given plane.

Through a given point there are six focal axes, viz. the generating lines of the three confocals through the point. Of these focal axes two are real and four imaginary.

The congruence of focal axes is therefore said to be of *class* 2 and *order* 6.

Ex. 1. Prove that the tangent-cones from a point P to the quadrics of a confocal system have a common system of principal planes, and form a system of confocal cones.

Ex. 2. Show that the six focal axes through any point P are the focal axes of the tangent-cones from P to the quadrics.

12·9. Deformable framework of generating lines of a quadric.

If the generating lines of a quadric are considered as thin wires pivoted at their points of intersection, the framework is deformable*.

Consider the hyperboloid of one sheet
$$x^2/a^2 + y^2/b^2 - z^2/c^2 = 1.$$
Freedom-equations in terms of two parameters λ, μ are
$$x/a = \lambda\mu + 1,$$
$$y/b = \lambda - \mu,$$
$$z/c = \lambda\mu - 1,$$
$$w = \lambda + \mu,$$

---

* This was discovered accidentally in 1873 by O. Henrici when he set his students at University College, London, to construct a model of a hyperboloid of one sheet by tying together a series of thin rods.

and the generating lines of the two systems are represented by $\lambda = $ const. and $\mu = $ const. Thus these values of x, y, z, w are the homogeneous coordinates of the point P of intersection of the generators λ and μ. Let P' be another point on the generator λ, and let μ' be the generator of the other system through P'; then the coordinates of P' are given by

$$x'/a = \lambda\mu' + \mathrm{I}, \text{ etc.}$$

We find then

$$PP'^2 = \{a^2(\lambda^2 - \mathrm{I})^2 + 4b^2\lambda^2 + c^2(\lambda^2 + \mathrm{I})^2\}(\mu - \mu')^2/(\lambda + \mu)^2(\lambda + \mu')^2.$$

Keeping λ, μ, μ' fixed we may vary a, b, c in such a way that PP' remains constant. If a, b, c are changed into a', b', c', PP' will remain unaltered if

$$a^2 + c^2 = a'^2 + c'^2, \text{ and } a^2 - 2b^2 - c^2 = a'^2 - 2b'^2 - c'^2,$$

hence if $a'^2 = a^2 + k$, then $c'^2 = c^2 - k$ and $b'^2 = b^2 + k$.

Hence without altering the distances between intersections along any generator the hyperboloid can be transformed continuously into the hyperboloid

$$\frac{x^2}{a^2 + k} + \frac{y^2}{b^2 + k} + \frac{z^2}{-c^2 + k} = \mathrm{I},$$

which is confocal with the given hyperboloid. When $k = c^2$ the framework flattens out in the plane $z = 0$ and the lines envelop the focal ellipse

$$x^2/(a^2 + c^2) + y^2/(b^2 + c^2) = \mathrm{I};$$

and (assuming $a > b$) when $k = -b^2$ it flattens out in the plane $y = 0$ and the lines envelop the focal hyperbola

$$x^2/(a^2 - b^2) - z^2/(b^2 + c^2) = \mathrm{I}.$$

12·91. In the case of the hyperbolic paraboloid

$$x^2/a^2 - y^2/b^2 = 2z/c$$

the freedom-equations are

$$x/a = \lambda + \mu,$$
$$y/b = \lambda - \mu,$$
$$z/c = 2\lambda\mu,$$

$\lambda = $ const. and $\mu = $ const. representing the two systems of generators. The distance between the two points $P \equiv [\lambda, \mu]$ and $P' \equiv [\lambda, \mu']$ is given by

$$PP'^2 = (a^2 + b^2 + 4c^2\lambda^2)(\mu - \mu')^2.$$

Keeping λ, μ, μ' constant and changing a, b, c into a', b', c' the distance PP' is unaltered if

$$a^2 + b^2 = a'^2 + b'^2, \quad c = c'.$$

Putting $a'^2 = a^2 + k$ we have $b'^2 = b^2 - k$ and on changing the origin we obtain the confocal paraboloid

$$\frac{x^2}{a^2 + k} + \frac{y^2}{-b^2 + k} = \frac{2z}{c} + \frac{k}{c^2}.$$

12·95. EXAMPLES.

1. Show that the foci of any plane section of a spheroid are the points of contact of the two spheres which can be drawn having ring-contact with the surface and touching the plane.

2. Find the locus of intersection of three mutually perpendicular planes which touch respectively three confocal central quadrics.

Ans. A sphere.

3. Show that the focal conics are the loci of umbilics of the confocal system.

4. Show that all the normals through a given point to the quadrics of a confocal system are generators of the same quadric cone.

5. Show that the locus of points on confocal central quadrics at which the normals are parallel is a rectangular hyperbola of which one asymptote is parallel to the normals.

6. Show that the normals to confocal quadrics at points the tangent-planes at which pass through a fixed line generate a hyperbolic paraboloid.

7. Prove that the axes of a tangent-cone to a quadric are the normals to the three confocals which pass through the point.

DEF. *Corresponding points* on two ellipsoids whose semi-axes are a, b, c and a', b', c' are points whose coordinates satisfy the equations $x/a = x'/a'$, $y/b = y'/b'$, $z/c = z'/c'$; and similarly for points on two hyperboloids of the same species.

8. Prove Ivory's Theorem: that the distance between points P and Q' on two confocal quadrics is equal to the distance between the corresponding points P' and Q.

9. If P, Q are two points on an ellipsoid, and P', Q' are the corresponding points on a confocal ellipsoid, prove that

$$OP^2 - OQ^2 = OP'^2 - OQ'^2.$$

10. If P and Q lie on one generator of a hyperboloid and P', Q' are the corresponding points of another hyperboloid of the same species, show that P' and Q' lie on one generator; and, if the hyperboloids are confocal, $PQ = P'Q'$.

11. Show that an umbilic on one ellipsoid is the corresponding point of an umbilic on a confocal ellipsoid.

12. A quadric has perpendicular directrices α, β corresponding to foci A, B respectively. Show that another quadric, with the same planes of circular section, has directrices α, β corresponding to foci B and A respectively.

(Math. Trip. II, 1913.)

Ans. If the given quadric is $ax^2 + by^2 + cz^2 = 1$, with foci $[X, 0, Z]$ and $[X', Y', 0]$, the other quadric is

$$ax^2 + by^2 + cz^2 - 1 = b\{(x-X)^2 + y^2 + (z-Z)^2\}$$
$$+ c\{(x-X')^2 + (y-Y')^2 + z^2\}.$$

CHAPTER XIII

LINEAR SYSTEMS OF QUADRICS

13·1. If $S = 0$ and $S' = 0$ are the point-equations of two quadrics, the equation

$$S - \lambda S' = 0$$

represents for all values of λ a quadric passing through all the points common to S and S'. This is called a *linear one-parameter system* or *pencil* of quadric loci, or a *point-system*.

13·11. The points common to two quadrics form a curve which has the property that it is cut by an arbitrary plane in four points. For a plane cuts the quadrics in two conics, and these intersect in four points. This curve, the *base-curve* of the pencil, is therefore of the fourth order. It will be considered in more detail in the following chapter.

13·12. *Through any point there passes just one quadric of the system,* for when the coordinates of the point are substituted in the equation we have a linear equation to determine λ.

13·13. *An arbitrary line is cut by the quadrics in pairs of points which form an involution.* For if P is any point on the line there is a unique quadric of the system through P and this cuts the line again in a unique point P'. Thus to P corresponds P' uniquely, and in the same way to P' corresponds P. Hence the points are connected by a symmetrical (1, 1) correspondence.

13·14. The involution has two double-points, and these are the points of contact of the quadrics of the system which touch the line. Hence in general *there are two quadrics of the system which touch a given line.* These may be real or imaginary. If they should be coincident the involution on the line is degenerate; the double-point D then corresponds to every point on the line and therefore D must be a point on the base-curve. Hence if the line cuts the base-curve in a point D there is just one quadric of the system which touches the line, and its point of contact is D. If the line cuts the base-curve in two distinct points D, D', a quadric of the system can be determined to pass through another point P of the line DD', and DD' is thus a generating line of this quadric; if D and D' coincide, so that DD' is a tangent to the base-curve, the line touches every quadric of the system at D, and there is, moreover, a quadric of the system having the line as a generator.

13·15. An arbitrary plane cuts the quadrics in a system of conics (pencil) which pass through the four fixed points A, B, C, D in which the plane cuts the base-curve. In this pencil of conics there are three line-pairs, viz. AD, BC; BD, CA; CD, AB, and the plane is a tangent-plane to the quadrics of which these are the sections. Hence in general *there are three quadrics of the system which touch a given plane.* Special cases may arise when coincidences occur among the four points A, B, C, D.

13·16. *Two quadrics have in general a unique common self-polar tetrahedron.* If the point $[X, Y, Z, W]$ has the same polar-plane with regard to the two quadrics the two equations

$$x\frac{\partial S}{\partial X}+y\frac{\partial S}{\partial Y}+z\frac{\partial S}{\partial Z}+w\frac{\partial S}{\partial W}=0,$$

$$x\frac{\partial S'}{\partial X}+y\frac{\partial S'}{\partial Y}+z\frac{\partial S'}{\partial Z}+w\frac{\partial S'}{\partial W}=0$$

are identical. Hence

$$\frac{aX+hY+gZ+pW}{a'X+h'Y+g'Z+p'W}=\frac{hX+bY+fZ+qW}{h'X+b'Y+f'Z+q'W}$$
$$=\frac{gX+fY+cZ+rW}{g'X+f'Y+c'Z+r'W}=\frac{pX+qY+rZ+dW}{p'X+q'Y+r'Z+d'W}.$$

Equating each of these to t, we have the four equations

$$(a-ta')X+(h-th')Y+(g-tg')Z+(p-tp')W=0,\text{ etc.};$$

and eliminating X, Y, Z, W,

$$\begin{vmatrix} a-ta' & h-th' & g-tg' & p-tp' \\ h-th' & b-tb' & f-tf' & q-tq' \\ g-tg' & f-tf' & c-tc' & r-tr' \\ p-tp' & q-tq' & r-tr' & d-td' \end{vmatrix}=0.$$

This is an equation of the fourth degree in t. Each root then determines one set of values of $[X, Y, Z, W]$, and we have four points forming a tetrahedron.

When this tetrahedron is taken as the tetrahedron of reference the equation of each quadric reduces to the form

$$ax^2 + by^2 + cz^2 + dw^2 = 0.$$

Special cases may arise when there are equalities among the roots, but we shall consider at present only the general case.

The equation of the system of quadrics can now be written

$$(ax^2 + by^2 + cz^2 + dw^2) - \lambda(a'x^2 + b'y^2 + c'z^2 + d'w^2) = 0.$$

Hence the same tetrahedron is self-polar with respect to all the quadrics of the system.

13·17. The discriminant of this equation is

$$(a - \lambda a')(b - \lambda b')(c - \lambda c')(d - \lambda d')$$

and if this vanishes the quadric becomes a cone. Hence *in a pencil of quadrics there are four cones, and their vertices are the vertices of the common self-polar tetrahedron.*

Ex. When the vertices of the four cones are real and distinct, show that the four cones are either all real or two real and two virtual.

13·2. If $\Sigma = 0$ and $\Sigma' = 0$ are the tangential equations of two quadrics, the equation

$$\Sigma - \lambda \Sigma' = 0$$

represents a system of quadrics which touch all the planes which are tangent to both Σ and Σ'. This is called a *linear tangential one-parameter system* or *pencil of quadric-envelopes*.

13·21. The assemblage of common tangent-planes forms a figure (one-dimensional envelope) which is dual to a curve in space (one-dimensional locus of points). It is called, for a reason that will appear later, a *developable*. It is to be distinguished from a surface, which is a two-dimensional envelope and is dual to a surface as a two-dimensional locus. The tangent-developable of two quadrics may in certain cases reduce to two cones, as in the case of two spheres, and a cone is a particular case of a developable. Reciprocally the curve of intersection of two quadrics may reduce to two conics (this also occurs in the case of two spheres), and we have already seen that a quadric cone is dual to a conic. The tangent-developable of two quadrics is said to be of the fourth *class* since there are four tangent-planes which pass through an arbitrary point, the common tangent-planes of the tangent-cones to the two quadrics with the given point as vertex. Developables will be treated at greater length in the following chapter.

13·22. Theorems, dual to those for a point-system, hold for the tangential system. Thus: *there is one and only one quadric of the system which touches a given plane.*

13·23. *The pairs of tangent-planes through a given line are in involution.* The double-planes of the involution being the tangent-planes to the quadrics which touch the given line it follows as for the point-system that *there are two quadrics of the system which touch a given line.*

13·24. With an arbitrary point P as vertex there is a system of tangent-cones to the quadrics, all touching four fixed planes α, β, γ, δ, the tangent-planes of the developable which pass through P. Three of these cones degenerate to line-pairs, viz. the intersections of the planes $\alpha\delta$, $\beta\gamma$; $\beta\delta$, $\gamma\alpha$; $\gamma\delta$, $\alpha\beta$; and P is then a point on the corresponding quadrics. Hence in general *there are three quadrics of the tangential system which pass through a given point.*

13·25. When the two quadrics Σ, Σ' are referred to their common self-polar tetrahedron their tangential equations are
$$\Sigma \equiv A\xi^2 + B\eta^2 + C\zeta^2 + D\omega^2 = 0,$$
$$\Sigma' \equiv A'\xi^2 + B'\eta^2 + C'\zeta^2 + D'\omega^2 = 0,$$
and the quadric $\Sigma - \lambda\Sigma' = 0$ will degenerate to a conic for four values of λ. Thus *in a tangential system of quadrics there are four which degenerate to conics, and their planes are the faces of the common self-polar tetrahedron.*

13·3. An example of a tangential system of quadrics is a confocal system. The above properties are then verified. The four degenerate quadrics are the three focal conics and the circle at infinity.

The orthogonal properties of the confocal system can be interpreted in the general tangential system, taking one of the four conics as absolute. Thus, considering the two quadrics which touch a given line, their tangent-planes at the points of contact are conjugate with regard to every quadric of the system, and therefore with regard to each of the four conics.

Considering the three quadrics through a point P: α, β, γ, δ being the tangent-planes through P to the tangent-developable, the tangent-planes at P to the three quadrics are the planes ξ, η, ζ containing the line-pairs $(\alpha\delta)$, $(\beta\gamma)$; $(\beta\delta)$, $(\gamma\alpha)$; $(\gamma\delta)$, $(\alpha\beta)$. These three planes are mutually conjugate with regard to any quadric which touches the four planes α, β, γ, δ; i.e. with regard to every quadric of the system and therefore in particular conjugate with respect to each of the four conics.

In a confocal system therefore the tangent-planes to the three quadrics which pass through any point P are mutually orthogonal and conjugate with respect to every quadric of the system.

13·31. Lines of curvature.

A given quadric of the system is cut by the other quadrics in curves, two passing through each point and therefore forming a network on the surface. If ξ is the tangent-plane to the given quadric at the point P, the tangents to the two curves through P are the lines $(\xi\eta)$ and $(\xi\zeta)$, while $(\alpha\delta)$ and $(\beta\gamma)$ are the generators through P of the given quadric. These two pairs are harmonic. In the confocal system $(\xi\eta)$ and $(\xi\zeta)$ are also at right angles and therefore the two systems of curves cut orthogonally; also $(\xi\eta)$ and $(\xi\zeta)$ are mutually polars with respect to the quadric. Let P' be a point on the curve of intersection of the quadric with η, very near to P, and let ξ' be the tangent-plane at P'. Then the line $(\xi\xi')$ is the polar of PP' and ultimately coincides with $(\xi\zeta)$ when $P' \to P$. The line $(\eta\zeta)$ is the normal at P, and the plane η which contains both P and P' is ultimately normal to $(\xi\xi')$ and therefore contains the normals at both P and P'. Hence the normals at P and P' ultimately intersect and the curve of intersection is a line of curvature. Hence *the quadrics of a confocal system intersect in lines of curvature.*

13·32.
The section of a quadric by a plane very near to the tangent-plane at a point P is a conic, and the limiting form of the section in the neighbourhood of P when the plane becomes the

tangent-plane is called the *indicatrix* at P. Its asymptotes are the generators through P, and its principal axes are the bisectors of the angles between the generators. At an elliptic point the indicatrix is the limiting form of an ellipse, the asymptotes and therefore the generators through the point being imaginary. A line of curvature has thus the property that the tangent at any point is one of the principal axes of the indicatrix at the point. At an umbilic the indicatrix is a point-circle and the directions of the lines of curvature appear to be indeterminate. The umbilics are in fact exceptional points. The four real umbilics of the ellipsoid $x^2/a^2 + y^2/b^2 + z^2/c^2 = 1$, where $a > b > c$, lie on the plane $y = 0$, and there is no other quadric of the system through one of these umbilics except the double-plane $y = 0$. The intersection of this plane with the ellipsoid passes through the four umbilics and is the only real line of curvature which passes through them. But the tangent-plane at the umbilic U cuts the surface in two imaginary lines UH, UK, where H, K are points on the circle at infinity, hence these lines are isotropic. Any plane UHT through UH is a tangent-plane, and the normal is the line joining U to the pole of HT with respect to Ω. But this point lies on the tangent to Ω at H. Hence the normals at points on UH all lie in one plane, the isotropic plane through U touching Ω at H. Similarly the normals at points of UK all lie in the isotropic plane through U touching Ω at K. These two lines are therefore lines of curvature. In addition to the real lines of curvature, which are the curves of intersection of the quadric with its confocals, there are therefore eight imaginary lines of curvature, the generators through the umbilics, each passing through one absolute point and three umbilics (cf. 10·71). These lines of curvature have the special property that the normals at all points lie in one plane; these planes are isotropic planes and are at the same time tangent-planes and normal planes.

13·33. The lines of curvature on an ellipsoid resemble confocal conics, the umbilics taking the place of foci. It can be shown, in fact, that if the lines of curvature are projected from the point at infinity on one of the axes (parallel projection) on to a plane of circular section which is not parallel to this axis the projections form a system of confocal conics.

Consider the ellipsoid

$$S \equiv x^2/a^2 + y^2/b^2 + z^2/c^2 - w^2 = 0 \qquad \ldots\ldots(1)$$

and its intersections with the confocal system

$$U \equiv \frac{x^2}{a^2-\lambda} + \frac{y^2}{b^2-\lambda} + \frac{z^2}{c^2-\lambda} - w^2 = 0. \qquad \ldots\ldots(2)$$

Taking the point $C \equiv [0, 0, 1, 0]$ as centre of projection we find the equations of a projecting cone (cylinder) by eliminating z between (1) and (2):

$$c^2 S - (c^2 - \lambda) U \equiv \frac{x^2(a^2-c^2)}{a^2(a^2-\lambda)} + \frac{y^2(b^2-c^2)}{b^2(b^2-\lambda)} - w^2 = 0.$$
$$\ldots\ldots(3)$$

The intersections of this system of cylinders with the plane $z = 0$ thus form a linear tangential system of conics, and therefore the cylinders (3) have four common tangent-planes, which intersect in three pairs of edges through C. For $\lambda = a^2$, b^2, or c^2, U degenerates to one of the focal conics, U_a, etc., and as $\lambda \to \infty$ it degenerates to the circle at infinity Ω. For $\lambda = a^2$ the cylinder degenerates as an envelope to two parallel straight lines, which form a pair of common chords of the focal conic U_a in $x = 0$ and the section of the ellipsoid S by this plane, and therefore pass through pairs of umbilics. These are two of the lines of intersection or edges of the common tangent-planes. Similarly for $\lambda = b^2$. As $\lambda \to \infty$ we get the third pair of edges which form a pair of common chords II', JJ' of the circle at infinity with the conic at infinity on the ellipsoid.

Hence a plane through one of the other common chords, say IJ, i.e. a plane of circular section not parallel to the axis of the cylinders, will cut the four common tangent-planes in two pairs of lines passing through I and J, and the cylinders in a system of conics touching these four lines and therefore forming a confocal system. The foci of the system are the intersections of the plane with the pairs of parallel edges through C, and are therefore the projections of the umbilics.

This may be proved also as follows. The central planes of real circular sections of the ellipsoid are $z = \pm x \tan\theta$, where

$$\tan\theta = \frac{c}{a}\left(\frac{a^2-b^2}{b^2-c^2}\right)^{\frac{1}{2}}.$$

Hence projecting from the plane of xy on to the plane $z = x \tan\theta$ with the axis of z as axis of projection, the equations of transformation are

$$x = x'\cos\theta, \quad y = y'.$$

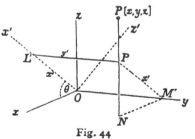

Fig. 44

The equations (3) then become

$$x'^2 \frac{a^2 - c^2}{a^2(a^2 - \lambda)} \frac{a^2(b^2 - c^2)}{b^2(a^2 - c^2)}$$

$$+ y'^2 \frac{b^2 - c^2}{b^2(b^2 - \lambda)} = 1,$$

i.e.

$$\frac{x'^2}{a^2 - \lambda} + \frac{y'^2}{b^2 - \lambda} = \frac{b^2}{b^2 - c^2},$$

which represents a system of confocal ellipses and hyperbolas.

13·34. Lines of curvature of a cone or a cylinder.

A cone or cylinder is a specialised quadric locus, and only a one-way envelope of planes; dually, a conic is a specialised quadric envelope, and only a one-way locus of points.

Instead of a developable, as the assemblage of common tangent-planes of a cone and the circle at infinity, we obtain just four imaginary planes through the vertex of the cone. If the point-equation of the cone is $x^2/A + y^2/B + z^2/C = 0$, the vertex being the origin, its tangential equations are

$$\omega = 0, \quad A\xi^2 + B\eta^2 + C\zeta^2 = 0,$$

and the circle at infinity is

$$\xi^2 + \eta^2 + \zeta^2 = 0.$$

Then the equations

$$\omega = 0, \quad A\xi^2 + B\eta^2 + C\zeta^2 - \lambda(\xi^2 + \eta^2 + \zeta^2) = 0$$

represent a system of cones touching the four common tangent-planes, and we can call this a system of confocal cones. The point-equation is

$$\frac{x^2}{A - \lambda} + \frac{y^2}{B - \lambda} + \frac{z^2}{C - \lambda} = 0.$$

If $A > B > C$ the cone is real if either $A > \lambda > B$ or $B > \lambda > C$, and through any point there pass just two real cones of the system, and these cut orthogonally. A triple orthogonal system can then be obtained by adjoining to this system the system of spheres with common centre at the vertex. The lines of intersection of pairs of cones (generating lines) and of a cone with a sphere then form the *lines of curvature*. For a cylinder the lines of curvature are the generators and the transverse plane sections.

13·35. The curve of intersection of a cone with a sphere whose centre is at the vertex is called a *sphero-conic*. These curves are also of course lines of curvature on the sphere. Lines of curvature on a sphere, however, are really indeterminate; any curve drawn on a sphere has the property that normals at consecutive points intersect, since all the normals to a sphere pass through the centre.

13·36. The system of sphero-conics on a given sphere, formed by the intersections with a system of confocal cones, have, however, in a special sense the properties of lines of curvature. They form, in fact, a concrete representation of confocal conics in *elliptic* plane geometry, in which the circular points or absolute are replaced by a virtual proper conic. Thus if x, y, z are homogeneous point-coordinates in a plane, and ξ, η, ζ the corresponding line-coordinates, the absolute may be represented by the point-equation $x^2 + y^2 + z^2 = 0$ and the line-equation

$$\xi^2 + \eta^2 + \zeta^2 = 0.$$

Then the equation

$$A\xi^2 + B\eta^2 + C\zeta^2 - \lambda(\xi^2 + \eta^2 + \zeta^2) = 0$$

represents a system of conic-envelopes touching the four common tangents of $A\xi^2 + B\eta^2 + C\zeta^2 = 0$ and $\xi^2 + \eta^2 + \zeta^2 = 0$. The foci are the intersections of these tangents. A sphero-conic or non-euclidean conic has one pair of real and two pairs of imaginary foci. In non-euclidean geometry the "distance" between two points P, Q is defined as $\frac{1}{2}i \log(PQ, XY)$, where (PQ, XY) denotes the cross-ratio (defined projectively) of the two points P, Q and the points X, Y in which PQ cuts the absolute. This is an extension of the well-known expression for the angle between two lines referred to the two tangents from the vertex to the absolute (or in particular the circular points). With this definition of distance we find that *the sum (or difference) of the distances of any point on a non-euclidean conic from a pair of foci is constant.* For a sphero-conic this means that *the sum (or difference) of the angles between any generator of a cone and a pair of focal lines is constant.*

Let the equation of the cone be

$$\frac{x^2}{a^2}+\frac{y^2}{b^2}-\frac{z^2}{c^2}=0 \quad (a>b).$$

The confocal system is

$$\frac{x^2}{a^2-\lambda}+\frac{y^2}{b^2-\lambda}-\frac{z^2}{c^2+\lambda}=0,$$

and the real focal lines are found by putting $\lambda=b^2$, viz.

$$y=0, \quad \frac{x^2}{a^2-b^2}-\frac{z^2}{c^2+b^2}=0,$$

i.e. $$[(a^2-b^2)^{\frac{1}{2}}, \ 0, \ \pm(c^2+b^2)^{\frac{1}{2}}].$$

Let $[\lambda, \mu, \nu]$ be the direction-cosines of any generator, and θ, θ' the angles which it makes with the two focal lines. Then

$$\lambda^2/a^2+\mu^2/b^2-\nu^2/c^2=0, \ \lambda^2+\mu^2+\nu^2=1,$$

and

$$\cos\theta=\frac{\lambda(a^2-b^2)^{\frac{1}{2}}+\nu(b^2+c^2)^{\frac{1}{2}}}{(a^2+c^2)^{\frac{1}{2}}}, \quad \cos\theta'=\frac{\lambda(a^2-b^2)^{\frac{1}{2}}-\nu(b^2+c^2)^{\frac{1}{2}}}{(a^2+c^2)^{\frac{1}{2}}}.$$

Eliminating μ we have

$$c^2(a^2-b^2)\lambda^2+a^2(b^2+c^2)\nu^2=a^2c^2.$$

Write $$\lambda(a^2-b^2)^{\frac{1}{2}}=a\cos\phi,$$

then $$\nu(b^2+c^2)^{\frac{1}{2}}=c\sin\phi;$$

write also $$a=(a^2+c^2)^{\frac{1}{2}}\cos\psi, \ c=(a^2+c^2)^{\frac{1}{2}}\sin\psi.$$

Then $$\cos\theta \ =\cos\phi\,\cos\psi+\sin\phi\,\sin\psi=\cos(\phi-\psi),$$

$$\cos\theta'=\cos\phi\,\cos\psi-\sin\phi\,\sin\psi=\cos(\phi+\psi),$$

whence either $\theta+\theta'$ or $\theta-\theta'=2k\pi\pm2\psi$, a constant angle.

There is a similar property for the lines of curvature on an ellipsoid: viz. the sum or difference of the geodesic distances from a pair of umbilics to any point on a given line of curvature is constant (see Salmon, *Analytic geometry of three dimensions,* 7th ed. § 400).

13·41. The polar of a point $P \equiv [x_1, y_1, z_1, w_1]$ with respect to a quadric of a linear point-system is

$$P - \lambda P' = 0,$$

where $P = 0$ represents the polar of P with respect to the quadric $S = 0$, and $P' = 0$ that with respect to $S' = 0$. Hence *the polars of a given point all pass through one line l.*

There is one quadric S_1 of the system which passes through P_1. Let α be the tangent-plane to S_1 at P_1. Then there are two other quadrics S_2 and S_3 which touch α; let P_2 and P_3 be their points of contact. Then the polar of P_1 with respect to S_1 is α. The polar of P_2 with respect to S_2 is also α and it passes through P_1, therefore the polar of P_1 with respect to S_2 passes through P_2. Similarly the polar of P_1 with respect to S_3 passes through P_3. But the polar of P_1 with respect to S_1 passes through both P_2 and P_3. Therefore the polars of P_1 with respect to all quadrics of the system pass through both P_2 and P_3. $P_2 P_3$ is therefore the line through which the polar planes of P_1 all pass.

13·42. *To every point P there corresponds in general one line l such that the polar planes of P with respect to all the quadrics of the system pass through l, and the whole assemblage of such lines l form a quadratic complex.* If l passes through a fixed point $Q \equiv [X, Y, Z, W]$, P lies on a fixed line, the line which corresponds to Q. We have then

$$a x_1 X + \ldots = 0,$$

$$a' x_1 X + \ldots = 0,$$

and, $[x, y, z, w]$ being any point on l,

$$a x_1 x + \ldots = 0,$$

$$a' x_1 x + \ldots = 0.$$

Eliminating x_1, y_1, z_1, w_1 between these four equations we obtain

$$\begin{vmatrix} ax & by & cz & dw \\ a'x & b'y & c'z & d'w \\ aX & bY & cZ & dW \\ a'X & b'Y & c'Z & d'W \end{vmatrix} = 0,$$

which represents a quadric cone through $[X, Y, Z, W]$. Hence the complex of lines is of class 2, or it is a quadratic complex. In a given plane there are two lines of the complex through any point, hence all the lines of the complex which lie in a given plane envelop a conic.

13·43. *The locus of points which correspond to lines of the complex which lie in a given plane is a cubic curve.*

Let $P \equiv [X, Y, Z, W]$ and let the given plane be $\alpha \equiv [l, m, n, p]$. Then the condition that the line of intersection of

$$aXx + bYy + cZz + dWw = 0$$

and
$$a'Xx + b'Yy + c'Zz + d'Ww = 0$$

should lie in the plane

$$lx + my + nz + pw = 0$$

is that the matrix
$$\begin{bmatrix} aX & bY & cZ & dW \\ a'X & b'Y & c'Z & d'W \\ l & m & n & p \end{bmatrix}$$

should be of rank 2. Hence

$$l(cd' - c'd)ZW + n(da' - d'a)WX + p(ac' - a'c)XZ = 0$$

and

$$m(cd' - c'd)ZW + n(db' - d'b)WY + p(bc' - b'c)YZ = 0.$$

The locus is therefore the intersection of these two quadrics which have the generating line $Z = 0 = W$ in common, and this locus is a cubic curve.

13·431. Since the polars of P with respect to S and S' intersect on the plane α, α is the polar plane of P with respect to one quadric of the system $S - \lambda S'$. That is *the locus of poles of a fixed plane with respect to the quadrics of the system is this cubic curve.* This may be shown also directly as follows. If $P \equiv [X, Y, Z, W]$ is the pole of the plane $[l, m, n, p]$ with respect to $S - \lambda S'$, then

$$(a - \lambda a')X/l = (b - \lambda b')Y/m = (c - \lambda c')Z/n = (d - \lambda d')W/p.$$

The locus of P is therefore represented by the freedom-equations

$$x = l(b - \lambda b')(c - \lambda c')(d - \lambda d'), \text{ etc.}$$

Ex. 1. Show that the polar quadratic complex of the system

$$(ax^2 + by^2 + cz^2 + dw^2) + \lambda\,(a'x^2 + b'y^2 + c'z^2 + d'w^2) = 0$$

is represented by any one of the equations

$$\frac{p_{01}p_{23}}{(da')\,(bc')} = \frac{p_{02}p_{31}}{(db')\,(ca')} = \frac{p_{03}p_{12}}{(dc')\,(ab')},$$

where (bc') stands for $bc' - b'c$, etc.

Ex. 2. Prove that the polar lines of the line

$$aXx + bYy + cZz + dWw = 0,$$

$$a'Xx + b'Yy + c'Zz + d'Ww = 0$$

with respect to the system

$$(ax^2 + \ldots) + \lambda\,(a'x^2 + \ldots) = 0$$

generate the quadric

$$\Sigma\,(bc')\,(ad')\,(yzXW + xwYZ) = 0.$$

13·44. Polar of a fixed line with respect to a pencil of quadrics.

If l is any line of the complex the polar planes of all points on it with respect to all the quadrics of the system pass through the corresponding point P. Hence the polar lines of l with respect to all the quadrics of the system are concurrent in P.

If l is an arbitrary line and (x_1), (x_2), two points upon it, the polar planes of (x_1) and (x_2) are $P_1 + \lambda P_1' = 0$ and $P_2 + \lambda P_2' = 0$. Eliminating λ we have $P_1P_2' - P_1'P_2 = 0$ which represents a quadric. Hence *the polar lines of an arbitrary fixed line generate a regulus.* The other regulus of this quadric consists of those lines of the polar complex which correspond to points on l.

13·45. The polar complex is an example of the *tetrahedral complex*, which is the complex of lines which are cut by four fixed arbitrary planes in a constant cross-ratio. The lines of the polar complex are in fact cut by the faces of the tetrahedron of reference in a constant cross-ratio.

A line of the complex is represented by

$$aXx + bYy + cZz + dWw = 0,$$

$$a'Xx + b'Yy + c'Zz + d'Ww = 0.$$

We obtain freedom-equations by taking the two points of reference on $z=0$ and $w=0$; thus

$$\left.\begin{array}{l} x = (bc')\,YZ + t(bd')\,YW \\ y = (ca')\,ZX + t(da')\,WX \\ z = (ab')\,XY \\ w = \qquad\quad t(ab')\,XY \end{array}\right\},$$

where (bc') stands for $bc' - b'c$, etc.

The line cuts the planes $x=0$, $y=0$, $z=0$, $w=0$ respectively where the parameter t has the values

$$-\frac{(bc')}{(bd')}\frac{Z}{W}, \quad -\frac{(ca')}{(da')}\frac{Z}{W}, \quad \infty, \quad 0.$$

The cross-ratio

$$(t_1 t_2,\ t_3 t_4) = t_1/t_2 = \frac{(bc')(da')}{(bd')(ca')},$$

which is independent of X, Y, Z, W.

13·5. The polar properties of the tangential system follow reciprocally. Thus if α is an arbitrary fixed plane, its poles with respect to the quadrics of a tangential system lie on one line l, and the assemblage of such lines, when the plane α is varied, form a tetrahedral complex. With certain exceptions, to every plane corresponds one line of the complex, and *vice versa*. If the plane α is a plane of the tangent-developable, so that it touches every quadric of the system, the points of contact are its poles and these lie on one line, which is a generating line of the developable. If α is one of the principal planes, its poles all coincide, and any line through a vertex of the fundamental tetrahedron belongs to the complex. If α passes through a vertex X, its poles lie in the opposite face and l lies in this face.

The assemblage of polar planes of a fixed point is a cubic developable, and this is also the assemblage of planes which correspond to lines of the polar complex which pass through the given point.

The polars of a fixed line l with respect to the quadrics of a tangential system generate one regulus of a quadric; the other regulus of this quadric consists of the lines of the polar complex which correspond to planes through l.

Ex. Show that the polar complex of the confocal system

$$\frac{x^2}{a^2-\lambda}+\frac{y^2}{b^2-\lambda}+\frac{z^2}{c^2-\lambda}=1$$

is represented by

$$a^2 p_{01} p_{23}+b^2 p_{02} p_{31}+c^2 p_{03} p_{12}=0.$$

13·61. A quadric is in general completely determined by nine points, provided these nine points do not lie on a quartic curve which is the intersection of two quadrics. If eight points are given, the system has one degree of freedom and the quadric is completely determined by one other point. Let S_1 and S_2 be two quadrics through the eight points. Then every quadric of the system $S_1+\lambda S_2=0$ passes through the eight points, and since λ can be determined so that this may represent a given quadric through the eight points, the general linear system is the system of quadrics through eight fixed points. All the quadrics which pass through eight arbitrary points have a quartic curve in common.

13·62. Eight points do not, however, always determine a one-parameter system. For any three quadrics S_1, S_2, S_3 have eight points in common, and then any quadric of the two-parameter system

$$S_1+\lambda S_2+\mu S_3=0$$

passes through these eight points. Eight points which form the basis of a two-parameter system are constituted in a particular way. The two-parameter system requires only seven points to determine it, for any seven of the base-points being taken a quadric of the system is determined by two other points. All the quadrics will then pass also through the eighth point. Hence *all quadrics which pass through seven given points will pass also through an eighth fixed point.* A group of eight points having this character is called a set of eight *associated points.*

Ex. Show that the vertices of any hexahedron form eight associated points.

When the eight points are divided into two sets of four, each set forms a tetrahedron self-polar with respect to one and the same quadric.

Thus the system

$$\lambda(yz + xw) + \mu(zx + yw) + \nu(xy + zw) = 0$$

passes through the eight points $[1, 0, 0, 0]$, $[0, 1, 0, 0]$, $[0, 0, 1, 0]$, $[0, 0, 0, 1]$, $[-1, 1, 1, 1]$, $[1, -1, 1, 1]$, $[1, 1, -1, 1]$, $[1, 1, 1, -1]$. The first four and also the last four form self-polar tetrahedra with respect to $x^2 + y^2 + z^2 + w^2 = 0$. For the general proof of this theorem see 15·212, Ex. 2.

13·7. As a particular case of the theorem 13·431 the locus of centres of a linear point-system of quadrics is a cubic curve. This cuts the plane at infinity in three points. Hence the system in general contains three paraboloids. This appears also otherwise, since there are three quadrics of the system which touch the plane at infinity. The directions of the axes of the three parabolas form a set of conjugate directions for each quadric of the system.

The locus of centres of a linear tangential system of quadrics is a straight line; but reduces to a fixed point if (as in a confocal system) the plane at infinity is a face of the common self-polar tetrahedron.

The condition for a rectangular hyperboloid ($w = 0$ being the plane at infinity) is $(a + b + c) - \lambda(a' + b' + c') = 0$. Hence there is in general one rectangular hyperboloid in a linear point-system. If there are two, then $a + b + c = 0$ and $a' + b' + c' = 0$, and every quadric of the system is a rectangular hyperboloid. As the condition $a + b + c = 0$ is linear in the coefficients, a rectangular hyperboloid is in general determined by eight points. But if the eight points form an associated group every quadric through seven of them will pass through the eighth. *All rectangular hyperboloids which pass through seven given points form a one-parameter system and have in common a quartic curve.* If six points are given there is a two-parameter system of rectangular hyperboloids, which is determined when three such hyperboloids are given. But these three intersect in eight points, which are common to all. Hence *all rectangular hyperboloids which pass through six given points form a bundle and pass through two other fixed points.*

The condition for an orthogonal hyperboloid is

$$(b - \lambda b')(c - \lambda c') + (c - \lambda c')(a - \lambda a') + (a - \lambda a')(b - \lambda b') = 0.$$

Hence there are in general two orthogonal hyperboloids in a linear system. If there are three, then $bc+ca+ab=0$, $b'c'+c'a'+a'b'=0$, $bc'+b'c+ca'+c'a+ab'+a'b=0$, and every quadric is an orthogonal hyperboloid. This requires

$$a'/a=b'/b=c'/c,$$

and the quadrics are then homothetic.

13·8. Classification of linear systems.

A linear point-system of quadrics, which is determined by two given quadrics, may be specialised in various ways, according to the nature of the base-curve.

13·81. [1111]*. In the general case the curve of intersection is a *quartic curve without singularities*. The two quadrics have a unique common self-polar tetrahedron and when this is taken as tetrahedron of reference the equation of each quadric is of the form

$$ax^2+by^2+cz^2+dw^2=0.$$

Further, by a suitable choice of unit-point we may obtain as canonical equations

$$S \equiv ax^2+by^2+cz^2+dw^2=0,$$

$$S' \equiv x^2+y^2+z^2+w^2=0.$$

The four cones of the system $S-\lambda S'=0$ are determined by the equation

$$(a-\lambda)(b-\lambda)(c-\lambda)(d-\lambda)=0.$$

There are three special cases when equalities occur among the roots.

13·811. [(11)11]. If $a=b$ two of the cones coincide and degenerate to two planes, so that the base-curve reduces to the two conics in which these planes cut all the quadrics. If the planes are $u=0$, $v=0$, the equation of the system is of the form

$$S-\lambda uv=0.$$

The planes of the two conics intersect in a line which cuts S in two points A, B which are points on each of the conics. (We assume that AB is not a generating line of S.) The two conics therefore intersect in two points. The base-curve thus consists of *two conics which intersect in two points*, and any two quadrics of the system have *double-contact* at A and B.

* These symbols are explained later (13·86).

Ex. 1. Show that the system $S - \lambda uv = 0$ has a single infinity of polar tetrahedra and two proper cones.

Ex. 2. If two conics intersect in two points, show that there are two centres from which one of the conics can be projected into the other.

13·812. [(111)1]. If $a = b = c$ the two planes u and v coincide and the base-curve becomes a *double-conic*. The equation of the system is of the form

$$S - \lambda u^2 = 0.$$

Any two quadrics of the system have *ring-contact* along this conic. An arbitrary plane cuts the system in a system of conics having double-contact at the points in which the plane cuts the double-conic. There is just one proper cone in the system and this is a tangent-cone to every quadric of the system. There are ∞^3 self-polar tetrahedra.

An example of a system of this type is a system of homothetic and concentric quadrics, in which case the double-conic is in the plane at infinity and the common tangent-cone is the asymptotic cone. Two concentric spheres have ring-contact along the circle at infinity.

13·813. [(11)(11)]. If $a = b$ and $c = d$, the base-curve is the intersection of two pairs of planes, i.e. a *skew quadrilateral*. The equation of the system can be written in the form

$$xy - \lambda zw = 0.$$

All the quadrics of the system have four common generating lines, two of each system, say a, b of the one system and a', b' of the other. The planes aa', ab', ba', bb' are tangent-planes to all, and any two quadrics of the system have *quadruple contact*. This system is self-dual and forms also a linear tangential system in which the tangent-developable reduces to four planes.

13·814. [(1111)]. The case $a = b = c = d$ is trivial, the two quadrics being coincident.

13·82. [211]. When the quadrics have *simple contact* at a point O every plane through O cuts the quadrics in conics which touch at O and therefore meets the curve of intersection in two coincident points at O. O is therefore a double-point on the curve;

the base-curve is a *nodal quartic*. In general the curve has two distinct tangents at O which lie in the tangent-plane to the quadrics at O. The lines joining O to the other points of the base-curve generate a cone which is cut by any plane through O in two lines; hence this is a quadric cone, and is, in fact, one of the cones of the linear system. The generators of this cone which lie in the tangent-plane at O are the tangents to the curve at its double-point.

Taking $z = 0$ as the common tangent-plane at $O \equiv [0, 0, 0, 1]$, and as $x = 0$ and $y = 0$ two planes through O which form with this plane a self-polar triad for the cone with vertex O, the equations of the two quadrics each reduce to the form

$$ax^2 + by^2 + cz^2 + 2rzw = 0.$$

The plane $cz + 2rw = 0$ may then be taken as the fourth plane of reference $w = 0$, thus making $c = 0$. Further by choosing the unit-point suitably we obtain as canonical equations of the two quadrics

$$S \equiv ax^2 + by^2 + z^2 + 2rzw = 0,$$

$$S' \equiv x^2 + y^2 + 2zw = 0.$$

The cones of the system $S - \lambda S' = 0$ are then found to correspond to the roots of the equation

$$(\lambda - r)^2 (\lambda - a)(\lambda - b) = 0.$$

Hence two of the cones coincide. $(a - r)x^2 + (b - r)y^2 + z^2 = 0$ is the cone which projects the base-curve from the node.

Here again there are three special cases when equalities occur among the roots.

13·821. [(21)1]. If $a = r$ the tangents at the double-point coincide and the quadrics are said to have *stationary contact*. The cone which projects the base-curve becomes two distinct planes, so that the base-curve reduces to *two conics which touch* at O.

13·822. [2(11)]. If $a = b$ the cone $S - aS' = 0$ reduces to two planes $z = 0$ and $z + 2(r - a)w = 0$, while $S - rS' = 0$ is a proper quadric cone. The plane $z = 0$ cuts all the quadrics in two fixed generators $z = 0$, $x \pm iy = 0$; the plane $z + 2(r - a)w = 0$ cuts all the quadrics in the same conic. Thus the base-curve consists of this *conic and two intersecting lines meeting the conic in distinct* *points* [±*i*, 1, 0, 0]. All the quadrics have the same tangent-planes at these two points and at the intersection of the common generators; they have therefore *triple contact*.

13·823. [(211)]. If $a = b = r$ the system is of the form $S - \lambda z^2 = 0$ and the base-curve is a double-conic, but this double-conic breaks up into *two double-lines*. All the quadrics of the system have *contact along these two lines*.

13·83. [22]. If the two quadrics S and S' have a single generator in common the remainder of their intersection is a space-cubic. An arbitrary plane through the common generator cuts each of the quadrics S, S' in a straight line, and the intersection of these lines gives one point on the space-cubic; hence the common generator must be a bisecant and thus the base-curve consists of a *space-cubic together with a bisecant*. The quadrics have *double contact* at the points, A and B, say, where the bisecant cuts the cubic curve. We assume that A, B are distinct points; the case in which they are coincident will be treated later (13·85). Let AC and BD be the other generators of S' through A and B respectively. There is one generator of S', of the system to which AB belongs, which is cut harmonically by AC, BD and the quadric S; let this be CD, and take $ABCD$ as tetrahedron of reference. Then by suitable choice of unit-point we obtain canonical equations of the two quadrics

$$S \equiv z^2 + w^2 + 2fyz + 2pxw = 0,$$
$$S' \equiv 2yz + 2xw = 0.$$

There are two distinct cones in the system $S - \lambda S' = 0$, determined by the equation

$$(\lambda - f)^2(\lambda - p)^2 = 0.$$

13·831. [(22)]. If $p = f$ the two quadrics have in common, in addition to the generator $z = 0 = w$ (twice), two generators $z \pm iw = 0$, $x \mp iy = 0$, and the base-curve consists of *a double-line and two lines mutually skew but each cutting the double-line*. In this case the quadrics *touch all along the double generator*.

13·84. [31]. In the case of stationary contact (13·821 above), which was derived as a special case of simple contact when the base-curve is a nodal quartic, the base-curve degenerated to two conics touching one another. Another case of *stationary contact* occurs when the base-curve becomes a *cuspidal quartic*. Let the point of stationary contact be $D[0, 0, 0, 1]$ and the tangent-plane $z = 0$, then the equation of each of the quadrics is of the form

$$ax^2 + by^2 + cz^2 + 2fyz + 2gzx + 2hxy + 2rzw = 0.$$

The plane $z = 0$ cuts the two quadrics in pairs of lines

$$ax^2 + by^2 + 2hxy = 0, \quad a'x^2 + b'y^2 + 2h'xy = 0,$$

and there is a pair of lines harmonic with regard to each pair. Taking these as DB $(z = 0 = x)$ and DA $(z = 0 = y)$ we have $h = 0 = h'$. The cone $r'S - rS' = 0$ projects the quartic curve from the cusp and is cut by the tangent-plane $z = 0$ in lines

$$(ar' - a'r)x^2 + (br' - b'r)y^2 = 0;$$

these are coincident, say $z = 0 = x$, therefore $br' = b'r$. We may choose the vertex C as a point on the quartic curve and the tangent-plane to S' at C as $w = 0$; then $c = 0$, $c' = 0$, $g' = 0$, $f' = 0$. The point C has still one degree of freedom and we can choose it so that the tangent at C to the quartic curve cuts DA; then $g = 0$. Lastly by a suitable choice of unit-point we obtain the canonical equations

$$S \equiv ax^2 + b(y^2 + 2zw) + 2yz = 0,$$
$$S' \equiv x^2 + (y^2 + 2zw) = 0.$$

There is a proper cone with vertex D which projects the quartic curve, viz. $(a - b)x^2 + 2yz = 0$.

13·841. [(31)]. If $a = b$ this cone degenerates to two planes. The two quadrics meet the tangent-plane $z = 0$ in the same two lines $x^2 + y^2 = 0$. These are common generators and form part of the base-curve; the remainder of the curve is the conic $y = 0$, $x^2 + 2zw = 0$, on which the two common generators intersect. The base-curve thus consists of a *conic and two lines intersecting each other on it.*

13·85. [4]. We consider now the case in which the base-curve is a *space-cubic together with a tangent.* The cubic curve which is represented by the parametric equations

$$x : y : z : w = t^3 : t^2 : t : 1$$

passes through the points A $[1, 0, 0, 0]$ and D $[0, 0, 0, 1]$, and the tangent at A is $z=o=w$ (for every plane $w=\mu z$ through this line meets the curve in two coincident points at A). The equation of a quadric containing the line $z=o=w$ is

$$cz^2+dw^2+2fyz+2gzx+2pxw+2qyw+2rzw=o.$$

Substituting $x=t^3$, $y=t^2$, $z=t$, $w=1$ and equating to zero the coefficients of the several powers of t we obtain $g=o$, $f+p=o$, $c+2q=o$, $r=o$, $d=o$. Then writing $f=\lambda c$ we have the pencil of quadrics

$$z^2-yw+2\lambda(yz-xw)=o.$$

The discriminant of this is λ^4, so that there is just one cone $z^2-yw=o$ in the system.

13·86. Invariant-factors.

These various cases are distinguished according to the nature of the discriminant $|S-\lambda S'|=o$, whose roots determine the four cones of the general system. In the general case the roots $\lambda_1, \lambda_2, \lambda_3, \lambda_4$ are all distinct, and different cases arise when there are equalities among the roots.

In general the matrix $[S-\lambda S']$ is of rank 4, but when λ is equal to one of the roots it is in general of rank 3. When $\lambda_1=\lambda_2$, $[S-\lambda_1 S']$ may still be of rank 3, but it will be of rank 2 if $\lambda-\lambda_1$ is a factor of each of the first minors.

Conversely, if $\lambda-\lambda_1$ is a factor of each of the first minors it will be a repeated factor of the determinant. If ∇ is the determinant formed from the first minors of a determinant Δ of order n, $\nabla=\Delta^{n-1}$; and if each element of ∇ contains the factor f, ∇ is divisible by f^n, but Δ^{n-1} could not be divisible by f^n unless Δ itself were divisible by f^2.

When $\lambda_1=\lambda_2=\lambda_3$, $[S-\lambda_1 S']$ may be of rank 3, 2, or 1; in the first case the first minors are not all divisible by $\lambda-\lambda_1$; in the last case the second minors are all divisible by $\lambda-\lambda_1$ and therefore the first minors are all divisible by $(\lambda-\lambda_1)^2$; in the second case the first minors may be divisible by $\lambda-\lambda_1$ or $(\lambda-\lambda_1)^2$ while the second minors are not divisible by $\lambda-\lambda_1$. These properties are of an invariant character and correspond to the geometrical characteristics of the system. They provide an exact method of distinguishing the different types of linear systems, and we shall explain briefly the notation which is used.

DEF. Let the determinant Δ contain the factor $\lambda-k$ to the power l_0, and let l_1, l_2,... be the indices of the powers of this factor which will divide all of the first, second, ... minors respectively. Then $l_0 > l_1 > l_2 > ....$ Write $l_0 - l_1 = e_1$, $l_1 - l_2 = e_2$,.... Then $(\lambda - k)^{e_1}$, $(\lambda - k)^{e_2}$, ... are called *invariant-factors* to the base $\lambda - k$.

$e_1, e_2, ...$ are each > 0 and their sum $= l_0$, hence the sum of the indices for all the invariant-factors, corresponding to all the factors of Δ, is equal to the order of the determinant. Further it can be proved (see Bromwich, *Quadratic forms and their classification by means of invariant-factors*, Cambridge Tracts, 1906) that $e_1 \geqslant e_2 \geqslant e_3 > ....$

If $e_1, e_2, ..., e_1', e_2', ...$ are the indices of the invariant-factors to the bases $(\lambda - k)$, $(\lambda - k')$, ..., the system is denoted by the *Segre characteristic*

$$[(e_1 e_2 ...)(e_1' e_2' ...) ...].$$

These characteristics have been attached to the various cases above.

13·87. There is one other type of pencil, for which there are no invariant-factors, and for which the discriminant $| S - \lambda S' |$ vanishes identically. This is called the Singular case. Every quadric of the system is a cone. We exclude the case in which the cones have a common vertex, as by taking this vertex as the point [0, 0, 0, 1] we have a system in three variables which is analytically equivalent to a pencil of conics.

Since the polar-planes of a given point with respect to the quadrics of a linear system form an axial pencil, and since in the case of a cone the polar-plane passes through the vertex, it follows that the vertices of all the cones lie on one line. Further, since the polar-plane of any point on this line passes through the point itself the line is a generator of every cone, and all the cones touch the same plane along this line. Thus part of the base-curve is a double-line; the remainder is a conic cutting the line in one point. To obtain canonical equations take A [1, 0, 0, 0] as vertex of S' and C [0, 0, 1, 0] as vertex of S and the common tangent-plane $y = 0$. Take B [0, 1, 0, 0] as a point on the conic and let the tangent at B cut $y = 0$ in D, so that BD is $x = 0 = z$ and the plane of the conic is $z = kx$. Then by suitable choice of unit-point the equations become

$$S \equiv w^2 + 2xy,$$

$$S' \equiv dw^2 + 2yz.$$

13·9. EXAMPLES.

1. Show that the locus of a point whose polar-planes with respect to two given quadrics are at right angles is another quadric.

2. If three quadrics have a common conic, show that the planes of their other conics of intersection have a common line.

3. If $ax^2 + by^2 + cz^2 = 1$ is an ellipsoid, show that as λ varies from $-\infty$ to $+\infty$ or the reverse, the quadric

$$\lambda(ax^2 + by^2 + cz^2 - 1) - (a'x^2 + b'y^2 + c'z^2 - 1) = 0$$

passes through one of the two series E, H_1, H_2, H_1, E, or E, H_1, H_2, Virtual, E.

4. Show that the surfaces

$$z^2 + 2xy + 2az = 0 \quad \text{and} \quad z^2 + 2xy + 2bz = 0$$

touch one another at three points, and have two generators common. Show also that they have a common enveloping cone whose equation is $\{z(a+b) + 2ab\}^2 + 8abxy = 0$.

(Math. Trip. II, 1915.)

5. Find the intersection of the pairs of surfaces:

(i) $x^2 - 2y^2 + yz + 2zx - xy - xw + 2yw = 0$,
$x^2 - y^2 + yz + zx - xw + yw = 0$.

(ii) $x^2 + 3z^2 - 2yz - 2yw + 4zw = 0$,
$x^2 + 3z^2 - 4yz + 4zx - 2yw = 0$.

(iii) $x^2 + y^2 - 2xz + 2yw = 0$, $3x^2 - y^2 - 2yz + 2xw = 0$.

(iv) $2x^2 - y^2 + z^2 + 2yz - 2xw = 0$,
$x^2 - 2y^2 - w^2 - 2zx - 2xy + 2yw + 2zw = 0$.

(v) $2x^2 - y^2 + w^2 - 3yz - 8zx + 4xw - 3zw = 0$,
$4x^2 - 5y^2 - w^2 + 2zx - 9xy - xw - 6yw = 0$.

(vi) $2x^2 + 2y^2 - 3z^2 + w^2 - 4yz - 4zx - 4xy - 2zw = 0$,
$x^2 + y^2 + 3z^2 + w^2 + 2yz + 2zx - 2xy + 4zw = 0$.

(vii) $6x^2 + y^2 + 6z^2 + 2yz + 16zx - 4xy - 2xw - 4zw = 0$,
$5x^2 + 5y^2 + 6z^2 - 6yz + 12zx - 4xy - 2xw - 4zw = 0$.

(viii) $-2x^2 + 8y^2 + w^2 + 8yz - 2zx + 6xy - 2xw - 4yw - 4zw = 0$,
$3x^2 - 5y^2 - z^2 + w^2 - 6yz + 2zx - 2xy + 4xw + 2yw + 2zw = 0$.

(ix) $2x^2 + y^2 - 2w^2 + 2yz + 2zx - 2yw - 2zw = 0$,
$x^2 + y^2 - z^2 - 3w^2 + 2yz + 2xw - 2yw = 0$.

(x) $3x^2 - 4y^2 + z^2 + w^2 - 8yz + 4zx - 4xw = 0$,
$x^2 + 2y^2 + 2yz + zx - 2xw - yw - 3zw = 0$.

(xi) $x^2 - z^2 - zx - xy + xw - 3yw - 2zw = 0$,

$\qquad 3x^2 - 2z^2 - 3w^2 - 4zx - 2xy + 4xw - 6yw - 10zw = 0$.

(xii) $x^2 + y^2 + z^2 + 5w^2 + 2xw - 2yw + 4zw = 0$,

$\qquad x^2 + 2y^2 + z^2 + 8w^2 + 2yz + 2zx + 4xw - 2yw + 6zw = 0$.

(xiii) $x^2 + y^2 + z^2 + w^2 - 2yz - 2zx + xy - xw + yw = 0$,

$\qquad x^2 - y^2 + yz - 2xy - 2xw + zw = 0$.

Ans. (i) $[(11)(11)]$ Two real lines, $x = 0 = y$, $z = 0 = x + y - w$; and two imaginary lines,

$\qquad (x + y) \pm i(x - y) = 0$, $x + y - w = \pm iz$.

(ii) $[(31)]$ Conic $(z = 0, \ x^2 - 2yw = 0)$, and two generators $2x - y - 2w = 0$, $(y - 2z - 2w)(y - 6z - 2w) = 0$ intersecting on conic.

(iii) $[2(11)]$ A conic in plane $x + y + z + w = 0$, and two lines $x = 0 = y$, $x = y = z - w$.

(iv) $[(111)1]$ Ring-contact in plane $x + y + z - w = 0$.

(v) $[(11)(11)]$ Four generators

$\qquad (x - 2y - z)(2x + y + w) = 0$,

$\qquad (3x + y + w)(y - 2z + w) = 0$.

(vi) $[(11)11]$ Double contact at $[1, 1, 0, 0]$ and $[0, 0, 1, -1]$.

(vii) $[(21)1]$ Two conics touching,

$\qquad (x + 2y)(x - 2y + 4z) = 0$.

(viii) $[2(11)]$ Conic $5x - 13y - z + 6w = 0$, and two generators $x + y + z = 0$, $w(2x + w) = 0$.

(ix) $[(211)]$ Two double-lines

$\qquad (x + z - w)^2 = 0$, $(y + 2z)(2x - y + 2z) = 0$.

(x) $[22]$ Cubic curve and a bisecant $(x + z = 0, \ 2y - w = 0)$.

(xi) $[(22)]$ A double-line $(z + 2w = 0, \ x + 3w = 0)$, and two single lines

$\qquad (x + z = 0, \ x + 3w = 0), (x - 2y - z = 0, \ x - 2z - w = 0)$.

(xii) $[31]$ Cuspidal quartic. Cusp at $[1, -1, 1, -1]$.

(xiii) $[4]$ Cubic curve and a tangent $(x = w, \ x + y = z)$ at $[1, -1, 0, 1]$.

6. When two quadrics cut in a conic and two straight lines passing through the same point A of this conic, the sections of the two quadrics by any plane through A have second order

contact at A; but if the plane passes through the tangent at A to the conic the contact is of the third order.

7. If two quadrics have two common intersecting generators and touch at all points of the one generator and at one point of the other, then they will touch at all points of the second generator also.

8. Show that $\rho x = t(t^2+1)$, $\rho y = t^2+1$, $\rho z = -2(f-p)t^2$, $\rho w = 2(f-p)t$ are freedom-equations of the cubic curve of intersection of the quadrics $z^2+w^2+2fyz+2pxw = 0$, $yz+xw = 0$.

9. Show that every quadric of the linear system determined by the two quadrics

$$y^2+z^2+w^2+yz+xy-xw-yw-3zw = 0,$$
$$2y^2+z^2+w^2+2xy-2xw-2yw-2zw = 0$$

is a cone. Find the locus of their vertices and the equation of their common conic.

Ans. $y = z = w$. Plane of conic $x+y-z = 0$. Common tangent-plane $y = w$.

10. At a point A of one given quadric U a tangent is drawn which touches another given quadric U' in a point B'. At B' a second tangent of U' is drawn to touch U in C. At C another tangent of U is drawn to touch U' in D'. Find the positions possible for A in order that AD' should touch U in A and U' in D'.   (Math. Trip. II, 1915.)

11. Prove (i) that if a fixed line through one vertex of the common self-polar tetrahedron of a linear system of quadric loci meet a variable surface of the system in P and P', the tangent-planes at P, P' touch a quadric cone; (ii) that the conics determined by the intersection of the surfaces of the system with a fixed tangent-plane of one of the four quadric cones of the system have all contact in two points. Obtain the properties of the confocal system which correspond to these two results respectively.   (Math. Trip. II, 1914.)

12. If $u = 0$, $u' = 0$; $v = 0$, $v' = 0$; $w = 0$, $w' = 0$ represent pairs of opposite planes of a hexahedron with quadrilateral faces, show that the general equation of a quadric which passes through the eight vertices is $\lambda uu' + \mu vv' + \nu ww' = 0$; deduce that the vertices of a hexahedron form a set of eight associated points.

CHAPTER XIV

CURVES AND DEVELOPABLES

14·1. A curve is a one-way locus of points. A plane (analytic) curve can always be represented by an equation $f(x, y) = 0$ connecting the cartesian coordinates x, y in its plane; when it is regarded as being in three dimensions we must supply another equation $z = 0$, representing the plane in which it lies, for a single equation in x, y, z represents a surface, and in particular the equation $f(x, y) = 0$, which does not contain z, represents a cylinder. The curve is thus represented as the intersection of a plane with a cylinder.

More generally, two equations in the cartesian coordinates x, y, z (or the homogeneous coordinates x, y, z, w) represent a curve in space as the intersection of two surfaces.

A curve may also be represented by parametric equations in which the coordinates x, y, z, w are expressed in terms of a single parameter.

Still more generally, we may have any number of equations connecting the coordinates x, y, z, w and any parameters p, q, ... If these are satisfied by a single infinity of sets of values of the coordinates (real or imaginary) they represent a curve. If the equations are all algebraic and involve the parameters rationally the curve is called an *algebraic curve*. In particular if the coordinates can be separately expressed as rational algebraic functions of one parameter the curve is called a *rational algebraic curve*.

For example, the intersection of two quadric surfaces is an algebraic curve, but, as we shall see later, it is not in general rational. The parametric equations $\rho x = t^3$, $\rho y = t^2$, $\rho z = t$, $\rho w = 1$ represent a rational algebraic curve. Eliminating t in different ways we find that it is the curve common to the three quadrics $xz = y^2$, $xw = yz$, $yw = z^2$.

14·11. An algebraic curve is cut by an arbitrary plane in a finite number of points (real or imaginary); this number is called the *order* of the curve. *The intersection of two surfaces of orders* *m and n is a curve of order mn*, for a plane cuts the two surfaces in plane curves of orders *m* and *n* and their *mn* points of intersection are the points of intersection of the plane with the given curve. Thus the intersection of two quadrics is in general a curve of the fourth order.

The rational curve

$$x : y : z : w = t^3 : t^2 : t : 1$$

is of the third order, since an arbitrary plane

$$lx + my + nz + pw = 0$$

cuts the curve in the three points whose parameters are the roots of the cubic $lt^3 + mt^2 + nt + p = 0$. In this case to every value of the parameter *t* there corresponds just one point, *and conversely*.

14·12. For any rational algebraic curve the homogeneous coordinates can be expressed as *polynomials* in a parameter *t* (with no common factor), for if they are expressed as algebraic fractions we can multiply each of these by the least common multiple of the denominators since only the ratios of the coordinates are significant. To every value of the parameter there corresponds then just one set of ratios of the coordinates and therefore one point. The converse, however, is not always true, as is seen at once from the simple example

$$x : y : z : w = u^6 : u^4 : u^2 : 1$$

which reduces to the above cubic by putting $u^2 = t$, so that to every point correspond two values of *u*, viz. $\pm\sqrt{t}$.

Provided there is a (1, 1) correspondence between the points and the parameters, a system of parametric equations of the form $\rho x = f(t)$, etc., where the functions $f(t)$, etc., are polynomials of degree *n*, represents a rational algebraic curve of order *n*.

14·13. Lüroth's Theorem.

The following theorem enables us to determine the order of a curve from its parametric equations and to obtain a (1, 1) correspondence between the points and the parameters.

Suppose a curve to be represented by the parametric equations

$$\rho x_i = f_i(t) \quad (i = 1, 2, 3, 4),$$

where f_i denote polynomials of degree *n* with no common factor, and suppose that to any point there correspond ν different values of the parameter, $t_1, t_2, ..., t_\nu$. Then it is always possible to find another parameter λ, a function of t, such that there is a $(1, 1)$ correspondence between the values of λ and points on the curve.

Since $t_1, ..., t_\nu$ all correspond to the same point, we have

$$f_1(t_r) : f_2(t_r) : f_3(t_r) : f_4(t_r) = f_1(t_s) : f_2(t_s) : f_3(t_s) : f_4(t_s)$$
$$(r, s = 1, 2, ..., \nu).$$

Form the three expressions

$$F_i(t) = f_i(t)f_4(t_1) - f_4(t)f_i(t_1) \quad (i = 1, 2, 3).$$

These are polynomials in t, of degree n, which all vanish for $t = t_1, ..., t_\nu$; they therefore have the common factor

$$(t - t_1)(t - t_2)...(t - t_\nu),$$

and no other common factor which contains t. They may also have a common factor which does not contain t; as it will involve t_1 denote the highest such factor by $\phi_0(t_1)$. Then the highest common factor of F_1, F_2, F_3 is

$$\phi_0(t_1)(t - t_1)...(t - t_\nu).$$

$t_2, ..., t_\nu$ are of course not contained in this explicitly, but only t and t_1, and we shall write it in the form

$$H \equiv \phi_0(t_1)(t - t_1)...(t - t_\nu)$$
$$\equiv \phi_0(t_1)t^\nu + \phi_1(t_1)t^{\nu-1} + ... + \phi_\nu(t_1).$$

F_1, F_2, F_3 are all skew symmetrical in t and t_1; the common factor $t - t_1$ is also skew symmetrical, and the remaining factor of each is symmetrical. Hence H is skew symmetrical in t and t_1, and as it is of degree ν in t, it is also of degree ν in t_1. At least one of the functions ϕ, say $\phi_i(t_1)$, is therefore of degree ν. Let $\phi_k(t_1)$ be any other of the coefficients which has no factor in common with $\phi_i(t_1)$ and write

$$\lambda = \phi_i(t)/\phi_k(t).$$

The coordinates can now be expressed rationally in terms of the new parameter λ. Since $(t - t_1)...(t - t_\nu)$ is symmetrical in $t_1, ..., t_\nu$ the ratios ϕ_i/ϕ_0 are symmetrical, and therefore λ is symmetrical. To every point correspond ν values of t, and to each of these corresponds the same point. Hence there is a $(1, 1)$ correspondence between the points and the values of the parameter λ.

In the parametric representation of a rational curve we shall assume generally this $(1, 1)$ correspondence.

Ex. 1. Show that the equations

$$\rho x = u^2 + 2,$$
$$\rho y = 2u^2 - u + 5,$$
$$\rho z = 3u^2 - 2u + 8,$$
$$\rho w = 4u^2 - 3u + 11,$$

represent a straight line, and find a parameter λ for a $(1, 1)$ correspondence.

$[\lambda = (u^2 + 2)/(u-1); \quad \rho x = \lambda, \rho y = 2\lambda - 1, \rho z = 3\lambda - 2, \rho w = 4\lambda - 3.]$

Ex. 2. Prove that the condition that the parametric equations $\rho x_i = a_i t^2 + b_i t + c_i$ $(i=1, 2, 3, 4)$ should represent a straight line is that the matrix

$$\begin{bmatrix} a_1 & a_2 & a_3 & a_4 \\ b_1 & b_2 & b_3 & b_4 \\ c_1 & c_2 & c_3 & c_4 \end{bmatrix}$$

should be of rank 2.

14·14. The complete intersection of two algebraic surfaces may consist of two or more distinct curves. Thus the intersection of two quadrics may break up into two conics. If two quadrics have a generating line in common the remaining part of the curve of intersection or *residual* is a cubic curve. In these cases the complete curve of intersection is said to be *reducible*. A plane curve $f(x, y) = 0$ is reducible simply when $f(x, y)$ breaks up into factors. To obtain a criterion for the reducibility of a curve in space we consider a cone which contains the curve. If we project the curve from the centre $O \equiv [0, 0, 0, 1]$, say, we obtain a cone whose homogeneous equation $f(x, y, z) = 0$ is found by eliminating w between the equations of the two surfaces. This cone, and therefore the curve, will be reducible if $f(x, y, z)$ breaks up into factors.

It is not always possible to represent an irreducible algebraic curve as the complete intersection of two algebraic surfaces. In the case of a cubic, for example, the only factors of 3 being 1 and 3, the curve cannot be the complete intersection of two algebraic surfaces unless it is a plane cubic. But two quadrics in general intersect in a curve of order 4 and if they have a common generator the residual is a curve of order 3, which cannot be a plane cubic since a plane cannot cut a quadric in a cubic curve. The cubic $x:y:z:w=t^3:t^2:t:1$ is the partial intersection of any two of the quadrics $xz=y^2$, $xw=yz$, $yw=z^2$, and can be represented exactly as the curve common to these three quadrics. To represent an algebraic curve exactly by the intersections of surfaces as many as four surfaces may be required.

14·21. A line does not in general meet a given curve in any point, but as ∞^2 lines pass through any point of the curve there are ∞^3 lines which meet the curve. These form a complex.

Again, there are ∞^2 lines, *bisecants* or chords, which cut a curve in two points, since each of the points has one degree of freedom. These lines form a congruence.

The lines which pass through a given point and cut a curve of order n form a cone of order n, and the lines which lie in a given plane and cut a given curve of order n form n plane pencils. The complex is therefore of degree n and can be represented by a single homogeneous equation of degree n in the line-coordinates, the line-equation of the curve.

The number of bisecants which lie in a given plane is $\frac{1}{2}n(n-1)$, the number of lines connecting the n points in which the plane cuts the curve. Hence the order of the congruence of bisecants is $\frac{1}{2}n(n-1)$. The class of the congruence is equal to the number of bisecants which pass through a given point O.

14·22. If the curve is projected on a plane from an arbitrary point O the projection is a curve of order n. Corresponding to a bisecant through O we have a double-point on the projection. Viewed from O the given curve has thus an *apparent double-point* there. The number of bisecants which pass through a given point O is thus equal to the number of double-points on the projection from O. This is a definite number for a given curve.

14·23. A plane curve of order n is determined by $\frac{1}{2}n(n+3)$ points, for this is one less than the number of coefficients in its equation, viz. $\frac{1}{2}(n+1)(n+2)$. The maximum number of double-points which it can possess is $\frac{1}{2}(n-1)(n-2)$, for if it had one more, then through these $\frac{1}{2}(n^2-3n+4)$ double-points and $n-3$ other points on the curve, i.e. altogether $\frac{1}{2}(n-2)(n+1)$, a curve of order $n-2$ is determined which would meet the given curve in $(n^2-3n+4)+(n-3)=n^2-2n+1$ points, whereas a curve of order $n-2$ cannot cut a curve of order n in more than $n(n-2)$ points.

14·231. *A plane curve which has its maximum number of double-points is rational,* for through the $\frac{1}{2}(n-1)(n-2)$ double-points and $n-2$ other points on the curve, i.e. altogether $\frac{1}{2}(n-2)(n+1)$, a curve of order $n-2$ is determined; and if only $n-3$ additional points are taken, then a linear system of curves of order $n-2$ is determined, each meeting the curve in

$$(n-1)(n-2)+(n-3)=n(n-2)-1$$

given points, and therefore meeting it in one variable point. If in particular these $n-3$ points are taken on a given straight line the variable curve of order $n-2$ meets this line in one other variable point. Hence there is a $(1, 1)$ correspondence between the points of the given curve of order n and the points of the straight line.

Ex. 1. The plane cubic curve $y^3+azx^2+bxyz=0$ has a double-point at $[0, 0, 1]$. The line $y=\lambda x$, which passes through the double-point, cuts the curve again where $\lambda^3 x+az+b\lambda z=0$. Hence we have the parametric equations

$$\left.\begin{array}{l} \rho x = a+b\lambda \\ \rho y = a\lambda + b\lambda^2 \\ \rho z = -\lambda^3 \end{array}\right\}.$$

Ex. 2. The cardioid, whose polar equation is $r=a\,(1+\cos\theta)$, and whose homogeneous cartesian equation is

$$(x^2+y^2-axz)^2=a^2\,(x^2+y^2)\,z^2,$$

has a cusp at the origin $[0, 0, 1]$, and cusps also at the circular points $[1, \pm i, 0]$. A conic through these three points and another fixed point on the curve, say $[2a, 0, 1]$, is a circle

$$(x-a)^2+(y-\lambda)^2=a^2+\lambda^2,$$

i.e. $$x^2+y^2=2\,(ax+\lambda y)\,z.$$

Eliminating z between the two equations we find, after removing the factors $(x^2+y^2)^2$ and y,

$$4a\lambda x=(a^2-4\lambda^2)\,y.$$

Hence we obtain the parametric equations

$$\rho x = 2a^3 (a^2 - 4\lambda^2),$$
$$\rho y = 8a^4\lambda,$$
$$\rho z = (a^2 + 4\lambda^2)^2$$

or, putting $2\lambda = a\mu$,

$$\left.\begin{aligned}
\rho x &= 2a (1 - \mu^2) \\
\rho y &= 4a\mu \\
\rho z &= (1 + \mu^2)^2
\end{aligned}\right\}.$$

14·232. A plane curve which has its maximum number of double-points, and is therefore rational, is said also to be *unicursal*, since all the real points of the curve (with the possible exception of certain isolated points, acnodes or double-points at which the tangents are imaginary) can be traced by the variation of a single parameter through all real values from $-\infty$ to $+\infty$. The number by which the number of double-points falls short of the maximum is called the *deficiency* or *genus*, usually denoted by p.

14·24. The maximum number of apparent double-points (for an arbitrary point O) for a space-curve of order n is therefore also $\frac{1}{2}(n-1)(n-2)$. If the point O lies on the curve, the projecting cone is of order $n-1$. An apparent double-point in this case corresponds to a trisecant, hence the maximum number of trisecants which pass through an arbitrary point of the curve is $\frac{1}{2}(n-2)(n-3)$. When the curve has the maximum number of apparent double-points for a given point O it is rational, for taking O as the point $[0, 0, 0, 1]$ and the plane of projection as $w = 0$ there is a $(1, 1)$ correspondence between a parameter λ and the points P' on the plane $w = 0$, and between these points and the points P of the curve there is also a $(1, 1)$ correspondence. It can be proved that the number by which the number of apparent double-points falls short of the maximum is the same from any general view-point, and this is called the *deficiency* or *genus* of the curve. This follows from Riemann's theorem that any two curves between which a $(1, 1)$ correspondence exists have the same genus, but it would be beyond the scope of this book to go further into it.

14·3. A line which meets a curve in two coincident points, i.e. the limiting case of a bisecant, is a *tangent-line*. The tangent-lines form a one-dimensional series like the points on the curve. The planes which pass through tangent-lines form a two-dimensional system like the tangent-planes of a surface, and can be represented by a single equation in plane-coordinates (ξ), the tangential equation of the curve. Thus an equation in (ξ) may represent, not a surface, but a curve.

A plane which meets the curve in three coincident points is called an *osculating plane*. At each point on the curve there is in general a unique osculating plane. These form a one-dimensional series, represented by two equations in plane-coordinates.

14·4. Thus there are three one-dimensional systems associated with a curve: (1) the points on the curve, (2) the tangent-lines, (3) the osculating planes. There is a symmetry among these systems. (3) is dual to (1), while (2) is self-dual, a tangent-line being as well the limiting line of intersection of two osculating planes as the limiting line joining two points. Any one of the three systems determines the other two. We have also the two-dimensional system of planes through tangent-lines, and dual to this we have the two-dimensional system of points on tangent-lines. But the latter is a sort of surface; it is represented by a single equation in (x) just as the system of tangent-planes is represented by a single equation in (ξ). Psychologically a locus of points has a concreteness which is much more difficult to attribute to assemblages of lines and planes, and this surface has been elevated to importance under the name *developable*, the significance of which will appear later. As a two-dimensional locus of points it is a sort of surface, but it differs from an ordinary surface just as essentially as a curve does. A surface as a locus is dual to a surface as an envelope, both being two-dimensional assemblages, the one of points, the other of planes. A curve as a locus of points is dual to a developable as an envelope of planes, both being one-dimensional assemblages of their respective elements; a curve as an envelope of planes is dual to a developable as a locus of points, both being two-dimensional assemblages; finally a curve as an assemblage of lines (tangents) is dual to a developable as an assemblage of lines (generating lines). From the last point of view a developable has the character of a ruled surface, the characteristic distinction being that the tangent-planes of a ruled surface form a two-dimensional system so that at each point of a generating line there is a different tangent-plane, while in the case of a developable the tangent-planes at all points of a given generator are the same and they form only a one-dimensional series.

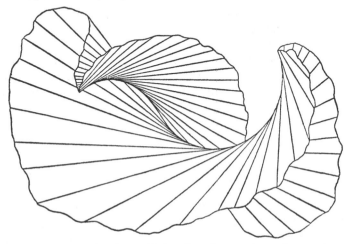

Fig. 45. A developable, with its edge of regression and generators

14·41. Starting with the developable as a one-dimensional series of planes depending on a single parameter, we obtain the generating lines as the limiting positions of the lines of inter-section of two planes which come into coincidence, and the locus of the limiting positions of points which lie in three ultimately coincident planes is the curve. The generating lines are tangents of the curve, and the planes are its osculating planes. From this point of view the curve is called the *edge of regression* of the developable.

14·42. The term "developable" refers to a characteristic property of the figure considered as a deformable surface or material sheet. By a succession of small successive rotations about the generating lines the surface can be laid flat or developed on to a plane, without any stretching or tearing. A cone has this property and is a developable, but its curve reduces to a single point—the vertex. A plane curve, which is dual to a cone, has its developable already flat—it is the plane of the curve.

14·43. Consider the section of a developable by a plane. The plane cuts the curve of the developable in n points (n being the order of the curve). It cuts each of the generating lines in a point, and the locus of these points is the curve of section C. Finally it cuts each of the planes of the developable in a line, and these lines are tangents to C. At an arbitrary point on C there is a unique plane of the developable and therefore a unique tangent to C; but at a point P on the edge of regression there are two coincident generators and the curve C has a double-point there; further there are two coincident planes of the developable passing through this generator, and the tangents to C at the double-point are coincident. Hence P is a cusp on the curve of section; that is, any plane in general cuts the developable in a curve having a cusp on the edge of regression. The edge of regression is thus a *cuspidal edge*. The developable considered as a surface consists of two sheets which meet at the cuspidal edge. The tangents to the curve are divided at the points of contact, the two parts belonging to the two separate sheets. In a similar way the assemblage of planes which pass through tangents of the curve, and produce the curve as an envelope, have the developable as a singularity or assemblage of double-planes. The envelope consists of two parts; each plane is divided by the tangent-line of the curve, and the two portions belong to the two parts of the envelope.

14·51. In addition to the order n of a curve there are other numbers which characterise it. It must always be borne in mind that we have to deal with not merely a curve but three one-dimensional systems, of points, lines and planes. The *order* is the number of points which the system of points has in common with an arbitrary plane field of points. Reciprocally, the *class m* is the number of planes which the system of planes has in common with an arbitrary bundle of planes, i.e. the number of osculating planes of the curve, or planes of the developable, which pass through an arbitrary point. This is clearly equal to the number of *apparent* inflexions of the curve, i.e. the number of inflexions of an arbitrary plane projection of the curve, or the number of inflexional or stationary tangent-planes of the projecting cone. Reciprocally, the order is equal to the number of stationary points or cusps on an arbitrary plane section of the developable, which agrees with what has been already observed. The class is also equal to the class of an arbitrary plane section of the developable. A third number is the order of the developable, i.e. the order of the plane curve obtained by taking a section of the system of planes by an arbitrary plane. This is called the *rank r*. Reciprocally it is also the class of the cone which projects the curve from an arbitrary point, or the class of an arbitrary projection of the curve.

14·52. For a plane algebraic curve the fundamental numbers: order n, class m, number of double-points δ, cusps κ, double-tangents τ, and inflexions ι, are connected by Plücker's equations, by means of which, any three of the numbers being given, the other three can be determined. These are

$$m = n(n-1) - 2\delta - 3\kappa,$$
$$3n(n-2) = \iota + 6\delta + 8\kappa,$$

and two others obtained by interchanging the pairs n, m; δ, τ; κ, ι. Now consider a space-curve of order n, rank r, class m, and having h apparent double-points, H actual double-points, K cusps, and I actual inflexions. Then for an arbitrary plane projection we have a plane curve for which

$$n' = n, \quad m' = r, \quad \delta' = h + H, \quad \kappa' = K, \quad \iota' = m + I.$$

Hence if n, h, H, K and I are given, the rank and class are determined by the equations

$$r = n(n-1) - 2(h+H) - 3K,$$
$$m = 3n(n-2) - 6(h+H) - 8K - I.$$

14·6. The space-cubic.

We shall consider now more particularly the curves of the third order or cubics.

14·61. If O is any point on the curve, the lines joining O to other points of the curve generate a quadric cone, since any plane through O cuts the curve in two other points and therefore the cone in two lines. A single infinity of quadric cones can therefore be passed through a space-cubic, each having as vertex a point of the curve.

Consider two such cones, vertices O_1 and O_2. O_1O_2 is a generating line of each cone, and each cone contains the whole curve. The intersection of the two cones then consists of the cubic curve and the line O_1O_2. If O_3 is a third point on the curve we obtain another quadric cone containing the whole curve. The three cones have in pairs the lines O_2O_3, O_3O_1, O_1O_2 in common, but these three lines have no common point, hence the cubic curve is represented as the complete intersection of three quadric cones.

14·62. *A line cannot cut a space-cubic in more than two points*, for suppose A, B, C to be three collinear points on the curve, and P any other point on the curve, then the plane $PABC$ meets the curve in more than three points. Hence there are no trisecants. All the bisecants which pass through a given point of the curve form a quadric cone.

Two bisecants cannot cut one another except at a point on the curve, for if they did they would determine a plane meeting the curve in more than three points.

14·631. Let AA', BB', CC' be three bisecants, and α any plane through AA'. α cuts the curve again in a single point P, and this determines with BB' a unique plane β and with CC' a unique plane γ. Hence we have three pencils of planes related in pairs in $(1, 1)$ correspondence. Conversely, the curve can be generated by the common point of a set of corresponding planes of three homographic pencils of planes whose axes have no point in common. As an example, let $ABCD$ be the tetrahedron of reference. Then the pencils with axes CD, AB and AD can be represented by the equations

$$x = \lambda y, \quad z = \mu w, \quad y = \nu z.$$

The simplest $(1, 1)$ correspondence is represented by

$$\lambda = \mu = \nu = t.$$

Then $$x/y = y/z = z/w = t,$$

and therefore the coordinates of the common point of corresponding planes are $[t^3,\ t^2,\ t,\ 1]$.

In a similar way it may be shown that for the general case of any three homographic pencils whose axes are not concurrent the coordinates of the common point of three corresponding planes are represented by polynomials of the third degree.

Now the equations
$$x=t^3,\ y=t^2,\ z=t,\ w=1$$
are freedom-equations of a space-cubic, for the plane
$$lx+my+nz+pw=0$$
cuts the curve in the three points whose parameters are the roots of the equation
$$lt^3+mt^2+nt+p=0.$$

The general space-cubic is therefore a rational curve.

14·632. The plane $x=0$ or BCD meets the curve in three coincident points at D and is the osculating plane at $D(t=0)$. Similarly the plane $w=0$ or ABC is the osculating plane at $A(t=\infty)$. Any plane $x=\lambda y$ through CD meets the curve in two coincident points at D; the line CD also meets the curve in two coincident points at D and is therefore the tangent at D. Similarly AB is the tangent at A.

14·64. $xz=y^2$ and $yw=z^2$ represent cones which contain the curve and have their vertices at D and A respectively. The equation $xw=yz$ represents a quadric, also containing the curve. This is generated by the line of intersection of the two homographic pencils
$$x=ty,\ z=tw,$$
and is therefore a quadric containing the bisecants CD and AB as generating lines. For all values of $\lambda,\ \mu,\ \nu$ the quadric
$$\lambda(y^2-zx)+\mu(yz-xw)+\nu(z^2-yw)=0$$
contains the curve. Hence through a space-cubic there are ∞^2 quadrics. The condition that a given quadric should contain a given space-cubic is therefore equivalent to seven linear conditions. An arbitrary quadric cuts a given cubic curve in $2\times3=6$ points; if it contains seven points of the curve it will therefore contain the whole curve.

14·641. When a conic and a cubic lie on the same quadric they intersect in three points, the points in which the plane of the conic meets the cubic curve. Hence, in particular, of two intersecting generators one meets the cubic in two points and the other in one point. It follows that each generator of one system meets the cubic in two points and each generator of the other system meets it in one point.

14·642. The bisecant generators of any quadric which contains the curve therefore determine an involution on the curve, since its points are thus made to correspond in pairs. Conversely an involution on the curve determines a regulus and therefore a quadric.

Analytically, the line joining two points t, t' on the cubic is given by

$$\rho x = ut^3 + vt'^3,$$
$$\rho y = ut^2 + vt'^2,$$
$$\rho z = ut + vt',$$
$$\rho w = u + v,$$

where u/v is a variable parameter. Substituting in the equation of the quadric

$$\lambda(y^2 - zx) + \mu(yz - xw) + \nu(z^2 - yw) = 0,$$

we obtain $\quad uv(t - t')^2 \{\lambda tt' + \mu(t + t') + \nu\} = 0.$

$u = 0$ and $v = 0$ give just the two points of the curve; the equation, however, is identically satisfied if

$$\lambda tt' + \mu(t + t') + \nu = 0.$$

When λ, μ, ν are given, this determines an involution; and when the involution is given, λ, μ, ν are determined, fixing the quadric.

14·643. If S and S' are two quadrics containing the curve they have in common a generating line l which is a bisecant of the curve. This is the line joining the pair of points common to the two involutions which determine the quadrics.

A unique quadric can be drawn to contain the curve and have two given (non-intersecting) bisecants as generators, for these two bisecants determine an involution on the curve.

14·651. *Through any point O, not on the curve, there passes one and only one bisecant.* Join O to any point R on the curve. If OR is not a bisecant a plane through OR will cut the curve in two other points P, P', and these pairs form an involution. The lines PP', which all intersect OR, form one regulus of a quadric, and the line of this regulus which passes through O is the bisecant required. From this result it follows that a cubic curve possesses one and only one apparent double-point. Every plane projection is a cubic having a double-point and is therefore rational.

14·652. A space-cubic cannot have an actual double-point, for if it had, then through this point and two other points on the curve a plane could be drawn which would have four points in common with the curve and would therefore contain the curve. The curve would then be a plane cubic. Further, a space-cubic cannot in general have an apparent cusp. A space-curve in fact can only have an apparent cusp viewed from an *arbitrary* point when it has an actual cusp; it will have an apparent cusp when the point of view lies on the developable.

The plane projection of the general cubic from an arbitrary point is thus in general a nodal cubic, which is of class 4 and has three inflexions. Hence the class of the space-cubic is also 3 and its rank is 4. It is a self-dual figure.

14·66. *A space-cubic can be made to pass through six arbitrary points,* no three of which are collinear and no four in one plane. For the five points B, C, D, E, F determine a quadric cone with vertex A, and the five points A, C, D, E, F a quadric cone with vertex B. The intersection of these cones is the line AB together with a definite space-cubic which passes through the six points.

14·671. On a given quadric surface U there are two systems of cubic curves; the one system has the λ-generators as bisecants, and the other the μ-system.

Consider two cubics C_1 and C_2 of the same system, and let V_1 and V_2 be two quadrics which contain them; these cut U further in two generators λ_1 and λ_2 of the same system, which are bisecants of both curves. The three quadrics have eight points in common; these all lie on either C_1 or λ_1, and on either C_2 or λ_2. Now λ_1

and λ_2 are both bisecants of both C_1 and C_2 and do not intersect one another; this accounts for four of the eight points and the remaining four must be common to C_1 and C_2.

If C_1 and C_2 belong to different systems on U, the quadrics V_1 and V_2 cut U in generators λ_1 and μ_2 of different systems. λ_1 cuts C_2 in one point, μ_2 cuts C_1 in one point, and λ_1 cuts μ_2 in one point. Thus just three of the eight points are accounted for and the two curves have five points in common.

14·672. *Through five arbitrary points on a given quadric there pass two space-cubics which lie entirely in the surface.* Let l, m be the generators of the surface through E. Then a definite quadric cone with vertex E is determined to contain the points A, B, C, D and the generator l. The intersection of this cone with the given quadric consists of the common generator l and a space-cubic which passes through A, B, C, D and E. A second space-cubic is determined by the cone with vertex E which has m as generator. The former has l and all the generators of this system as bisecants, the latter has m and all the generators of the other system as bisecants. Besides these two there are no others, for two cubics of the same system can have only four points in common.

Ex. If a conic cuts a cubic curve in three points, show that there is a unique quadric surface which contains them both.

14·681. The tangent at t is the limiting position of the chord joining t and t' when $t' \to t$. The line-coordinates of the chord tt' are given by the matrix

$$\begin{bmatrix} t^3 & t^2 & t & 1 \\ t'^3 & t'^2 & t' & 1 \end{bmatrix},$$

and are therefore (cancelling the factor $t-t'$)

$$[t^2+tt'+t'^2, \ t+t', \ 1; \ -tt', \ tt'(t+t'), \ -t^2t'^2].$$

The coordinates of the tangent are therefore

$$[3t^2, \ 2t, \ 1; \ -t^2, \ 2t^3, \ -t^4].$$

These can also be obtained from the matrix

$$\begin{bmatrix} t^3 & t^2 & t & 1 \\ 3t^2 & 2t & 1 & 0 \end{bmatrix},$$

the second line being obtained by differentiating the first.

14·682. The osculating plane at t is that which meets the curve in three coincident points at t, and its equation is therefore

$$x - 3ty + 3t^2z - t^3w = 0.$$

The osculating plane at $t = 0$ is $x = 0$, and at $t = \infty$ is $w = 0$. These three osculating planes intersect at $[0, t, 1, 0]$. The plane through the three points t, 0, ∞ is $y = zt$, and this passes through the common point of the three osculating planes. Hence *the osculating planes at any three points intersect on the plane containing the three points*.

14·683. The plane π $(lx + my + nz + pw = 0)$ cuts the curve in three points A_1, A_2, A_3 whose parameters t_1, t_2, t_3 are the roots of the equation

$$lt^3 + mt^2 + nt + p = 0.$$

The osculating planes at these points are

$$x - 3t_iy + 3t_i^2z - t_i^3w = 0 \quad (i = 1, 2, 3).$$

The coordinates of the point P common to these are

$$[3t_1t_2t_3, \ \Sigma t_1t_2, \ \Sigma t_1, \ 3],$$

i.e.

$$[-3p/l, \ n/l, \ -m/l, \ 3]$$

or

$$[3p, \ -n, \ m, \ -3l].$$

There is thus a correlation between the planes π and the points P, forming a null system since each point lies on the corresponding plane, and the self-corresponding lines of the system form a linear complex. The self-corresponding lines, or lines of the complex, are those which pass through any given point and lie in the corresponding (polar) plane. Among these lines are the tangents to the cubic curve, for the osculating plane α at a point A contains the tangent a at A; any plane π which contains a cuts the curve in A and two other points B, C, and the osculating planes at A, B, C intersect on π at the point P on a which is the pole of the plane π. Thus the tangent a lies in π and passes through P.

Since the coordinates of the tangent at t are

$$p_{01} = 3t^2, \ p_{02} = 2t, \ p_{03} = 1; \ p_{23} = -t^2, \ p_{31} = 2t^3, \ p_{12} = -t^4,$$

there is just one *linear* relation connecting them, viz.

$$p_{01} + 3p_{23} = 0.$$

This represents a linear complex which contains all the tangents, and is the only linear complex in which they are contained. It is therefore the linear complex determined above. This follows also from the equations of the correlation or null system, viz.

$$\rho\xi_0 = x_1, \quad \rho\xi_1 = -x_0, \quad \rho\xi_2 = 3x_3, \quad \rho\xi_3 = -3x_2,$$

i.e. $a_{01} = 1$, $a_{23} = 3$, while the other coefficients are zero.

The polar of any point on the curve, with regard to the linear complex, is the osculating plane at that point, and in general the polar of any point P is the plane through the points of contact of the three osculating planes which pass through P.

14·69. Projectively, all space-cubics, like all conics, are equivalent. Metrically, space-cubics are classified according to their relation to the plane at infinity and the circle at infinity. In general the plane at infinity cuts the curve in three distinct points, and then the curve goes to infinity in three different directions; the tangents at the three points at infinity are *asymptotes*, and the osculating planes are *asymptotic planes*. The asymptotes are all mutually skew. The asymptotic planes intersect in a point on the plane at infinity, and therefore form a triangular prism.

There are four types of curves:

(1) *The cubical hyperbola.* Three real and distinct points at infinity; three finite asymptotes and asymptotic planes. The cone containing the curve and having its vertex at one of the points at infinity becomes a cylinder; and as there are other real points at infinity it is a hyperbolic cylinder. The curve is therefore the intersection of two hyperbolic cylinders which have a common line at infinity, i.e. having one asymptotic plane of one cylinder parallel to one of the other.

(2) *The cubical ellipse.* One real point at infinity and two conjugate imaginary points; one real asymptote. The curve is the intersection of an elliptic cylinder and a cone.

(3) *The cubical hyperbolic parabola.* Two coincident points at infinity, and one single point; one finite asymptote. The curve is the intersection of a hyperbolic and a parabolic cylinder which have a common line at infinity, i.e. having one asymptotic plane of the hyperbolic cylinder parallel to the axial plane of the parabolic cylinder.

(4) *The cubical parabola.* Three coincident points at infinity. The plane at infinity is an osculating plane. The curve is the intersection of a parabolic cylinder and a cone, the tangent-plane to the cone along the common generator being parallel to the axial plane of the cylinder.

Ex. 1. Show that the equations

$$x : y : z : w = t(t-1) : t(t+1) : t^2 - 1 : t(t^2-1)$$

represent in cartesian coordinates a cubical hyperbola, and that the equations of the hyperbolic cylinders which contain it are

$$-yw + yz + zw = 0,$$
$$xw - zw + xz = 0,$$
$$xw - wy + 2xy = 0.$$

Ex. 2. Show that the intersection of the cone $y^2 + z^2 - zx = 0$ and the elliptic cylinder $y^2 + z^2 - yw = 0$ is a cubical ellipse which can be represented by the parametric equations

$$x : y : z : w = t(t^2+1) : t^2 : t : t^2+1.$$

Ex. 3. Show that the intersection of the parabolic cylinder $z^2 - wy = 0$ with the hyperbolic cylinder $zw + wx - xz = 0$ is a cubical hyperbolic parabola

$$x : y : z : w = t : t^2(t-1) : t(t-1) : t-1.$$

Ex. 4. Show that a cubical parabola can be represented by the equations

$$x : y : z : w = t^3 : t^2 : t : 1,$$

and that it is the intersection of the cone $y^2 = zx$ with the parabolic cylinder $z^2 = wy$.

14·7. Quartic curves.

A curve of order 4, or space-quartic, is met by an arbitrary plane in four points, and by an arbitrary quadric surface in eight points. Through nine arbitrary points on the curve a definite quadric is determined which must contain the whole curve provided it does not degenerate. Hence at least one quadric $S = 0$ can be determined to contain a given quartic curve. If a second quadric $S' = 0$ can be drawn containing the curve, the curve is the intersection of these two quadrics, and every quadric of the linear system $S + \lambda S' = 0$ contains the curve.

We have to distinguish then two different species of space-quartics:

First Species: those which are the intersection or base of a pencil of quadrics,

Second Species: those through which only one quadric passes.

14·71. Space-quartics of the First Species.

A space-quartic of the First Species is the complete intersection of two quadrics which do not have a common generator; as we have just seen, it lies also on every quadric of the linear system determined by these two.

14·711. *Every generator of a quadric S which contains the curve cuts it in two points.* Let l be any generating line of S. A plane through l cuts S in another generator l' as well, and cuts any other quadric S' of the linear system in a conic C. But C cuts l and l' each in two points which, being common to S and S', lie on the curve.

Conversely, *if a quartic curve K on a quadric P cuts every generator in two points it is a quartic of the First Species*, i.e. a second quadric can be drawn to contain the curve. Take eight arbitrary points $A_1, ..., A_8$ on K and let O be any other point. These nine points determine a quadric Q which cuts P in a quartic curve K' of the First Species also passing through the eight points A, since these are common to P and Q. We have to prove that, with the given conditions, K' coincides with K. With centre A_1 project K and K' on to an arbitrary plane α. The projections are two plane cubics C, C', which, if they are distinct, intersect in nine points; seven of these are the projections $B_2, ..., B_8$ of the common points $A_2, ..., A_8$. Further, the two generators of P through A_1 are by hypothesis bisecants of K, and they are also bisecants of K', since K' is a quartic curve of the First Species; hence the remaining two points of intersection of the two plane cubics are the points, G and G', say, where these two generators cut α. Now a plane cubic is in general uniquely determined by nine points, and through eight given points a linear system of ∞ cubics can be drawn. Any two cubics which pass through the eight points intersect in a ninth point which is common to all the cubics of the linear system; this point is then determined when the eight other points are given. Thus nine points uniquely determine a cubic unless they are associated in a particular way. But the nine points $B_2, \ldots, B_8, G, G'$ are not so associated, since $B_2, \ldots, B_8$ are all independently variable, hence the cubic through these is uniquely determined and therefore C and C' coincide. As the projections of K and K' from any other point A_i must similarly coincide, these two quartics must themselves coincide.

14·712. A quartic curve of either species can be considered as the partial intersection of a quadric with a cubic surface. The complete intersection is a curve of order 6, and the *residual* is a conic, two skew lines or a double-line of the cubic surface. If the residual is a conic, any generator of the quadric meets this in just one point, the point in which it meets the plane of the conic; the two other intersections of this generator with the cubic surface must therefore be on the quartic curve. Thus every generator of the quadric is a bisecant, and the quartic is of the First Species.

14·713. A quartic curve of the First Species, being the complete intersection of two quadrics, has no trisecants, for if the line l cut the curve in three points A, B, C, a plane through l would cut the two quadrics in conics intersecting in three collinear points A, B, C, which is impossible. Hence, viewed from any point on the curve, it has no apparent double-points. The cubic curve which is the plane projection from a point on the curve is therefore not unicursal but with deficiency 1. It follows that the quartic curve is not rational, i.e. its coordinates cannot be expressed rationally and algebraically in terms of a parameter.

14·714. Assuming that the two quadrics do not touch, they have a common self-polar tetrahedron. By a suitable choice of unit-point the equations can be expressed in the forms

$$S \equiv x^2 + y^2 - z^2 - w^2 = 0, \quad S' \equiv ax^2 + by^2 + cz^2 - w^2 = 0.$$

The coordinates of a point on S can be expressed in terms of parameters t, u

$$\rho x = t + u, \quad \rho y = 1 - tu, \quad \rho z = 1 + tu, \quad \rho w = t - u.$$

Substituting in the equation of the other quadric we have an equation which is of the second degree in both t and u. It is not possible to eliminate one of the parameters and express x, y, z, w rationally and algebraically in terms of one of them.

The linear system $\lambda S + S' = 0$ contains four cones. Suppose, by another choice of unit-point, the equations of two of these are

$$x^2 + y^2 - w^2 = 0, \quad k^2 x^2 + z^2 - w^2 = 0.$$

The parametric equations of the former can be written

$$\rho x = \sin\theta, \quad \rho y = \cos\theta, \quad \rho w = 1,$$

and substituting in the second we have

$$z^2 = 1 - k^2 \sin^2\theta.$$

Write $\theta = \mathrm{am}\,(u, k)$ (an elliptic function), then

$$\rho x = \mathrm{sn}\ u, \quad \rho y = \mathrm{cn}\ u, \quad \rho z = \mathrm{dn}\ u, \quad \rho w = 1.$$

Thus the coordinates of a point on the quartic are expressed parametrically in terms of elliptic functions. On this account this type of quartic curve is called *elliptic*. It can be shown generally that a curve is elliptic when its genus $= 1$. For higher genera other functions (hyperelliptic) are required for a parametric representation.

14·715. If the quadrics touch, the curve of intersection has a double-point there. This is a particular case of the general theorem for any two surfaces. At an ordinary point of intersection P the tangent-planes are distinct and determine one definite tangent to the curve of intersection, and any plane through this *line* cuts the surfaces in curves which touch at P. But when the surfaces touch at P every plane through this *point* cuts the surface in curves which touch at P.

14·7151. To express most simply the equations of two quadrics which touch, take the point of contact $O \equiv [0, 0, 0, 1]$ and the common tangent-plane $z = 0$. Then the equations of the quadrics are of the form

$$ax^2 + by^2 + cz^2 + 2fyz + 2gzx + 2hxy + 2zw = 0.$$

Since the quadrics do not have a common generating line the pairs of generating lines in the common tangent-plane are distinct and there is a unique pair of lines which are harmonic conjugates with regard to each pair. Choosing these as axes of x and y, $h = 0 = h'$. We can then choose the vertex $A \equiv [1, 0, 0, 0]$ of the tetrahedron of reference so that its polar-planes with respect to the two quadrics coincide, and as this plane must pass through OB we can choose it as the plane $x=0$. Then $g=0=g'$. $B \equiv [0, 1, 0, 0]$ can similarly be chosen so that its polar-plane with respect to each is $y=0$. Then $f=0=f'$. The line OC is then determined as the intersection of these planes, and C may be chosen arbitrarily, as also the scales on the three axes. The equations of the two quadrics can thus be written

$$U \equiv ax^2 + by^2 + cz^2 + 2zw = 0,$$
$$U' \equiv a'x^2 + b'y^2 + c'z^2 + 2zw = 0.$$

In the linear system $U - \lambda U' = 0$ there are three cones, the values of the parameter being $\lambda = a/a'$, b/b' and 1 (twice). $\lambda = 1$ gives

$$(a-a')x^2 + (b-b')y^2 + (c-c')z^2 = 0.$$

Adjusting the scales on the three axes we may reduce this equation to
$$x^2 + y^2 - z^2 = 0.$$

Freedom-equations for this are $\rho x = 2t$, $\rho y = t^2 - 1$, $\rho z = t^2 + 1$. Substituting in U we get for w a biquadratic in t divided by $2(t^2+1)$. Hence putting $\rho = \frac{1}{2}\rho'/(t^2+1)$ we have

$$\rho'x = 4t(t^2+1), \quad \rho'y = 2(t^4-1), \quad \rho'z = 2(t^2+1)^2$$
and $\qquad \rho'w = $ a quartic polynomial in t.

The nodal quartic curve is therefore rational.

14·716. The common tangent-plane $z=0$ cuts the cone with vertex O in the two lines $(a-a')x^2 + (b-b')y^2 = 0$, which are the tangents to the quartic curve at the double-point. They may be real or imaginary. If they are coincident the two quadrics have *stationary contact*, and the quartic curve has a *cusp*. This will be the case if $b=b'$. We cannot, however, now take $f=f'$, for then the cone with vertex O would break up into two planes and the curve of intersection would reduce to two conics. Keeping $g=g'=0$, the equations of two quadrics of the system are

$$U \equiv ax^2 + by^2 + cz^2 + 2fyz + 2zw = 0,$$
$$U' \equiv a'x^2 + by^2 + c'z^2 + 2f'yz + 2zw = 0,$$

and the cone with vertex O is
$$(a-a')x^2 + (c-c')z^2 + 2(f-f')yz = 0.$$

With scales adjusted this can be simplified to
$$x^2 + z^2 - 2yz = 0.$$

Freedom-equations of this are $\rho x = 2t$, $\rho y = t^2 + 1$, $\rho z = 2$, and substituting in U we obtain again for w a quartic polynomial in t of the form $\rho w = pt^4 + qt^2 + r$.

14·717. Through a quartic curve of the first species there pass a linear system of quadrics, of which the curve forms the base. All the generating lines of these quadrics are bisecants, or chords of the curve, and conversely every bisecant of the curve is a generating line of a quadric of the system. All the bisecants form a congruence. Since there is just one quadric of the system through a given point, and through this point there are two generators of the quadric, there are two bisecants through any arbitrary point which does not lie on the curve. Also since there are three quadrics which touch a given plane, and each is met by the plane in two generators, there are six bisecants in any plane. The congruence of bisecants is therefore of order 2 and class 6. This agrees with the general result in 14·21.

If the curve is projected from an arbitrary point, the projecting lines form a quartic cone and the projection is a plane quartic curve with just two double-points. This is not a rational curve, since a rational plane quartic has three double-points. But when the space-quartic has itself a double-point, its projection has three double-points and is rational.

From the equations in 14·52 we find the rank and class of the quartic curves of the First Species to be as follows:

	n	r	m
The general elliptic quartic curve	4	8	12
Nodal quartic	4	6	6
Cuspidal quartic...	4	5	4

Ex. 1. Show that the general quartic curve of the first species is uniquely determined by eight arbitrary points, of which no three are collinear, and no five lie in one plane.

Ex. 2. Show that through eight associated points on a quadric surface there pass a linear system of quartic curves, one through each point on the surface, and two touching each generating line.

Ex. 3. Show that all chords of a given quartic which cut a given chord are generators of one quadric.

14·718. The figure which is dual to the quartic curve of intersection of two quadrics is the assemblage of common tangent-planes of two quadrics. This is a developable, an important example of which being the focal developable determined by a given quadric and the circle at infinity. It is in general of class 4, order 12, and rank 8. To a bisecant of the quartic curve corresponds a line lying in two common tangent-planes, a "line-in-two-planes" or *axis*. The axes form a congruence of order 6 and class 2; in each plane there are two axes and through any point there are six axes, and these axes are all generators of quadrics of the same linear tangential system. In the case of the focal developable the axes are the focal axes (cf. 12·21 and 12·81).

14·72. Space-quartics of the Second Species.

A space-quartic of the Second Species is the partial intersection of a quadric surface with a cubic surface when the residual consists either of two skew lines or of a line which counts double on the cubic surface.

14·721. Let U be a quadric surface and K a quartic curve of this species lying on it and therefore such that there is no other quadric but U which contains the curve. If the curve is projected from a point O on itself, projecting lines form a cubic cone and the plane projection is a cubic curve. In the case of a quartic of the first species without double-point this cubic curve has no double-point, and the cubic cone has no double-line; the quadric and the cone have two intersecting generators (a degenerate conic) in common. If the quartic curve (still of first species) has a double-point or a cusp, the cone has a double-line or a cuspidal edge passing through O, but this is not a generating line of the quadric; the quadric and the cone have still only two single generators common. If, however, the cubic cone has a double-line and this coincides with a generator of the quadric, the projection from O of the quartic curve of intersection is a rational cubic, and the quartic curve is rational. This is also the case for the intersection of a quadric surface with any cubic surface which has a double-line coinciding with a generator of the quadric.

14·722. Again, consider a quadric and a cubic surface with two non-intersecting lines λ and λ' in common, belonging therefore to the same λ-set of generators of the quadric. Every other generator of the quadric cuts the cubic surface in three points, and in the case of the λ-generators these three points all belong to the quartic curve. But a μ-generator cuts both λ and λ', and therefore cuts the quartic curve in just one point. On a given quadric surface there are therefore two distinct systems of quartic curves of the second species: one cutting each of the λ-generators in three points and each of the μ-generators in one point, the other cutting the μ-generators in three points and the λ-generators in one point.

14·723. A quartic curve of the second species has thus a single infinity of trisecants, which are the generators of one system of the quadric surface which contains the curve. There can be no trisecants which are not generators of this quadric since a line cannot cut a quadric in three points. Through each point on the curve passes one trisecant, and therefore the projection of the curve from any point on itself is a cubic having one double-point and therefore is rational.

A quartic curve of the second species can have no actual double-point, for if it had, the projection from an arbitrary point on the curve would be a cubic with two double-points and the curve would degenerate. Further, as the curve is rational, its projection from an arbitrary point is a plane quartic with three double-points. Hence *through an arbitrary point there pass three bisecants*.

14·724. Parametric equations of the general quartic curve of the second species can always be expressed by first forming parametric equations of the quadric which contains the curve in terms of two parameters λ and μ such that the generators are expressed by $\lambda = \text{const.}$ and $\mu = \text{const.}$, and then connecting λ, μ by a (3, 1) relation of the form

$$\mu(a_0\lambda^3 + 3a_1\lambda^2 + 3a_2\lambda + a_3) + (b_0\lambda^3 + 3b_1\lambda^2 + 3b_2\lambda + b_3) = 0.$$

14·725. Three essentially different types are found according to the existence of *stationary* or *inflexional tangents*, i.e. tangents which meet the curve in three coincident points. The cubic curve and the quartics of the first species, since they have no trisecants, can have no inflexional tangents. The conditions that the cubic equation in λ (14·724) should have three equal roots are

$$\frac{\mu a_1 + b_1}{\mu a_0 + b_0} = \frac{\mu a_2 + b_2}{\mu a_1 + b_1} = \frac{\mu a_3 + b_3}{\mu a_2 + b_2}.$$

These give two quadratics in μ, and eliminating μ we find a relation connecting the coefficients. In general therefore (First Type) there are no stationary tangents. If the condition is satisfied there is (Second Type) one stationary tangent corresponding to the common root of the two quadratics in μ. If further these two quadratics have both their roots in common we have (Third Type) two stationary tangents.

The class and rank of the three types of quartics of the Second Species are therefore as follows:

	n	r	m
First Type (no inflexions) ...	4	6	6
Second Type (one inflexion) ...	4	6	5
Third Type (two inflexions) ...	4	6	4

14·8. Number of intersections of two curves lying on a quadric surface.

Let C_1 denote a λ-generator, C_1' a μ-generator, C_2 a conic, C_3 a cubic which cuts a λ-generator in one point and a μ-generator in two points, C_3' a cubic of the other system, C_4 a quartic of the first species, K_4 a quartic of the second species which cuts a λ-generator in one point and a μ-generator in three points, K_4' a quartic of the other system. Then the following table shows the number of intersections of each pair*:

	C_1	C_1'	C_2	C_3	C_3'	C_4	K_4	K_4'
C_1	0	1	1	1	2	2	1	3
C_1'		0	1	2	1	2	3	1
C_2			2	3	3	4	4	4
C_3				4	5	6	5	7
C_3'					4	6	7	5
C_4						8	8	8
K_4							6	10

* The general result, from another point of view, is given in 14·95, Ex. 6.

Some of these results have been already proved (see 14·641, 14·671, 14·711, 14·722). The others can be left as an exercise to the reader. The method can be illustrated in one case; for example, to prove that $\{K_4 K_4'\} = 10$. Let U be the given quadric, Q and Q' cubic surfaces containing K_4 and K_4' respectively. Then the intersection of U and Q consists of K_4 and two generators C_1' and K_1'; the intersection of U and Q' consists of K_4' and two generators C_1 and K_1. The three surfaces U, Q, Q' have $2 \times 3 \times 3 = 18$ points in common, and we can write symbolically

$$\{UQQ'\} = 18 = \{K_4 K_4'\} + \{K_4 C_1\} + \{K_4 K_1\} + \{C_1' K_4'\}$$
$$+ \{C_1' C_1\} + \{C_1' K_1\} + \{K_1' K_4'\} + \{K_1' C_1\} + \{K_1' K_1\}.$$

But each of the last eight terms $= 1$, hence $\{K_4 K_4'\} = 10$.

14·9. As an example of a quartic curve of the second species we shall consider the line of striction of a quadric surface. Two generators of the same system have a unique common perpendicular, and in the limiting case, when the generators come to coincide, the foot of the common perpendicular is a unique point on the generator. The locus of this point is called the *line of striction* for this system of generators.

Consider the hyperboloid

$$x^2/a^2 + y^2/b^2 - z^2/c^2 = w^2.$$

Its freedom-equations are

$$\rho x/a = 1 - \lambda\mu, \quad \rho y/b = \lambda + \mu, \quad \rho z/c = 1 + \lambda\mu, \quad \rho w = \lambda - \mu,$$

and the generating lines correspond to $\lambda = $ const. and $\mu = $ const. The equations of the generator $\lambda = $ const. are

$$x/a - z/c = \lambda(w - y/b),$$
$$\lambda(x/a + z/c) = w + y/b,$$

and its direction-cosines are proportional to

$$a(1 - \lambda^2), \quad 2b\lambda, \quad c(1 + \lambda^2).$$

The direction-cosines of the line joining the points (λ, μ) and $(\lambda + \delta\lambda, \mu + \delta\mu)$ are proportional to

$$a\{(1 - \mu^2)\delta\lambda - (1 - \lambda^2)\delta\mu\},$$
$$2b(\mu\delta\lambda - \lambda\delta\mu),$$
$$c\{(1 + \mu^2)\delta\lambda - (1 + \lambda^2)\delta\mu\}.$$

The condition that this line should be perpendicular to the generator λ is

$$a^2(1-\lambda^2)\{(1-\mu^2)\delta\lambda-(1-\lambda^2)\delta\mu\}+4b^2\lambda(\mu\delta\lambda-\lambda\delta\mu)$$
$$+c^2(1+\lambda^2)\{(1+\mu^2)\delta\lambda-(1+\lambda^2)\delta\mu\}=0.$$

Expressing also the condition that it should be perpendicular to the generator $\lambda+\delta\lambda$, and subtracting, we obtain

$$-a^2\lambda\{(1-\mu^2)\delta\lambda-(1-\lambda^2)\delta\mu\}+2b^2(\mu\delta\lambda-\lambda\delta\mu)$$
$$+c^2\lambda\{(1+\mu^2)\delta\lambda-(1+\lambda^2)\delta\mu\}=0.$$

Eliminating $\delta\lambda$ and $\delta\mu$ between the last two equations we obtain, after cancelling the factor $\lambda-\mu$,

$$a^{-2}(1-\lambda^2)(1-\lambda\mu)+2b^{-2}\lambda(\lambda+\mu)+c^{-2}(1+\lambda^2)(1+\lambda\mu)=0,$$

or $$\mu\lambda(\lambda^2+A)+A\lambda^2+1=0,$$

where $$A\equiv(2b^{-2}+c^{-2}-a^{-2})/(a^{-2}+c^{-2}).$$

Thus we have λ, μ connected by a $(3, 1)$ relation, and the locus is a quartic curve of the second species which cuts every λ-generator in one point and every μ-generator in three points. Substituting for μ in terms of λ in the freedom-equations of the quadric we obtain the parametric equations of the curve

$$\rho x/a=(1+A)\lambda(\lambda^2+1),\ \rho y/b=\lambda^4-1,$$
$$\rho z/c=(1-A)\lambda(\lambda^2-1),\quad \rho w=\lambda^4+2A\lambda^2+1.$$

The curve passes through the vertices of the principal sections of the quadric. In certain cases the line of striction degenerates.

Ex. 1. Show that for the paraboloid $x^2/a^2-y^2/b^2=zw$ the lines of striction are the parabolas in which the surface is cut by the two planes $b^3x\pm a^3y=0$.

Ex. 2. Show that for the paraboloid $x^2-y^2=zw$ the lines of striction are the two generating lines in which the surface is cut by the plane $z=0$ and the two lines at infinity $w=0$, $x^2=y^2$.

Ex. 3. Show that for a hyperboloid of revolution the two lines of striction coincide with the principal section (a circle).

Ex. 4. Show that the curve of striction of the quadric

$$ax^2+by^2+cz^2=1$$

has in general no inflexion, but has two inflexions if

$$(2a-b-c)(2b-c-a)(2c-a-b)=0.$$

14·95. EXAMPLES.

1. Show that parametric equations of the cubic curve which passes through the vertices of the tetrahedron of reference and the points $[1, 1, 1, 1]$ and $[a, b, c, d]$ are

$$x : y : z : w = a/(t-a) : b/(t-b) : c/(t-c) : d/(t-d).$$

2. Find the equations of the six cones which pass through the six points $[1, 0, 0, 0]$, $[0, 1, 0, 0]$, $[0, 0, 1, 0]$, $[0, 0, 0, 1]$ $[1, 1, 1, 1]$, $[a, b, c, d]$ and have one of these points as vertex; and show that they have one cubic curve in common.

3. Prove that a quadric surface can be drawn through a cubic curve in space to contain two assigned chords of the curve.

Show that if A, B, C, A', B', C' be six assigned points of the cubic, the three quadrics containing the curve and, respectively, the three pairs of chords AA', BC'; BB', CA'; CC', AB', have all a common generator.   (Math. Trip. II, 1915.)

4. Show that the problem of finding a polygon of n sides whose corners lie on a twisted cubic curve and whose sides belong to a linear complex is poristic. (Math. Trip. II, 1914.)

5. Prove that any four points on a twisted cubic curve and the osculating planes at the points are the vertices and faces of two tetrahedra each of which is inscribed in the other.

A variable tetrahedron has its vertices on a cubic. The cubic is given by equations $x/t^3 = y/t^2 = z/t = w$, and the parameters of the vertices $u + \lambda v = 0$, where u and v are quartics in t, and λ is variable. Show that the faces of the tetrahedron are osculating planes of another cubic curve and that the tetrahedron is self-polar with respect to a fixed quadric.   (Math. Trip. II, 1914.)

6*. If the quadric $xw - yz = 0$ is represented by the parametric equations $x : y : z : w = \lambda\mu : \lambda : \mu : 1$, show that an algebraic equation $\phi(\lambda, \mu) = 0$ in λ, μ, of degree p in λ and q in μ $(p + q = n)$ represents a curve, to be denoted by (p, q), lying on the quadric and having the following properties:

(i) It is met by an arbitrary plane in n points, and is therefore of order n.

---

* See Cayley, *Coll. Math. Papers*, vol. v, pp. 70–2.

(ii) It meets every λ-generator in q points and every μ-generator in p points.

(iii) For a given value of n there are, according as n is even or odd, $\frac{1}{2}n$ or $\frac{1}{2}(n-1)$ essentially different species of curves without singularities.

(iv) If $p=q$ show, by substituting $\lambda=y/w$, $\mu=z/w$ in the equation $\phi(\lambda, \mu)=0$, that the curve is the complete intersection of the quadric with a surface of order p.

(v) If $p>q$ show that we can derive from the equation

$$w^p \phi(\lambda, \mu) f(\mu, 1)=0,$$

where $f(\mu, 1)$ is any polynomial in $\mu:1$ of degree $p-q$, by substituting for λ, μ in terms of the coordinates, an equation of the form $Pf(x, y)+Qf(z, w)=0$, where P and Q are homogeneous polynomials in x, y, z, w of degree q; and deduce that the curve (p, q) is the partial intersection of the quadric and a surface of order p, the remaining intersection consisting of $p-q$ μ-generators which may be arbitrarily selected.

(vi) Show that the equation $\phi(\lambda, \mu)=0$ represents in the (λ, μ)-plane a curve passing p times through the point $[0, 1]$ and q times through the point $[1, 0]$. Deduce that two curves (p, q) and (p', q') on the quadric intersect one another in $pq'+p'q$ points.

(vii) From the number of constants in the equation

$$\phi(\lambda, \mu)=0$$

show that a curve (p, q) on a quadric is determined by $pq+p+q$ points on the quadric.

(viii) If the quadric has no real generating lines show that the only real curves which lie upon it are curves of even order, of the type (p, p).

7. Show that the coordinates of any point on the developable, which is the envelope of the polar-planes of a fixed point $[x', y', z']$ with respect to quadrics confocal with

$$x^2/a+y^2/b+z^2/c=1,$$

may be written in the form

$$xx' = \frac{(a+\lambda)^2(a+\mu)}{(a-b)(a-c)}, \quad yy' = \frac{(b+\lambda)^2(b+\mu)}{(b-a)(b-c)}, \quad zz' = \frac{(c+\lambda)^2(c+\mu)}{(c-a)(c-b)},$$

where λ, μ are parameters. Show that the equation $\lambda = $ const. defines a generator, and that the equation $\mu = $ const. defines a parabola, which together with a generator makes up the complete intersection of the surface by the polar-plane with respect to the confocal of parameter μ.   (Trinity, 1914.)

8. Show that the constant-number of a conic in space is 8; of a space-cubic 12; of a quartic curve of the first species 16.

CHAPTER XV

INVARIANTS OF A PAIR OF QUADRICS

15·1. We have seen that a single quadric has only one projective invariant, the discriminant Δ, whose vanishing is the condition that the quadric should be specialised as a cone. We now consider a system determined by two quadrics

$$S \equiv \Sigma\Sigma a_{rs} x_r x_s$$

and

$$S' \equiv \Sigma\Sigma a_{rs}' x_r x_s.$$

In the linear system

$$\lambda S + S' = 0$$

we have seen that there are in general four cones, corresponding to the roots of the equation

$$\begin{vmatrix} \lambda a_{00} + a_{00}' \dots \lambda a_{03} + a_{03}' \\ \dots\dots\dots\dots\dots\dots\dots\dots \\ \lambda a_{30} + a_{30}' \dots \lambda a_{33} + a_{33}' \end{vmatrix} = 0.$$

This is a quartic equation in λ, and we shall write it in the form

$$\Delta(\lambda) \equiv \Delta\lambda^4 + \Theta\lambda^3 + \Phi\lambda^2 + \Theta'\lambda + \Delta' = 0.$$

Δ and Δ' are the discriminants of S and S', of the fourth degree in the respective coefficients; Θ, Φ, Θ' are functions of the coefficients of both S and S', of degrees 3 and 1, 2 and 2, 1 and 3 respectively.

If the quadrics are referred to any other tetrahedron of reference, i.e. if their equations are subjected to a linear transformation, the values of λ for which $\lambda S + S' = 0$ represents a cone will remain the same, provided the coefficients of S and S' are not multiplied separately by different factors. The roots of the quartic equation are therefore unaltered, and therefore the ratios of the coefficients are invariants.

But if M is the modulus of the transformation it has been shown that the discriminant Δ is transformed into Δ_1, where

$$\Delta_1 = M^2\Delta,$$

and since S' has been subjected to the same transformation

$$\Delta_1' = M^2\Delta'.$$

Hence also $\quad \Theta_1 = M^2\Theta, \ \Phi_1 = M^2\Phi, \ \Theta_1' = M^2\Theta'.$

Δ and Δ' are the projective invariants of S and S' separately; Θ, Φ, Θ' are *simultaneous invariants* of the two quadrics.

15·11. The vanishing of one of these simultaneous invariants represents some projective relationship between the two quadrics. By a particular choice of the frame of reference the forms of these invariants may be simplified and their meanings arrived at.

Choose a tetrahedron of reference self-polar with respect to S', and a suitable unit-point. Then we can write

$$S' \equiv x^2 + y^2 + z^2 + w^2 = 0,$$
$$S \equiv ax^2 + by^2 + cz^2 + dw^2 + 2fyz + 2gzx + 2hxy$$
$$+ 2pxw + 2qyw + 2rzw = 0.$$

The invariants then become

$$\Theta' \equiv a + b + c + d,$$
$$\Phi \equiv (bc - f^2) + (ca - g^2) + (ab - h^2) + (ad - p^2)$$
$$+ (bd - q^2) + (cd - r^2),$$
$$\Theta \equiv A + B + C + D,$$

where A is the cofactor of a in the determinant Δ.

15·2. If S is circumscribed about the tetrahedron of reference, a, b, c, d, all vanish, and therefore $\Theta' = 0$; that is, *if there is a tetrahedron inscribed in the quadric S and self-polar with respect to the quadric S', the projective invariant $\Theta' = 0$.*

This result is somewhat remarkable, for we might expect it to be possible always to find a tetrahedron self-polar with respect to one quadric and inscribed in another, without any relation whatever between the two quadrics. For the number of conditions required in order that a given tetrahedron should be self-polar with respect to a given quadric is 6, each pair of vertices being conjugate; and the number of conditions required in order that a given tetrahedron should be inscribed in a given quadric is 4. But a tetrahedron can in general be constructed to satisfy twelve conditions since each vertex has three degrees of freedom. What we should expect then is that there should be a double infinity of tetrahedra satisfying the given conditions. Actually no tetrahedron at all exists unless the two quadrics are related in a particular way. We shall show now, conversely, that if this condition, $\Theta' = 0$, is satisfied there is a triple infinity of such tetrahedra. A problem of this sort is said to be *poristic*.

15·21. Assuming the condition $\Theta' = 0$, choose the tetrahedron of reference with one vertex A on S. The polar-plane of A with respect to S' cuts S in a conic; choose the second vertex B on this conic. The polar-line of AB with respect to S' cuts S in two points; choose one of these as the third vertex C. Then choose D as the fourth vertex of the self-polar tetrahedron with respect to S'. Three of its vertices A, B, C lie on S, and therefore $a = 0$, $b = 0$, $c = 0$; and since $\Theta' = 0$ it follows that $d = 0$ as well, hence D also lies on S.

Hence $\Theta' = 0$ *is the necessary and sufficient condition that there should be one tetrahedron (and therefore a triple infinity of tetrahedra) inscribed in S and self-polar with respect to S'.*

If capital letters denote the cofactors of the corresponding small letters in the determinant Δ', the tangential equation of S' in the general case is

$$\Sigma' \equiv A'\xi^2 + \dots + 2F'\eta\zeta + \dots + 2P'\xi\omega + \dots = 0,$$

and $$\Theta' \equiv aA' + \dots + 2fF' + \dots + 2pP' + \dots,$$

i.e. Θ' is linear in the coefficients of S and Σ'. Following Baker, we say that the quadric locus S is *outpolar* to the quadric envelope Σ', S being circumscribed to a tetrahedron which is self-polar with respect to Σ'.

15·211. Again, if we choose a tetrahedron of reference self-polar with respect to S, so that f, g, h, p, q, r all vanish, Θ' will vanish if A', B', C', D' all vanish, i.e. if the tetrahedron is circumscribed about Σ'; and conversely. We say that the quadric envelope Σ' is *inpolar** to the quadric locus S, Σ' being inscribed in a tetrahedron which is self-polar with respect to S. Hence $\Theta' = 0$ *is also the necessary and sufficient condition that there should be one tetrahedron (and therefore a triple infinity of tetrahedra) circumscribed to Σ' and self-polar with respect to S.*

Since Θ' is linear in the coefficients of S and Σ' it follows that if two quadrics S_1 and S_2 are both outpolar to Σ' all quadrics of the linear system $S_1 + \lambda S_2$ are outpolar to Σ'.

---

* The term *apolar* is also used for both outpolar and inpolar.

15·212. *Ex.* 1. *The vertices of two self-polar tetrahedra of the same quadric form eight associated points.*

We may take one of the tetrahedra as frame of reference and the equation of the quadric

$$S \equiv x^2 + y^2 + z^2 + w^2 = 0.$$

Let $S' \equiv a'x^2 + b'y^2 + c'z^2 + d'w^2 + \ldots$ be any quadric passing through the vertices of the second tetrahedron and through three vertices A, B, C of the first. Then since the second tetrahedron is self-polar with respect to S and is inscribed in S', S' is outpolar to S and therefore $\Theta = 0$, i.e.

$$a' + b' + c' + d' = 0.$$

But $a' = 0$, $b' = 0$, and $c' = 0$, therefore $d' = 0$ and S' passes also through the fourth vertex D. Hence every quadric which passes through seven of the vertices passes also through the eighth.

Ex. 2. Conversely, *if eight associated points are divided in any way into two sets, the two sets of four points form self-polar tetrahedra with respect to the same quadric.*

Choose the four points A, B, C, D as frame of reference. Any quadric for which $ABCD$ is self-polar is represented by

$$S \equiv ax^2 + by^2 + cz^2 + dw^2 = 0.$$

The ratios of the coefficients will be determined by the three conditions that A', B', C' are mutually conjugate. Let D_1 be the pole of the plane $A'B'C'$ with respect to S when thus determined. Then $ABCD$ and $A'B'C'D_1$ are both self-polar with respect to S, hence they form a set of eight associated points. But since $ABCDA'B'C'D'$ also form such a set and since the eighth point is uniquely determined by the other seven, D' coincides with D_1.

15·22. To determine the meaning of the vanishing of Φ we note that $\Phi = 0$ if all the six terms $(bc - f^2), \ldots, (ad - p^2), \ldots$ vanish. But $bc - f^2 = 0$ is the condition that the edge $y = 0 = z$ of the tetrahedron of reference should touch S, and similarly for the other conditions. Hence $\Phi = 0$ *when there is a tetrahedron self-polar with respect to S' and having all its edges touching S.*

To construct a tetrahedron, self-polar with respect to one quadric and having its edges touching another quadric, requires twelve conditions, just the right number to determine a tetrahedron in general, but the above result shows that $\Phi = 0$ is a necessary condition. Hence no tetrahedron exists satisfying the given conditions unless the quadrics are suitably related, and then there will be a single infinity. As Φ is symmetrical as regards the coefficients of S and S' it follows that if $\Phi = 0$ there will be an infinity of tetrahedra self-polar with respect to S and having all their edges touching S'.

15·31. The invariant Φ involves the coefficients of the line-equations of the two quadrics. The point-equation of a quadric being

$$S \equiv \Sigma\Sigma a_{rs} x_r x_s = 0,$$

the line-equation is

$$\Psi = \Sigma\Sigma\Sigma\Sigma (a_{ij} a_{kl} - a_{ik} a_{jl}) p_{il} p_{jk}.$$

Changing the notation so that

$$S \equiv ax^2 + \ldots + 2fyz + \ldots + 2pxw + \ldots = 0,$$

and writing the line-coordinates

$$p_{23} = u_1, \quad p_{31} = u_2, \quad p_{12} = u_3,$$

$$p_{01} = v_1, \quad p_{02} = v_2, \quad p_{03} = v_3,$$

we find that the line-equation is a homogeneous quadratic in the six variables $u_1, \ldots, v_1, \ldots$, viz.

$$\Psi \equiv c_{11} u_1^2 + \ldots + C_{11} v_1^2 + \ldots + k_{11} u_1 v_1 + \ldots + K_{11} v_1 u_1 + \ldots$$
$$+ 2c_{23} u_2 u_3 + \ldots + 2C_{23} v_2 v_3 + \ldots + 2k_{23} u_2 v_3 + \ldots$$
$$+ 2K_{23} v_2 u_3 + \ldots = 0,$$

where

$$c_{11} = bc - f^2, \quad C_{11} = ad - p^2, \quad k_{11} = K_{11} = gq - hr,$$
$$c_{23} = gh - af, \quad C_{23} = fd - qr, \quad k_{23} = gr - cp, \quad K_{23} = bp - hq,$$

the other expressions being written down by the simultaneous permutations (123), (abc), (fgh), (pqr); corresponding small and capital letters represent complementary minors of the determinant Δ.

Then the invariant Φ of two quadrics is the bilinear expression

$$\Phi \equiv c_{11} C_{11}' + \ldots + k_{11} K_{11}' + \ldots + 2c_{23} C_{23}' + \ldots + 2k_{23} K_{23}' + \ldots$$
$$+ c_{11}' C_{11} + \ldots + k_{11}' K_{11} + \ldots + 2c_{23}' C_{23} + \ldots$$
$$+ 2k_{23}' K_{23} + \ldots.$$

15·32. Invariants for the reciprocal system.

If we take the linear tangential system of quadrics

$$\lambda\Sigma + \Sigma' = 0,$$

where $\quad \Sigma \equiv A\xi^2 + \ldots + 2F\eta\zeta + \ldots + 2P\xi\omega + \ldots = 0,$

we obtain the discriminant

$$\nabla(\lambda) \equiv \Delta^3\lambda^4 + \Delta^2\Theta'\lambda^3 + \Delta\Delta'\Phi\lambda^2 + \Delta'^2\Theta\lambda + \Delta'^3;$$

for the determinant

$$\nabla \equiv \begin{vmatrix} A & H & G & P \\ H & B & F & Q \\ G & F & C & R \\ P & Q & R & D \end{vmatrix} = \Delta^3,$$

as is found by multiplying together $\nabla.\Delta.$

Also

$$\Delta \begin{vmatrix} B & F & Q \\ F & C & R \\ Q & R & D \end{vmatrix} = \begin{vmatrix} a & h & g & p \\ h & b & f & q \\ g & f & c & r \\ p & q & r & d \end{vmatrix} \cdot \begin{vmatrix} 1 & 0 & 0 & 0 \\ H & B & F & Q \\ G & F & C & R \\ P & Q & R & D \end{vmatrix}$$

$$= \begin{vmatrix} a & 0 & 0 & 0 \\ h & \Delta & 0 & 0 \\ g & 0 & \Delta & 0 \\ p & 0 & 0 & \Delta \end{vmatrix} = \Delta^3 a,$$

and $\quad \Delta(AB - H^2) = \begin{vmatrix} a & h & g & p \\ h & b & f & q \\ g & f & c & r \\ p & q & r & d \end{vmatrix} \cdot \begin{vmatrix} A & H & G & P \\ H & B & F & Q \\ 0 & 0 & 1 & 0 \\ 0 & 0 & 0 & 1 \end{vmatrix}$

$$= \begin{vmatrix} \Delta & 0 & g & p \\ 0 & \Delta & f & q \\ 0 & 0 & c & r \\ 0 & 0 & r & d \end{vmatrix} = \Delta^2(cd - r^2).$$

Hence $\quad \begin{vmatrix} B & F & Q \\ F & C & R \\ Q & R & D \end{vmatrix} = \Delta^2 a, \quad \begin{vmatrix} A & H \\ H & B \end{vmatrix} = \Delta \begin{vmatrix} c & r \\ r & d \end{vmatrix},$ etc.

If

$$BC - F^2 \equiv \gamma_{11}, \ AD - P^2 \equiv \Gamma_{11}, \ GQ - HR \equiv \kappa_{11} \equiv \mathrm{K}_{11},$$

$$GH - AF \equiv \gamma_{23}, \ FD - QR \equiv \Gamma_{23}, \ GR - CP \equiv \kappa_{23}, \ BP - HQ \equiv \mathrm{K}_{23},$$

then $\qquad \Delta\Delta'\Phi \equiv \gamma_{11}\Gamma_{11}' + \text{etc.}$

15·33. Absolute invariants.

The five invariants Δ, Θ, Φ, Θ', Δ' are relative. To form an absolute invariant it is necessary to take a function of these which is not only of zero dimensions in the five invariants, i.e. involves only their ratios, but is also of zero dimensions in the coefficients of both quadrics. Such expressions, of the form $\Delta^\alpha \Theta^\beta \Phi^\gamma$, can be formed with any three of the five invariants. To determine α, β, γ we express that the dimensions in the coefficients of the two quadrics are each zero. Hence, for this example, we have $4\alpha + 3\beta + 2\gamma = 0$ and $\beta + 2\gamma = 0$, which give also $\alpha + \beta + \gamma = 0$. Hence $\alpha : \beta : \gamma = -1 : 2 : -1$. Ten absolute invariants such as this can be obtained, but only three of these are independent. If we write

$$P \equiv \frac{\Phi^2}{\Theta\Theta'}, \quad Q \equiv \frac{\Theta^2}{\Delta\Phi}, \quad Q' \equiv \frac{\Theta'^2}{\Delta'\Phi},$$

we can eliminate any two of the quantities Δ, Θ, Φ, Θ', Δ' between these equations; thus any invariant equation involving the five invariants can be expressed in terms of P, Q, Q'.

It is proved further in text-books on invariant-theory[*] that the five invariants form a *complete system* in the sense that any polynomial simultaneous invariant of the two quadrics can be expressed in terms of these.

15·34. A general procedure in finding an invariant equation corresponding to a projective relationship between two quadrics is as follows. First choose a convenient frame of reference so as to simplify the equations of the quadrics. Express the five invariants and form P, Q, Q', or some other set of independent absolute invariants. If a single invariant equation exists, P, Q, Q' can be expressed in terms of two variable parameters alone, and by eliminating these we get an equation connecting the three,

---

[*] See, for example, Turnbull, *The theory of determinants, matrices, and invariants* (London: Blackie, 1928), p. 304.

which can then be transformed to a homogeneous equation in the five invariants.

Ex. 1. Condition that there should be a tetrahedron inscribed in one quadric S and having two pairs of opposite edges as generators of another quadric S'.

Taking the tetrahedron as frame of reference we have

$$S \equiv 2fyz + 2gzx + 2hxy + 2pxw + 2qyw + 2rzw = 0,$$
$$S' \equiv 2f'yz + 2p'xw = 0.$$

We find
$$\Delta' \equiv f'^2 p'^2, \quad \Theta' \equiv 2f'p'\,(f'p + fp'),$$
$$\Phi \equiv (fp' + f'p)^2 + 2f'p'\,(fp - gq - hr),$$
$$\Theta \equiv 2\,(fp' + f'p)\,(fp - gq - hr).$$

Write
$$f'p' = \alpha, \quad fp' + f'p = \beta, \quad fp - gq - hr = \gamma,$$

then
$$\Delta' \equiv \alpha^2, \quad \Theta' \equiv 2\alpha\beta, \quad \Phi \equiv \beta^2 + 2\alpha\gamma, \quad \Theta \equiv 2\beta\gamma.$$

Without using the absolute invariants we see that as these four expressions are homogeneous in α, β, γ we can eliminate the latter and obtain an equation homogeneous in Δ', Θ', Φ, Θ. The result is

$$4\Delta'\Theta'\Phi = \Theta'^3 + 8\Delta'^2\Theta.$$

Ex. 2. Condition that there should be a tetrahedron whose six edges touch two given quadrics.

We have here the exact number of conditions required to determine a tetrahedron, but we shall find that the problem is again poristic.

Taking the tetrahedron as frame of reference we find that the coefficients of the quadric

$$S \equiv ax^2 + \dots + 2fyz + \dots + 2pxw + \dots = 0$$

satisfy the six equations

$$bc - f^2 = 0, \quad ad - p^2 = 0, \text{ etc.}$$

Writing $a = \alpha^2, b = \beta^2, c = \gamma^2, d = \delta^2$, we have $f = \pm\beta\gamma$, etc. There are sixty-four possible combinations of sign, but it can be verified that the only ones which do not make Δ vanish are that in which the signs are all negative and those obtained from this by changing the signs of α, β, γ, or δ. It is therefore quite general to take the signs all negative.

Choose the other quadric S' similarly with α', β', γ', δ' instead of α, β, γ, δ. Then we find

$$\Delta \equiv -16\alpha^2\beta^2\gamma^2\delta^2, \qquad \Delta' \equiv -16\alpha'^2\beta'^2\gamma'^2\delta'^2,$$
$$\Theta \equiv -4\Sigma\,(\alpha'^2\beta^2\gamma^2\delta^2 + 2\beta'\gamma'\,.\,\alpha^2\beta\gamma\delta^2),$$
$$\Theta' \equiv -4\Sigma\,(\alpha^2\beta'^2\gamma'^2\delta'^2 + 2\beta\gamma\,.\,\alpha'^2\beta'\gamma'\delta'^2),$$
$$\Phi \equiv -8\Sigma\alpha^2\beta\gamma\,.\,\beta'\gamma'\delta'^2.$$

If we write $\alpha' = p\alpha$, $\beta' = q\beta$, $\gamma' = r\gamma$, $\delta' = s\delta$ and $-4\alpha^2\beta^2\gamma^2\delta^2 = k$, the invariants are expressed as homogeneous polynomials in p, q, r, s. Then writing $p = \lambda s$, $q = \mu s$, $r = \nu s$, we obtain

$$\Delta \equiv 4k, \quad \Theta \equiv ks^2(\Sigma\lambda + 1)^2, \quad \Theta' \equiv ks^6(\Sigma\mu\nu + \lambda\mu\nu)^2,$$

$$\Delta' \equiv 4ks^8\lambda^2\mu^2\nu^2, \quad \Phi \equiv 2ks^4\{(\Sigma\lambda + 1)(\Sigma\mu\nu + \lambda\mu\nu) - 4\lambda\mu\nu\}.$$

Finally, writing $\Sigma\lambda + 1 = 2u$, $\Sigma\mu\nu + \lambda\mu\nu = 2v$, $\lambda\mu\nu = w$, we have

$$\Delta \equiv 4k, \quad \Theta \equiv 4ks^2u^2, \quad \Theta' \equiv 4ks^6v^2,$$

$$\Delta' \equiv 4ks^8w^2, \quad \Phi \equiv 8ks^4(uv - w).$$

Then we have the absolute invariants

$$\Theta\Theta'/\Delta\Delta' \equiv u^2v^2/w^2, \quad \Phi^2/\Delta\Delta' \equiv 4(uv - w)^2/w^2.$$

Eliminating uv/w we have

$$\Phi^2 = 4\{(\Theta\Theta')^{\frac{1}{2}} - (\Delta\Delta')^{\frac{1}{2}}\}^2,$$

or rationalising

$$(4\Delta\Delta' + 4\Theta\Theta' - \Phi^2)^2 = 64\Delta\Delta'\Theta\Theta'.$$

15·35. Contact of quadrics.

The complete conditions for the various sorts of contact of two quadrics require invariant factors, as we have seen in 13·8, but certain conditions can be expressed in terms of the invariants Δ, Θ, Φ, Θ', Δ'. These are the conditions which depend only on equalities among the roots of the quartic equation $\Delta(\lambda) = 0$. Let us write this equation in the form

$$\Delta(\lambda) \equiv a_0\lambda^4 + 4a_1\lambda^3 + 6a_2\lambda^2 + 4a_3\lambda + a_4 = 0.$$

Then it may be transformed by the substitution $a_0\lambda + a_1 = \mu$ to the form

$$\mu^4 + 6H\mu^2 + 4G\mu + (a_0^2 I - 3H^2) = 0,$$

where
$$H \equiv a_0 a_2 - a_1^2,$$
$$G \equiv 2a_1^3 - 3a_0 a_1 a_2 + a_0^2 a_3,$$
$$I \equiv a_0 a_4 - 4a_1 a_3 + 3a_2^2.$$

15·351. The condition for *simple contact* (denoted by [211] in invariant-factor notation) is that two roots of the equation $\Delta(\lambda) = 0$ should be equal. This is expressed by the vanishing of the discriminant $I^3 - 27J^2 = 0$, where

$$J \equiv a_0 a_2 a_4 - a_0 a_3^2 + 2a_1 a_2 a_3 - a_1^2 a_4 - a_2^3.$$

This discriminant is called a *tact-invariant*.

The same condition is satisfied in the case of *double contact* [(11)11], and the two cases have to be distinguished by considering the rank of the matrix [$\Delta(\lambda)$]. In the case of double contact the curve of intersection of the two quadrics has two double-points and breaks up into two conics. For some value of λ, therefore, $\lambda S + S' = 0$ represents two planes. The conditions for this are that for some value of λ the matrix [$\Delta(\lambda)$] should be of rank 2.

15·352. The conditions for three equal roots λ_1 are $I = 0, J = 0$. This happens in the cases of *ring-contact* and the two forms of *stationary contact*. In the first form of stationary contact [31], when the curve of intersection is a cuspidal quartic, the matrix [$\Delta(\lambda_1)$] is of rank 3; when the curve of intersection consists of two conics touching one another [(21)1] the matrix is of rank 2; and in the case of ring-contact [(111)1] it is of rank 1.

15·353. The conditions for two pairs of equal roots λ_1 and λ_2 are $G = 0$, $12H^2 = a_0^2 I$. (This condition is symmetrical; with $G = 0$ the second condition is equivalent to

$$2a_3{}^3 - 3a_2 a_3 a_4 + a_1 a_4{}^2 = 0.)$$

In this case the quadrics have at least one generator in common. When the matrices [$\Delta(\lambda_1)$] and [$\Delta(\lambda_2)$] are both of rank 3 there is just one common generator, the remainder of the curve of intersection being a space-cubic of which the generator is a bisecant, and the quadrics have *double contact* [22]. When [$\Delta(\lambda_1)$] is of rank 3 and [$\Delta(\lambda_2)$] is of rank 2, there are two common intersecting generators and the quadrics have *triple contact* [2(11)]. When [$\Delta(\lambda_1)$] and [$\Delta(\lambda_2)$] are both of rank 2, the quadrics have four common generators and have *quadruple contact* [(11)(11)].

15·354. The conditions for four equal roots λ_1 are

$$a_1/a_0 = a_2/a_1 = a_3/a_2 = a_4/a_3.$$

When [$\Delta(\lambda_1)$] is of rank 3 the curve of intersection consists of a space-cubic and a tangent, and the quadrics have stationary contact at the point of contact [4]. When [$\Delta(\lambda_1)$] is of rank 2, there are two cases according as $\lambda - \lambda_1$ or $(\lambda - \lambda_1)^2$ is a factor of each of the first minors. In the former case the curve of intersection consists of a conic and two lines intersecting upon it; the two quadrics touch at this point [(31)]. In the latter case the base-curve consists of a double-line and two mutually skew lines both cutting it; the quadrics touch along the double-line [(22)]. When $[\Delta(\lambda_1)]$ is of rank 1 the base-curve consists of two inter-secting double-lines, and the quadrics have contact along them both [(211)].

Ex. 1. Show that if two quadrics have simple contact and the generating lines through the point of contact are harmonic, $\Theta\Theta' = 4\Delta\Delta'$ in addition to the condition for simple contact.

Ex. 2. In a pencil of quadrics having double contact show that each quadric is paired with another, such that the generating lines at either point of contact are harmonic.

15·36. In interpreting the meanings of the simultaneous in-variants we have assumed that neither of the quadrics is a cone. Let us consider the case in which $\Delta' = 0$, so that S' is a cone, say

$$S' \equiv x^2 + y^2 + z^2 = 0,$$

the vertex being $D \equiv [0, 0, 0, 1]$, and the edges through D forming a self-conjugate trihedron. Assume S to be general:

$$S \equiv ax^2 + \dots + 2fyz + \dots + 2pxw + \dots = 0.$$

Forming the linear system $S + \lambda S' = 0$ the discriminant equation is

$$\Delta(\lambda) \equiv d\lambda^3 + \lambda^2 \Sigma (ad - p^2) + \lambda \Sigma A + \Delta = 0,$$

so that $\Theta' \equiv d$, $\Phi \equiv \Sigma(ad - p^2)$, $\Theta \equiv \Sigma A$.

$\Theta' = 0$ when the vertex of S' lies on S.

The invariants Θ and Φ can be interpreted by considering the tangent-cone T from D to S. This is

$$T \equiv d(ax^2 + by^2 + cz^2 + 2fyz + 2gzx + 2hxy) - (px + qy + rz)^2 = 0.$$

T and S' are homogeneous in x, y, z, and with $w = 0$ represent two conics C and C'. We can form the simultaneous invariants of these two conics. The discriminant equation of the linear system $T + \lambda S' = 0$ is

$$\Delta_1'\lambda^3 + \Theta_1'\lambda^2 + \Theta_1\lambda + \Delta_1 = 0,$$

where $\Delta_1' \equiv 1$, $\Theta_1' \equiv \Sigma(ad - p^2) \equiv \Phi$,

$\Theta_1 \equiv \Sigma\{(bd - q^2)(cd - r^2) - (fd - qr)^2\} \equiv d\Theta$,

$\Delta_1 \equiv d^2\Delta$.

(The last result is obtained by multiplying together the two determinants

$$\begin{vmatrix} d & o & o & -p \\ o & d & o & -q \\ o & o & d & -r \\ o & o & o & 1 \end{vmatrix} \cdot \begin{vmatrix} a & h & g & p \\ h & b & f & q \\ g & f & c & r \\ p & q & r & d \end{vmatrix} \cdot)$$

If $\Theta = o$, so that $\Theta_1 = o$, the conic C is inpolar to C'. Hence there is an infinity of trihedra with vertex D, self-conjugate with respect to the cone S' and having their faces touching the quadric S.

If $\Phi = o$, so that $\Theta_1' = o$, there is an infinity of trihedra with vertex D, self-conjugate with respect to the cone S' and having their edges touching the quadric S.

15·37. Reciprocally, if one of the quadrics Σ' reduces as an envelope to a conic C' in the plane $w = o$, its tangential equation referred to a self-polar triangle is

$$\Sigma' \equiv \xi^2 + \eta^2 + \zeta^2 = o.$$

The general quadric S cuts this plane in a conic C.

The discriminant equation for the linear system $\Sigma + \lambda\Sigma' = o$ is (cf. 15·32)

$$D\lambda^3 + \lambda^2 D\Sigma(bc - f^2) + \lambda\Delta^2\Sigma a + \Delta^3 = o,$$

so that $\Theta' \equiv \Sigma a$, $\Delta'\Phi \equiv \Sigma(bc - f^2)$, $\Delta'^2\Theta \equiv D$.

If $\Theta = o$ the plane of the conic C' touches the quadric S.

If $\Phi = o$ there is an infinity of triangles in the plane of the conic self-conjugate with respect to the conic Σ' and circumscribed about C.

If $\Theta' = o$ there is an infinity of triangles self-conjugate with respect to Σ' and inscribed in C.

Ex. 1. If S and S' are both cones, not having a common vertex, show that $\Theta = o$ is the condition that the vertex of S lies on S', and if $\Phi = o$ the tangent-planes to the two cones through the line joining their vertices are harmonic.

Ex. 2. If S' breaks up into two planes show that $\Theta = o$ is the condition that the two planes should be conjugate with respect to S, and that $\Phi = o$ is the condition that the line of intersection of the two planes should touch S.

Ex. 3. If every quadric of the linear system $S+\lambda S'=0$ is a cone (not with common vertex) show that all the cones pass through a fixed conic C and have their vertices on a fixed line l which cuts C.

Ex. 4. If $\Sigma+\lambda\Sigma'=0$ represents a system of conics (not all in one plane) show that all the conics lie on a fixed cone C and that their planes all pass through a fixed line l which touches C.

15·4. Metrical applications.

A metrical invariant of a quadric is a function of the co-efficients which is unaltered by an *orthogonal transformation*. The orthogonal transformation, equivalent to transformation from one set of rectangular cartesian coordinates to another, is a particular case of the general linear transformation, and is characterised by the property that it preserves the circle at infinity unaltered.

15·41. The general linear or projective transformation transforms the point-coordinates (x_i) into (x_i')

$$x_i' = \sum_{r=0}^{3} l_{ir}x_r \quad (i=0, 1, 2, 3),$$

and the inverse transformation by which (x_i) are expressed in terms of (x_i') is

$$Lx_i = \sum_{r=0}^{3} L_{ri}x_r' \quad (i=0, 1, 2, 3),$$

where L is the determinant of the coefficients l_{ij}, and L_{ij} is the cofactor of l_{ij}. L must not vanish.

If $x_0=0$ represents the plane at infinity, after the transformation this becomes

$$\Sigma L_{r0}x_r' = 0.$$

In order that the plane at infinity should be unaltered, i.e. represented by $x_0'=0$ after the transformation, $L_{10}=0$, $L_{20}=0$, $L_{30}=0$; and as $x_0'=0$ is transformed by the inverse transformation into $x_0=0$, $l_{01}=0$, $l_{02}=0$, $l_{03}=0$. Taking $l_{00}=1$, so that $x_0=x_0'$, we can take $x_0=1$, and the equations assume the form

$$x_1' = l_{11}x_1 + l_{12}x_2 + l_{13}x_3 + l_{10},$$
$$x_2' = l_{21}x_1 + l_{22}x_2 + l_{23}x_3 + l_{20},$$
$$x_3' = l_{31}x_1 + l_{32}x_2 + l_{33}x_3 + l_{30},$$

which are the general equations of transformation of cartesian coordinates, the plane at infinity remaining unaltered. This is called the general *affine* transformation.

15·42. The circle at infinity is represented by the tangential equation $\xi_1^2 + \xi_2^2 + \xi_3^2 = 0$, and we have to find the equations of transformation of the plane-coordinates ξ_0, ξ_1, ξ_2, ξ_3.

The plane $\Sigma \xi_r x_r = 0$ is transformed into the plane

$$\sum_r \{\xi_r (\sum_i L_{ir} x_i')\} = 0 \quad \text{or} \quad \Sigma \xi_i' x_i' = 0.$$

Hence

$$\xi_i' = \sum_r L_{ir} \xi_r,$$

and, inversely,

$$L^3 \xi_i = \Sigma l_{ri} \xi_r'.$$

Now if $\xi_1^2 + \xi_2^2 + \xi_3^2 = 0$ is transformed into $\xi_1'^2 + \xi_2'^2 + \xi_3'^2 = 0$ we have

$$\sum_{i=1}^{3} l_{1i}^2 = \Sigma l_{2i}^2 = \Sigma l_{3i}^2$$

and

$$\Sigma l_{2i} l_{3i} = \Sigma l_{3i} l_{1i} = \Sigma l_{1i} l_{2i} = 0.$$

But these are just the conditions that the transformation should be orthogonal.

15·43. The metrical properties of a quadric should therefore be expressible in terms of the simultaneous invariants of the quadric and the circle at infinity considered as a specialised quadric. Now taking the circle at infinity as $\xi^2 + \eta^2 + \zeta^2 = 0$, which is equivalent to assuming rectangular cartesian coordinates, we found in 15·37 the simultaneous invariants

$$D, \quad \Sigma(bc - f^2), \quad \text{and} \quad \Sigma a,$$

and these are exactly the metrical invariants which were found in 8·82.

These are really simultaneous invariants of the circle at infinity Ω and the conic at infinity C on the quadric, and we shall denote them by Δ_0, Θ_0 and Θ_0'.

15·44. Thus $\Sigma a = 0$, or $\Theta_0' = 0$, is the condition that there should be an infinity of triangles inscribed in C and self-polar with respect to Ω. But since two lines are at right angles when their points at infinity are conjugate with respect to Ω, the quadric has an infinity of sets of three mutually rectangular generators of each system (*rectangular hyperboloid*).

When $\Sigma(bc - f^2) = 0$, or $\Theta_0 = 0$, there is an infinity of triangles circumscribed about C and self-polar with respect to Ω, and hence the asymptotic cone of the quadric has an infinity of sets of three mutually orthogonal tangent-planes (*orthogonal hyperboloid*).

15·45. There is another special form of hyperboloid to which the name orthogonal was given by Schröter, which we shall now explain.

In general a quadric possesses no generating line which is perpendicular to a plane of circular section, but if it has one pair of parallel generators per-pendicular to one set of circular sections, there is a pair of parallel generators perpendicular to the com-plementary circular sec-tions also. This follows from a well-known theorem for the conic. Let Ω be the circle at infinity, AB and $A'B'$ two chords, C

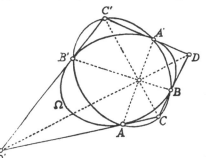

Fig. 46

and C' their poles. Then the inscribed quadrangle $ABA'B'$ and the circumscribed quadrilateral $CDC'D'$ formed by the tangents at A, B, A', B' have the same harmonic triangle, and therefore AA', BB', CC' and DD' are concurrent. Hence by the converse of Pascal's theorem the six points $AA'C'B'BC$ lie on one conic.

Let S be any quadric which contains this conic. The planes through AB and $A'B'$ are complementary planes of circular section; the two generators through C are perpendicular to the former, and those through C' to the latter. When $ABCA'B'C'$ is a proper conic and the planes of circular section, and the corresponding generators, are real, the quadric is either a cone or a hyperboloid of one sheet. We shall call such quadrics *orthocyclic* cones and hyperboloids. If AB and $A'B'$ are con-jugate lines with respect to Ω, so that AB passes through C' and $A'B'$ through C, the conic $ABCA'B'C'$ reduces to these two lines, and the quadric is either a *rectangular hyperbolic paraboloid*, a rectangular hyperbolic cylinder, or two orthogonal planes.

There are no proper circular sections, however, in this case; a circle reduces to a generating line and a line at infinity lying in the same plane. In the case of the paraboloid all the generators of one system are parallel to one plane and there is one generator of the other system perpendicular to this plane. In the case of the cylinder the only generator through C is the line at infinity $CA'B'$, and the property in question fails altogether.

If C denotes the conic at infinity and Σ' the circle at infinity, the invariant equation corresponding to this property is

$$\Theta_0'^3 + 8\Delta_0'^2\Delta_0 = 4\Delta_0'\Theta_0\Theta_0'.$$

Ex. 1. Show that the condition that the quadric

$$ax^2 + by^2 + cz^2 = 1$$

should be orthocyclic is $(-a+b+c)(a-b+c)(a+b-c)=0$.

Ex. 2. Prove that the locus of a point whose distances from two skew lines have a constant ratio $k\,(\neq 1)$ is an orthocyclic hyperboloid; and that if $k=1$ the locus is a rectangular hyperbolic paraboloid. If the two lines intersect, the locus is a cone, or (if $k=1$) two orthogonal planes.

Ex. 3. If the squares of the distances of a variable point from two fixed lines satisfy a linear equation, show that the locus is an orthocyclic quadric, which may be an ellipsoid, a hyperboloid of one or of two sheets, or a hyperbolic paraboloid. (If the two lines are conjugate imaginaries the locus may also be an elliptic paraboloid.)

15·46. The theorem in plane geometry which is reciprocal to the theorem on which the property of orthocyclic quadrics is based is that if AB and $A'B'$ are the chords of contact of tangents drawn from any two points C and C' to a given conic the six sides of the two triangles ABC and $A'B'C'$ all touch one conic.

Let the given conic be the circle at infinity and let S be any quadric which contains the conic which touches the six lines, O the vertex of its asymptotic cone. Then OC and OC' are focal lines of the cone, since the tangent-planes through these lines touch the absolute. Also OAB is a tangent-plane of the cone and is perpendicular to OC. Hence the cone has the property that one pair of focal lines are perpendicular to tangent-planes. In general a cone does not possess a focal line which is perpendicular to a tangent-plane. The above theorem shows that if it possesses one then it possesses a pair. These focal lines are then generators of the reciprocal cone. We shall call such a cone *orthofocal*, and likewise any quadric of which it is the asymptotic cone. The reciprocal cone passes through $ABCA'B'C'$ and is therefore *orthocyclic*. As the equation of the reciprocal cone is

$$Ax^2 + \ldots + 2Fyz + \ldots = 0,$$

where $A \equiv bc - f^2, \ldots, F \equiv gh - af, \ldots,$

the condition for an orthofocal cone is

$$(-A + B + C)(A - B + C)(A + B - C) = 0.$$

The relation between the reciprocal cones is brought out more closely by considering their intersections with a sphere whose centre is at the vertex, *sphero-conics*. In the case of an orthocyclic cone the ratio of the distances of any point from two fixed intersecting lines is constant. Hence on the sphere we have two fixed points M, N and a variable point P such that

$$\sin MOP : \sin NOP = \text{const.}$$

For an orthofocal cone we have reciprocally two fixed great circles on the sphere and a variable great circle which is such that the ratio of the sines of the angles which it makes with the fixed great circles is constant. Hence an orthofocal cone with vertex O can be generated by a plane through O which moves so that the ratio of the sines of the angles which it makes with two fixed planes through O is constant.

Ex. Prove that the envelope of a plane through O which moves so that the ratio of the sines of the angles which it makes with the two fixed planes $y = \pm \mu x$ is constant is

$$\xi^2 + \mu^2 \eta^2 + (1 + \mu^2)\zeta^2 + 2k\xi\eta = 0.$$

15·51. Contravariants.

An arbitrary plane $\xi x + \eta y + \zeta z + \omega w = 0$ cuts a linear system of quadrics $S + \lambda S' = 0$ in a linear system of conics $C + \lambda C' = 0$. The projective properties of this system are the same as those of any system into which it may be projected. Thus if we eliminate w between the equation of the plane and the equation of the quadric we obtain a homogeneous equation in x, y, z which represents the projection from the vertex W upon the plane $w = 0$. The coefficients are expressions of the second degree in ξ, η, ζ, ω, but linear in the coefficients of the quadric.

The discriminant of the conic $C + \lambda C' = 0$, which expresses that the plane (ξ) should touch the quadric $S + \lambda S' = 0$, is of the form

$$\Sigma + \tau\lambda + \tau'\lambda^2 + \Sigma'\lambda^3 = 0,$$

where Σ, τ, τ', Σ' are expressions of the second degree in ξ, η, ζ, ω, and of degrees 3, 0; 2, 1; 1, 2; 0, 3 in the coefficients of S and S' respectively. Thus if

$$S \equiv ax^2 + by^2 + cz^2 + dw^2, \quad S' \equiv x^2 + y^2 + z^2 + w^2,$$

we find, after dividing by w^4,

$$\Sigma \equiv bcd\xi^2 + acd\eta^2 + abd\zeta^2 + abc\omega^2,$$

$$\tau \equiv (bc + bd + cd)\xi^2 + (ac + ad + cd)\eta^2 + (ab + ad + bd)\zeta^2 + (bc + ca + ab)\omega^2,$$

$$\tau' \equiv (b + c + d)\xi^2 + (a + c + d)\eta^2 + (a + b + d)\zeta^2 + (a + b + c)\omega^2,$$

$$\Sigma' \equiv \xi^2 + \eta^2 + \zeta^2 + \omega^2.$$

$\Sigma = 0$ is the condition that the plane (ξ) should touch the quadric S, and is the tangential equation of S; similarly $\Sigma' = 0$ is the tangential equation of S'. $\tau = 0$ is the condition that the conic C should be outpolar to the conic C', and is the tangential equation of a quadric projectively associated with the two given quadrics. Similarly $\tau' = 0$ is the tangential equation of a quadric which is the envelope of planes cutting S and S' in conics C and C' such that C' is outpolar to C. These envelopes, which are thus associated with S and S', are called simultaneous *contravariants* of S and S'. Any equation homogeneous in Σ, τ, τ', Σ' and involving the invariants of the two quadrics, homogeneous in the coefficients of both quadrics, is a simultaneous contravariant and represents some envelope projectively associated with the two quadrics.

15·52. Tangential equation of the curve of intersection of two quadrics.

As an example let us find the condition that the plane (ξ) should contain a tangent to the curve of intersection (SS') of the two quadrics. The four points of intersection of the conics C and C' are the points in which the plane (ξ) cuts the curve (SS'). If C and C' have simple contact the plane (ξ) contains a tangent-line of (SS'). The condition for this, which is the discriminant of the cubic

$$\Sigma + \tau\lambda + \tau'\lambda^2 + \Sigma'\lambda^3 = 0,$$

viz. $$(9\Sigma\Sigma' - \tau\tau')^2 = 4(3\Sigma\tau' - \tau^2)(3\Sigma'\tau - \tau'^2),$$

is therefore the tangential equation of the quartic curve (SS'), an equation of the eighth degree in (ξ) and of the sixth degree in the coefficients of each of the quadrics.

Ex. Show that the osculating planes of the curve of intersection of the two quadrics satisfy the equations

$$3\Sigma\tau' = \tau^2, \quad 3\Sigma'\tau = \tau'^2.$$

15·53. Covariants.

Reciprocally, the tangent-cones from an arbitrary point (x) to a linear tangential system of quadrics $\Sigma + \lambda\Sigma' = 0$ form a linear system which is projectively the same as that represented by the equation $\omega = 0$ together with the equation obtained by eliminating ω between the equation of the point $x\xi + \ldots = 0$ and the tangential equation of the quadric. We thus get an equation $K + \lambda K' = 0$ in ξ, η, ζ. Taking the canonical forms

$$\Sigma \equiv bcd\xi^2 + acd\eta^2 + abd\zeta^2 + abc\omega^2, \quad \Sigma' \equiv \xi^2 + \eta^2 + \zeta^2 + \omega^2,$$

we find, after dividing by w^4, the equation

$$\Delta^2 S + \Delta T\lambda + \Delta'T'\lambda^2 + \Delta'^2 S'\lambda^3 = 0,$$

where
$$S \equiv ax^2 + by^2 + cz^2 + dw^2,$$
$$T \equiv a(b+c+d)x^2 + \ldots,$$
$$T' \equiv a(bc+bd+cd)x^2 + \ldots,$$
$$S' \equiv x^2 + y^2 + z^2 + w^2.$$

$\Delta^2 S$, ΔT, $\Delta'T'$, $\Delta'^2 S'$ are of degrees 9, 0; 6, 3; 3, 6; 0, 9 respectively in the coefficients of S and S'.

S, T, T', S' are *covariants* of the two quadrics, S and S' being of course just the quadrics themselves.

Ex. Show that the point-equation of the circumdevelopable of the two quadrics is

$$(9\Delta\Delta'SS' - TT')^2 = 4(3\Delta'ST' - T^2)(3\Delta S'T - T'^2),$$

an equation of degree 8 in x, y, z, w, and degree 10 in the coefficients of each of the quadrics.

15·54. The point-, line-, and tangential equations of a quadric can from a certain point of view be considered as invariants. The tangential equation, for instance, is a simultaneous invariant of the quadric and a plane

$$L \equiv \xi x + \eta y + \zeta z + \omega w = 0,$$

involving the coefficients of S in the third degree and those of L in the second, viz.

$$\begin{vmatrix} a & h & g & p & \xi \\ h & b & f & q & \eta \\ g & f & c & r & \zeta \\ p & q & r & d & \omega \\ \xi & \eta & \zeta & \omega & 0 \end{vmatrix} = 0,$$

and its vanishing is the condition that L should touch S. Similarly the point-equation is an invariant of the quadric S and a point $P \equiv x\xi + y\eta + z\zeta + w\omega = 0$, involving the coefficients of Σ in the third degree and x, y, z, w (the coefficients of P) in the second, and its vanishing is the condition that P should lie on S. The line-equation $\Psi = 0$ is a simultaneous invariant of S and two planes, involving the coefficients of S and those of each of the two planes L, L' all in the second degree; and it is also a simultaneous invariant of S and two points P, P'; its vanishing is the condition that the line of intersection of L and L' or the line joining P and P' should touch S. Any covariant or contravariant can in fact be considered as an invariant; it is a question of view-point.

15·61. Reciprocal of one quadric with respect to another.

The locus of the pole, with respect to a given quadric S_0, of a variable tangent-plane of another quadric S is a quadric.

Let
$$S \equiv ax^2 + by^2 + cz^2 + dw^2 = 0,$$
$$S_0 \equiv x^2 + y^2 + z^2 + w^2 = 0.$$

The tangential equation of the pole P of the plane (ξ') with respect to $\Sigma_0 \equiv \xi^2 + \eta^2 + \zeta^2 + \omega^2$ is

$$\xi'\xi + \eta'\eta + \zeta'\zeta + \omega'\omega = 0,$$

i.e. the coordinates of P are $[\xi', \eta', \zeta', \omega']$. But the plane (ξ') is a tangent-plane to Σ, therefore

$$bcd\xi'^2 + acd\eta'^2 + abd\zeta'^2 + abc\omega'^2 = 0,$$

i.e. P lies on the quadric

$$S' \equiv bcdx^2 + acdy^2 + abdz^2 + abcw^2 = 0.$$

Similarly it may be shown that the envelope of the polar with respect to S_0 of a variable point on S is the quadric whose tangential equation is

$$\Sigma' \equiv a\xi^2 + b\eta^2 + c\zeta^2 + d\omega^2 = 0,$$

but this is just the tangential equation of S'.

The two quadrics S and S' are symmetrically related with respect to S_0, and each is said to be the *reciprocal* of the other with respect to S_0.

15·62. The reciprocal of S with respect to S' is a simultaneous covariant of S and S', and can be expressed in terms of the covariants T, T' and the simultaneous invariants of S and S'.

Let $\quad S \equiv ax^2 + by^2 + cz^2 + dw^2$, $\quad S' \equiv x^2 + y^2 + z^2 + w^2$.

Then the reciprocal of S with respect to S' is

$$R \equiv bcdx^2 + acdy^2 + abdz^2 + abcw^2 = 0.$$

Now $\qquad\qquad T' \equiv (acd + abd + abc)x^2 + \dots$

$$\equiv \Sigma abc . \Sigma x^2 - \Sigma bcdx^2.$$

Therefore $\qquad\qquad R \equiv \Theta S' - T'.$

We verify that the expression on the right is homogeneous in the coefficients of each quadric, the degrees being 3 and 2 respectively.

Similarly the reciprocal of S' with respect to S is $\Theta'S - T = 0$.

Ex. Prove that the tangential equation of the reciprocal of S with respect to S' is

$$\Theta'\Sigma' - \Delta'\tau' = 0.$$

15·7. The harmonic complex of two quadrics.

The line-equation of the quadric S is (see 15·31)

$$\Psi \equiv c_{11}u_1^2 + \dots + C_{11}v_1^2 + \dots + k_{11}u_1v_1 + \dots + K_{11}v_1u_1 + \dots$$
$$+ 2c_{23}u_2u_3 + \dots + 2C_{23}v_2v_3 + \dots + 2k_{23}u_2v_3 + \dots$$
$$+ 2K_{23}v_2u_3 + \dots = 0;$$

this represents the quadratic complex of lines touching the quadric. Replacing S by $S + \lambda S'$ we have

$$\Psi_\lambda \equiv \Psi + \lambda \Upsilon + \lambda^2 \Psi',$$

where $\qquad \Psi' \equiv c_{11}' u_1^2 + \ldots \equiv (b'c' - f'^2) u_1^2 + \ldots$

and $\qquad \Upsilon \equiv (bc' + b'c - 2ff') u_1^2 + \ldots.$

Υ is obtained from Ψ by the usual "polarising" process represented by

$$\Upsilon \equiv \Sigma \left(a' \frac{\partial}{\partial a} \Psi \right),$$

the summation extending over all the coefficients of S.

$\Upsilon = 0$ represents a quadratic complex associated with the two quadrics. It is *the complex of lines which are cut harmonically by the two quadrics.* Let the line (p) be determined by the two points (x_1) and (x_2), so that

$$u_1 \equiv p_{23} \equiv y_1 z_2 - y_2 z_1, \text{ etc.}$$

A variable point on the line is represented by $x = x_1 + \lambda x_2$. Substituting in $S = 0$ we have

$$S_1 + 2\lambda (a x_1 x_2 + \ldots) + \lambda^2 S_2 = 0,$$

and the roots of this quadratic in λ determine the two points of intersection with S. A similar equation determines its intersections with S'. The condition that the two pairs of points should be harmonic is

$$(a x_1^2 + \ldots)(a' x_2^2 + \ldots) + (a x_2^2 + \ldots)(a' x_1^2 + \ldots)$$
$$= 2 (a x_1 x_2 + \ldots)(a' x_1 x_2 + \ldots),$$

and this reduces to

$$(bc' + b'c - 2ff')(y_1 z_2 - y_2 z_1)^2 + \ldots = 0,$$

i.e. $\Upsilon = 0$. This is called the *Harmonic Complex* or Complex of Battaglini.

Similarly the assemblage of lines through which the tangent-planes to S and S' are harmonic is another harmonic complex

$$(BC' + B'C - 2FF') p_{01}^2 + \ldots = 0.$$

A particular case of the last complex is obtained when we take Σ' as the circle at infinity. We have then the complex of lines through which the tangent-planes to the quadric S are orthogonal.

15·81. Line-equation of the curve of intersection of two quadrics.

An arbitrary line is cut in involution by the quadrics of a linear system, the double-points of the involution being the points of contact of the two quadrics of the system which touch the line. The line-equations of the two quadrics S and S' being $\Psi = 0$ and $\Psi'' = 0$ the line-equation of $S + \lambda S' = 0$ is

$$\Psi + \lambda \Upsilon + \lambda^2 \Psi'' = 0.$$

If the line-coordinates of an arbitrary line l are substituted in this equation we have a quadratic in λ which determines the two quadrics which touch l, and their points of contact are the double-points of the involution on l. But if l cuts the curve of intersection (SS') of the two quadrics, the involution degenerates, for then one point of each pair is this point of intersection. The two quadrics which touch l then coincide and

$$4\Psi\Psi'' = \Upsilon^2.$$

This is therefore the line-equation of the curve (SS'). It represents a quartic complex of lines.

15·82. If the line l is a tangent to (SS') it touches every quadric of the system and we have $\Psi = 0$, $\Psi'' = 0$, $\Upsilon = 0$, three equations in line-coordinates determining a line-series, in fact the assemblage of tangents to the curve. These tangents are also of course the generating lines of the developable belonging to the curve.

A curve and its developable can be represented in six different ways:

(1) The curve as a one-way locus of points is represented by two equations in point-coordinates $(S = 0,\ S' = 0)$.

(2) The curve as a one-way assemblage of its tangent-lines is represented by three equations in line-coordinates

$$(\Psi = 0,\ \Psi'' = 0,\ \Upsilon = 0).$$

The same equations represent the developable as a one-dimensional assemblage of its generating lines.

(3) The curve as the complex of lines passing through its points is represented by a single equation in line-coordinates $(4\Psi\Psi'' = \Upsilon^2)$. The same equation represents the developable as the complex of lines lying in its generating planes.

(4) The curve as a two-dimensional envelope of planes is represented by a single equation in tangential coordinates (15·52).

(5) The developable as a one-dimensional envelope of planes is represented by two equations in tangential coordinates

$$(3\Sigma\tau' = \tau^2, \ 3\Sigma'\tau = \tau'^2).$$

(6) The developable as a two-dimensional locus of points is represented by a single equation in point-coordinates.

We have still to find this last representation.

15·83. Point-equation of the developable belonging to the curve of intersection of two quadrics.

If P is any point on a tangent-line to the curve (SS') its polar-plane with respect to any quadric of the linear system passes through the point of contact of the tangent-line; all such planes pass through one line, which therefore cuts the curve. If we express the condition that the line of intersection of the polar-planes of P with respect to S and S' cuts (SS') we shall obtain an equation involving the coordinates of P, and this will be the point-equation of the developable.

Taking $\qquad S \equiv \Sigma a_r x_r^2, \ S' \equiv \Sigma x_r^2,$

and $P \equiv [x_0', x_1', x_2', x_3']$, we have the line of intersection of the two planes $\qquad \Sigma a_r x_r' x_r = 0, \ \Sigma x_r' x_r = 0,$

viz. $\qquad \varpi_{ij} = p_{kl} = (a_i - a_j) x_i' x_j'.$

Then $\quad \Psi \equiv \Sigma\Sigma a_i a_j p_{ij}^2, \ \Psi' \equiv \Sigma\Sigma p_{ij}^2, \ \Upsilon \equiv \Sigma\Sigma(a_i + a_j) p_{ij}^2,$

and the equation required is

$$\Upsilon^2 - 4\Psi\Psi'' = 0.$$

This has now to be expressed in terms of the point-coordinates x_r', and we can do this by means of S, S' and the covariants T, T'. Thus

$$T \equiv \Sigma a_i (a_j + a_k + a_l) x_i^2 \equiv \Sigma a_i . \Sigma a_i x_i^2 - \Sigma a_i^2 x_i^2,$$
$$T' \equiv \Sigma (a_i a_j a_k + a_i a_j a_l + a_i a_k a_l) x_i^2$$
$$\equiv \Sigma a_i a_j a_k . \Sigma x_i^2 - \Sigma a_j a_k a_l x_i^2.$$

Therefore $\qquad \Sigma a_i^2 x_i^2 \equiv \Theta' S - T,$

$$\Sigma a_j a_k a_l x_i^2 \equiv \Theta S' - T'.$$

Now

$$\Psi \equiv \Sigma\Sigma a_k a_l (a_i - a_j)^2 x_i^2 x_j^2$$
$$\equiv \Sigma\Sigma a_k a_l (a_i^2 + a_j^2) x_i^2 x_j^2 - 2a_0 a_1 a_2 a_3 \Sigma\Sigma x_i^2 x_j^2$$
$$\equiv \Sigma a_i x_i^2 . \Sigma a_j a_k a_l x_i^2 - \Delta (\Sigma x_i^2)^2,$$

therefore $\Psi \equiv S(\Theta S' - T') - \Delta S'^2.$

Similarly $\Psi'' \equiv S'(\Theta' S - T) - \Delta' S^2.$

Again, $\Upsilon \equiv \Sigma\Sigma (a_k + a_l)(a_i - a_j)^2 x_i^2 x_j^2.$

Υ is symmetrical and of degree 3 in the coefficients of each of the quadrics, and by comparing dimensions we see that it can be expressed only in terms of $(ST + S'T')$, $\Phi SS'$, and $(\Theta S'^2 + \Theta' S^2)$. Comparing coefficients we find

$$\Upsilon \equiv \Phi SS' - (ST + S'T').$$

Hence the point-equation of the developable is

$$(\Phi SS' - ST - S'T')^2$$
$$= 4(\Theta SS' - ST - \Delta S'^2)(\Theta' SS' - S'T - \Delta' S^2).$$

15·91. Conjugate generators of a quadric.

We shall consider the generators of the quadric S which are conjugate with respect to another quadric S'.

Let $S \equiv ax^2 + by^2 + cz^2 + dw^2 = 0$

and $S' \equiv x^2 + y^2 + z^2 + w^2 = 0.$

The generators of one system of S are

$$\sqrt{b}.y + \sqrt{-c}.z = \lambda(\sqrt{-a}.x + \sqrt{d}.w),$$
$$\lambda(\sqrt{b}.y - \sqrt{-c}.z) = \sqrt{-a}.x - \sqrt{d}.w.$$

The line-coordinates are thus

$$p_{23} = 2\lambda \sqrt{(-ad)}, \qquad p_{01} = -2\lambda \sqrt{(-bc)},$$
$$p_{31} = -(\lambda^2 + 1)\sqrt{(bd)}, \qquad p_{02} = (\lambda^2 + 1)\sqrt{(ca)},$$
$$p_{12} = (\lambda^2 - 1)\sqrt{(-cd)}, \qquad p_{03} = -(\lambda^2 - 1)\sqrt{(-ab)}.$$

The polar of

$$[p_{23}, \ldots, p_{01}, \ldots]$$

with respect to S' is

$$[p_{01}, \ldots, p_{23}, \ldots],$$

and the condition that two lines (p), (p') should be conjugate, or that (p) should intersect the polar of (p'), is

$$p_{01}p_{01}' + \ldots + p_{23}p_{23}' + \ldots = 0,$$

i.e.

$$-4(bc+ad)\lambda\lambda' + (ca+bd)(\lambda^2+1)(\lambda'^2+1)$$
$$-(ab+cd)(\lambda^2-1)(\lambda'^2-1) = 0.$$

Write for shortness $bc+ad=l$, $ca+bd=m$, $ab+cd=n$, then we have a $(2, 2)$ symmetrical relation between λ, λ'

$$(m-n)\lambda^2\lambda'^2 + (m+n)(\lambda+\lambda')^2$$
$$-2(m+n+2l)\lambda\lambda' + (m-n) = 0.$$

If we put $\lambda'=\lambda$ we get an equation of the fourth degree in λ

$$(m-n)(\lambda^4+1) + 2(m+n-2l)\lambda^2 = 0.$$

Hence *there are four self-conjugate generators*.

In general there are two generators conjugate to a given generator λ', corresponding to the roots of the quadratic equation in λ

$$\{(m-n)\lambda'^2 + (m+n)\}\lambda^2 - 4l\lambda'\lambda + \{(m+n)\lambda'^2 + (m-n)\} = 0.$$

If the two generators which are conjugate to λ' are conjugate to one another, the roots λ_2, λ_3 of this equation must be connected by the same equation, i.e.

$$(m-n)\lambda_2^2\lambda_3^2 + (m+n)(\lambda_2+\lambda_3)^2 - 2(m+n+2l)\lambda_2\lambda_3 + (m-n) = 0.$$

On substituting the values of the symmetric functions $\lambda_2+\lambda_3$ and $\lambda_2\lambda_3$ we obtain the equation

$$\{(m-n)(\lambda'^4+1) + 2(m+n-2l)\lambda'^2\}(mn+nl+lm) = 0.$$

If $\Sigma mn \neq 0$ we obtain again the quartic equation giving the four self-conjugate generators, and there are no sets of three generators mutually conjugate. But if $\Sigma mn = 0$ the equation is identically satisfied and there is an infinity of sets of three mutually conjugate generators. The condition $\Sigma mn = 0$ is now easily identified with $\Theta\Theta' - 4\Delta\Delta' = 0$, for

$$\Sigma mn = \Sigma a^2 bc = \Sigma a \cdot \Sigma bcd - 4abcd.$$

Hence, since the condition is symmetrical, $\Theta\Theta' - 4\Delta\Delta' = 0$ *is the necessary and sufficient condition that each of the quadrics S, S' should have an infinity of triads of generators of each system mutually conjugate with regard to the other quadric.*

The cross-ratio of the four generators of a quadric S which are self-conjugate with respect to a quadric S' is evidently a simultaneous invariant. In particular the condition that it should have the value -1 is the vanishing of the invariant J of the quartic equation*. This gives

$$(m+n-2l)(n+l-2m)(l+m-2n)=0,$$

which is equivalent to

$$2\Phi^3 - 9\Theta\Theta'\Phi - 72\Delta\Delta'\Phi + 27(\Delta\Theta'^2 + \Delta'\Theta^2) = 0.$$

If the roots of the quartic form an equianharmonic tetrad, with cross-ratio a complex cube root of unity, the invariant $I=0$. This gives

$$\Sigma(l^2 - mn) = 0,$$

which is equivalent to

$$\Phi^2 - 3\Theta\Theta' + 12\Delta\Delta' = 0.$$

15·92. The results of the last section can be interpreted in an interesting way in non-euclidean geometry in which the circle at infinity (a degenerate quadric) is replaced by a proper quadric $S' \equiv x^2 + y^2 + z^2 + w^2 = 0$. Two lines which are conjugate with respect to S' are perpendicular, and we have the result that in the general quadric there are four lines (imaginary) which are self-orthogonal, and when a certain condition is satisfied the quadric contains an infinity of triads of mutually rectangular generators, the non-euclidean rectangular hyperboloid.

In euclidean geometry the general quadric has four self-orthogonal generators of each system. These are the (imaginary) generators which pass through the four points of intersection of the conic at infinity with the circle at infinity.

If the four points of intersection of the circle at infinity with the conic at infinity form a harmonic set on the conic at infinity, one pair of common chords of the conic and circle at infinity are conjugate with respect to the conic. In this case one pair of complementary planes of circular section are conjugate with respect to the quadric. If the four points form a harmonic

---

* See *Analytical Conics*, chap. XIX, § 19.

tetrad on the circle at infinity, one pair of complementary planes of circular section are orthogonal.

Ex. If Δ_0, Θ_0, Θ_0', Δ_0' are the simultaneous invariants of the conic at infinity C and the circle at infinity C', show that the condition that the four self-orthogonal generators of S should be harmonic is

$$2\Theta_0{}^3 - 9\Delta_0\Theta_0\Theta_0' + 27\Delta_0{}^2\Delta_0' = 0.$$

15·95. EXAMPLES.

1. If two quadrics have a common generator, show that

$$\Delta\Theta'^2 = \Delta'\Theta^2.$$

2. If two quadrics have in common two generators of one system and one generator of the other system, show that

$$\Delta\Theta'^2 = \Delta'\Theta^2$$

and

$$(\Theta\Theta' + 8\Delta\Delta')^2 = 16\Phi^2\Delta\Delta'.$$

3. If two quadrics touch along a generator and have also two generators of the other system in common, show that

$$4\Delta/\Theta = 3\Theta/2\Phi = 2\Phi/3\Theta' = \Theta'/4\Delta'.$$

4. If a hexahedron, which is a projection of a parallelepiped, can be inscribed in the quadric S' and circumscribed about the quadric S, prove that

$$16\Delta^3\Delta' - 8\Delta^2\Theta\Theta' + 4\Delta\Theta^2\Phi - \Theta^4 = 0.$$

5. If two spheres are orthogonal show that $4\Delta\Delta' = \Theta\Theta'$, and conversely if $4\Delta\Delta' = \Theta\Theta'$ either the spheres are orthogonal or $d^2 = 3(r^2 + r'^2)$, where r and r' are their radii and d the distance between their centres.

6. Prove that the volume of an ellipsoid is $\frac{4}{3}\pi(-\Delta^3/D^4)^{\frac{1}{2}}$.

7. Show that

$$x^2 + y^2(1 - t^2) \pm 2(tyz + ax) = 0$$

represents two hyperboloids of revolution having contact along the generator $x = 0 = y$; and that if $t = 1$ they have contact along two generators. [This arrangement affords a system of bevel gearing, the axes of rotation being inclined at the angle $2\tan^{-1} t$.]

CHAPTER XVI

LINE GEOMETRY

16·11. A line is determined uniquely by either two points or two planes, and in either case we have a four-by-two matrix, the coordinates of the two points and the coordinates of the two planes

$$\begin{bmatrix} x_0 & x_1 & x_2 & x_3 \\ y_0 & y_1 & y_2 & y_3 \end{bmatrix} \quad \text{and} \quad \begin{bmatrix} \xi_0 & \xi_1 & \xi_2 & \xi_3 \\ \eta_0 & \eta_1 & \eta_2 & \eta_3 \end{bmatrix}.$$

We denote the six determinants $x_i y_j - x_j y_i$ by p_{ij}, and the six determinants $\xi_i \eta_j - \xi_j \eta_i$ by ϖ_{ij}. From the fundamental incidence relation

$$\xi_0 x_0 + \xi_1 x_1 + \xi_2 x_2 + \xi_3 x_3 = 0$$

which expresses that the point (x) lies on the plane (ξ) we derive the relations (2·523)

$$\varpi_{01} : \varpi_{02} : \varpi_{03} : \varpi_{23} : \varpi_{31} : \varpi_{12} = p_{23} : p_{31} : p_{12} : p_{01} : p_{02} : p_{03},$$

so that the ratios of one set of numbers (p) determine the ratios of the other set (ϖ).

Further, since the determinant

$$\begin{vmatrix} x_0 & x_1 & x_2 & x_3 \\ y_0 & y_1 & y_2 & y_3 \\ x_0 & x_1 & x_2 & x_3 \\ y_0 & y_1 & y_2 & y_3 \end{vmatrix}$$

is identically zero, we have the identical relation

$$\omega(p_{23}, \ldots, p_{01}, \ldots) = p_{01}p_{23} + p_{02}p_{31} + p_{03}p_{12} = 0.$$

The set of six numbers (p) connected by this identical relation are Plücker's coordinates of the line; the numbers (ϖ) are sometimes, to distinguish them, called the *axial coordinates*, and (p) the *ray coordinates*.

16·12. Two lines have an incidence relation when they have a point in common and therefore also a common plane. Expanding the determinant

$$\begin{vmatrix} x_0 & x_1 & x_2 & x_3 \\ y_0 & y_1 & y_2 & y_3 \\ x_0' & x_1' & x_2' & x_3' \\ y_0' & y_1' & y_2' & y_3' \end{vmatrix},$$

which vanishes identically, we have

$$p_{01}p_{23}' + p_{02}p_{31}' + p_{03}p_{12}' + p_{23}p_{01}' + p_{31}p_{02}' + p_{12}p_{03}' = 0,$$

which is
$$\Sigma p_{23}' \frac{\partial\omega}{\partial p_{23}} = 0.$$

Conversely, if (p) and (p') are the coordinates of two lines which are connected by this equation, the lines intersect.

16·13. A single homogeneous equation in (p) represents a three-dimensional assemblage of lines or *complex*; two equations represent a two-dimensional assemblage or *congruence*; three equations determine a *line-series* such as the regulus of a quadric, or the tangents to a curve; and four equations determine a finite number of lines.

16·14. If a line is given to pass through a fixed point it is deprived of two degrees of freedom. The conditions that the line (p) should pass through the point (x) are any two of the linear equations
$$x_i p_{jk} + x_j p_{ki} + x_k p_{ij} = 0,$$
where i, j, k are given any three of the values 0, 1, 2, 3.

Similarly, if a line is given to lie in a fixed plane it is deprived of two degrees of freedom. The conditions that the line (ϖ) should lie in the plane (ξ) are any two of the linear equations
$$\xi_i \varpi_{jk} + \xi_j \varpi_{ki} + \xi_k \varpi_{ij} = 0.$$

16·15. The lines of a complex which pass through a given point form a one-dimensional assemblage or cone. Let the equation of the complex be $f(p_{23}, \ldots, p_{01}, \ldots) = 0$, homogeneous and of the nth degree in (p). If we substitute $p_{ij} = x_i y_j - x_j y_i$ in the equation we obtain an equation which is homogeneous and of degree n in the coordinates (x) and also in the coordinates (y). If (y) is fixed, the equation then represents a cone of order n. Similarly the lines of a complex which lie in a given plane envelop a curve. If we substitute $\varpi_{ij} = \xi_i \eta_j - \xi_j \eta_i$ we obtain an equation of degree n in (ξ) and also in (η), and if (η) is fixed the equation represents a curve of class n lying in this plane.

The number n is called the *degree* of the complex and is thus both the order of the cone formed by all the lines through an arbitrary fixed point and the class of the curve formed by all the lines lying in an arbitrary fixed plane.

16·16. The linear complex.

A complex of degree 1 is called a *linear complex*. All the lines through a given point P lie in a plane π, and all the lines in a given plane π pass through a point P, in each case forming a plane pencil. The point and plane so associated are called *pole* and *polar* with respect to the complex.

The general equation of the linear complex is

$$\Sigma a_{ij}p_{ij}=0$$

and thus depends upon the ratios of six numbers a_{ij}. There is no loss of generality in making the convention that

$$a_{ij}=-a_{ji}, \quad a_{ii}=0.$$

With this convention the equation when written in terms of the coordinates of two points (x), (y) is

$$\Sigma a_{ij}x_iy_j=0.$$

If (y) is fixed this equation represents the polar-plane of (y). Since the equation is skew symmetrical in (x) and (y) it follows that if the polar-plane of P passes through Q, that of Q passes through P. If (y) belongs to a linear range of points $(y+\lambda z)$ the polar-planes form an axial pencil

$$\Sigma a_{ij}x_iy_j+\lambda\Sigma a_{ij}x_iz_j=0.$$

Thus to a line l as a range of points corresponds a line l' as the axis of a pencil of planes. If P and Q are points on l, and P' and Q' points on l', the polar-planes of P' and Q' both pass through P and Q, hence the relation between l and l' is symmetrical; each line is called the *polar* of the other.

Two polar lines do not intersect, for if l cuts l', and P is any point on l, the polar-plane of P contains P and the line l' and is therefore the fixed plane (Pl') or (ll'), unless l and l' coincide; and if l and l' coincide, the polar-plane of any point on l contains l and therefore l belongs to the complex.

Any line of the complex which meets a line l meets also its polar l', for if p cuts l in P the polar of P contains p and l'; thus p and l' lie in one plane and therefore intersect.

Ex. Prove that the polar of the line (q) with respect to the linear complex $\Sigma a_{ij}p_{ij}=0$ is

$$q_{ij}'=a_{kl}\Sigma a_{rs}q_{rs}-q_{ij}\omega(a_{rs}).$$

16·2. Line-geometry is perhaps best studied in connection with geometry of higher dimensions. The geometry of ordinary space is said to be of three dimensions because in it a point has three degrees of freedom. A plane has also three degrees of freedom, and if the plane is taken as the element we have again a three-dimensional geometry, dual to point-geometry. But the line has four degrees of freedom, and a geometry in which the line is the element should be of four dimensions.

16·21. Four-dimensional geometry.

There is no difficulty in extending analytical point-geometry to four and more dimensions. For a space of four dimensions S_4 we define a point as a set of values of the ratios of five numbers $x_0, \ldots, x_4$ taken in a fixed order; these are the homogeneous coordinates of the point. If the numbers are all multiplied by the same factor k, not zero, they will continue to represent the same point. If

$$P' \equiv [x_0', \ldots, x_4'] \quad \text{and} \quad P'' \equiv [x_0'', \ldots, x_4'']$$

are two given points, the coordinates

$$\rho x_i = x_i' + \lambda x_i''$$

represent for all values of λ a point on the line $P'P''$. Similarly

$$\rho x_i = x_i' + \lambda x_i'' + \mu x_i'''$$

represents points on the plane determined by three fixed points, and

$$\rho x_i = x_i^{(1)} + \lambda x_i^{(2)} + \mu x_i^{(3)} + \nu x_i^{(4)}$$

represents a point having three degrees of freedom and lying in the space determined by four given points. In the last case, eliminating λ, μ, ν and ρ between the five equations we obtain an equation of the first degree and homogeneous in $x_0, \ldots, x_4$, the equation of the space. This is of the form

$$\xi_0 x_0 + \ldots + \xi_4 x_4 = 0,$$

and involves the ratios of five numbers $\xi_0, \ldots, \xi_4$ which may be regarded as homogeneous tangential coordinates of the space. With two simultaneous linear equations we can express x_i linearly in terms of two parameters. Thus two equations determine a plane as the intersection of two spaces. Three equations determine a line, and four equations a point. A space cuts a space, a plane, and a line respectively in a plane, a line, and a point. A plane cuts another plane in a point, and does not in general cut a given line.

16·22. A homogeneous equation of the second degree

$$\Sigma\Sigma a_{rs}x_r x_s = 0$$

represents a three-dimensional assemblage of points which is cut by an arbitrary line in two points. We shall call this a *quadric variety* or simply a quadric, and denote it by the symbol $V_3{}^2$. A quadric in three dimensions is $V_2{}^2$. The theory of pole and polar can be extended in an obvious way. Two points (x) and (y) are conjugate with respect to the quadric

$$\Sigma\Sigma a_{rs}x_r x_s = 0$$

when $\Sigma\Sigma a_{rs}x_r y_s = 0.$

If (y) is fixed and (x) is variable, the equation

$$\Sigma\Sigma a_{rs}y_r x_s = 0$$

represents a three-dimensional space or *three-flat*, the *polar three-flat* of (y). If (y) lies on the quadric this is the *tangent three-flat* at (y).

If two points P and Q are conjugate and each lies on the quadric, all points of the line PQ lie on the quadric. (Proof as in 8·22.)

16·23. As in three dimensions, therefore, it appears that a quadric possesses generating lines. If P lies on the quadric the locus of points conjugate to P is a three-dimensional tangent-space α, and this meets the quadric in a two-dimensional quadric surface; if Q is any point on this two-dimensional quadric the line PQ, since it joins two conjugate points both lying on the quadric, is a generating line of the original quadric and therefore a generating line of the two-dimensional quadric. The latter is therefore a quadric cone with vertex P. Through every point P on the quadric there is thus a cone of generating lines. On this cone take a point Q. Then the polar of Q is a three-dimensional tangent-space β which cuts α in a plane, the polar-plane of the line PQ. This plane cuts the quadric in a conic consisting in part of the line PQ, and such that every line in its plane cuts PQ in two coincident points; hence the conic consists of the line PQ counted twice.

16·24. If the quadric variety V_3^2 contains a plane α, let P be any other point on the variety, then the polar of P is a tangent-space which cuts α in a line p. All points on p are conjugate to P and lie on the quadric, therefore all the lines joining these points to P lie on the quadric. The quadric therefore contains also the plane (Pp). If Q is another point on the quadric we obtain similarly the plane (Qq). p and q both lie in the plane α and intersect in a point O. The planes (Pp) and (Qq) both pass through O, but have no other point in common, for if they had another point in common they would have a line in common and the three planes would all be contained in one three-dimensional space S. The intersection of this space with the quadric variety would then be a quadric V_2^2 containing these three planes; this is impossible unless the quadric variety contains the whole of S and degenerates to two three-flats, S and another.

Assuming then that the quadric variety does not degenerate, the point O is conjugate to all points in either of the planes (Pp) and (Qq), and therefore to any point on any line joining two points, one on each of these planes, i.e. to any point in S_4. Then if R is any point of the quadric, and r the line in which the tangent-space at R cuts α, the plane (Rr) passes through O. The quadric variety is in this case a *hypercone* with vertex O. Its section by a space not passing through O is a quadric V_2^2, and the hypercone is generated by lines joining O to the points of V_2^2. The planes determined by O and the generating lines of V_2^2 all lie on the hypercone. Thus *a hypercone in S_4 has two singly infinite systems of planes all passing through the vertex O. Two planes of the same system have only the point O in common, two planes of different systems have a line through O in common. A quadric variety in S_4, not specialised, has no planes but a triple infinity of lines, through every point a cone of lines.*

16·25. The generating lines of a quadric in S_4 are not divided into two systems like those of a hyperboloid. If we project stereographically from any point O on the quadric on to a space α, the tangent-space τ at O cuts α in a plane and this plane cuts the cone of generators through O in a conic C. The generating lines through O are then projected into points of this conic. If l is any other generating line it cuts the tangent-space τ in a point P which lies on the quadric and therefore on the cone;

P is therefore projected into a point P' lying on C, and l into a line through P'. Thus all the other generators of the quadric are projected into lines in α which meet the conic C. The generators form a three-dimensional assemblage and are projected into the points of the conic C and the complex of lines which meet C.

16·3. Five-dimensional geometry.

It is possible to represent the lines of S_3 by points of S_4, but the most symmetrical representation requires space of five dimensions, taking the six homogeneous coordinates of a line in S_3 as homogeneous coordinates of a point in S_5.

16·31. We shall therefore sketch further the geometry of five dimensions so far as it is required for this representation. In S_5 we have points, lines, planes, three-dimensional spaces or three-flats (S_3), and four-flats (S_4), represented respectively by freedom-equations in o, 1, 2, 3, and 4 parameters, and by 5, 4, 3, 2, and 1 linear equations respectively in $x_0, \ldots, x_5$. The dual elements are point and four-flat, line and three-flat, plane and plane.

16·32. A quadric variety $V_4{}^2$ in S_5 is cut by a four-flat in a quadric variety $V_3{}^2$ of one dimension fewer, by a three-flat in an ordinary quadric, by a plane in a conic, and by a line in two points. If P and Q are conjugate points with respect to $V_4{}^2$ and both lie on the quadric variety, all points of the line PQ lie on $V_4{}^2$. The tangent four-flat α at P meets $V_4{}^2$ in a hypercone with vertex P. If Q is a point on this hypercone the tangent four-flat β at Q cuts α in a three-flat and this three-flat meets $V_4{}^2$ in a quadric which has the line PQ as a double-line and therefore consists of two planes which lie entirely in the quadric variety. Thus *a quadric variety in S_5 possesses not only lines but also planes, two through each line.*

16·33. The planes of a $V_4{}^2$, like the lines of a quadric in S_3, are separated into two systems, but two planes of the same system always intersect in a point, while two planes of different systems either intersect in a line or not at all.

Project the $V_4{}^2$ from a point O upon it on to a four-flat π. The tangent four-flat τ at O cuts π in a three-flat and meets the quadric variety in a hypercone $V_3{}^2$ with vertex O. This hypercone possesses both lines and planes, and these are projected into the points and lines of a quadric V_2^2 in the space of intersection of π and τ. Thus the planes of V_4^2 through O, which are also the planes of the hypercone, are separated into two systems corresponding to the two systems of generators of V_2^2; two planes of the same system have only the point O in common, two planes of different systems have a line in common. Through each point O there is a single infinity of planes of each system.

Let α be any plane of V_4^2 not passing through O. α cuts τ in a line l which is projected into a line l' of V_2^2, and α is projected into a plane α' in π, which passes through l'. The planes of π in general cut the space of V_2^2 in lines which cut V_2^2 in pairs of points, those which are the projections of planes of the quadric V_4^2 pass through lines of V_2^2. Also lines of π which are projections of lines of V_4^2 pass through points of V_2^2.

Let α, β be two planes of V_4^2 of the same system, not passing through O. Their projections α', β' on π cut the space of V_2^2 in two generating lines of V_2^2 of the same system and therefore non-intersecting. α' and β' do not then intersect in a line but have just one point C' in common. As C' is not on V_2^2 it is the projection of just one point C on V_4^2 and this point is common to α and β. Hence two planes of the same system have always one point in common, and we have proved above that they have only one point in common.

Let now α, β be two planes of V_4^2 of different systems. If they have one point in common we have proved above that they have a line in common. We shall now show that they may have no point in common. Let l and m be two generators of V_2^2 of different systems, C their point of intersection. Let A be a point in π, not in the space of V_2^2 and call the plane (Al) α'. Let B be a point in π, not in the space of V_2^2 nor in the three-flat (Alm), and call the plane (Bm) β'. Then α' and β' are the projections of two planes α and β of V_4^2 not passing through O, and since α' and β' do not lie in the same three-flat they have no line in common, therefore α and β have no line in common and therefore no point in common.

A V_4^2 in S_5 has ∞^4 points, ∞^5 lines, and two systems of ∞^3 planes. Through each line there are two planes, one of each system. Through each point there are ∞^2 lines and two systems of ∞ planes belonging to a hypercone V_3^2.

16·34. A four-flat cuts a V_4^2 in a V_3^2 which has in general no planes. A four-flat which contains a plane of V_4^2 meets V_4^2 in a hypercone and is therefore a tangent four-flat, the vertex of the hypercone being the point of contact; the four-flat then contains ∞ planes of V_4^2.

A three-flat cuts V_4^2 in a quadric. If the three-flat contains three concurrent lines of V_4^2 it meets V_4^2 in a cone and is a tangent three-flat, the vertex of the cone being the point of contact. If the three-flat contains a plane of V_4^2 it meets V_4^2 in two planes and is a tangent three-flat along the line of intersection of the two planes. Similarly a plane may touch a V_4^2 at a point or along a line, or lie entirely in the V_4^2.

A line l cuts a V_4^2 in two points P, Q. The tangent four-flats at P and Q intersect in a three-flat L which is the polar of the given line. A plane through PQ and containing a generating line through P is a tangent-plane, hence through PQ there are ∞^2 tangent-planes to V_4^2. As the point of contact is the intersection of a generator through P with a generator through Q, all the points of contact lie in the polar three-flat and their locus is a quadric. Through the three-flat L there are just two tangent four-flats, their points of contact being the points in which the polar line l cuts the quadric. If the line l is a tangent at P, its polar three-flat L is also a tangent at P and all the tangent-planes through l lie in L.

The tangent four-flats at the points in which V_4^2 is cut by a plane p all pass through another plane p', the polar of p, and the tangent four-flats which pass through p all have their points of contact in p'.

16·4. We take now the six homogeneous coordinates $p_{23}, \dots$ of a line as homogeneous coordinates of a point in S_5. The identical relation

$$p_{01}p_{23} + p_{02}p_{31} + p_{03}p_{12} = 0,$$

being homogeneous and of the second degree, represents a quadric variety V_4^2. We shall denote this by ω and the function on the left by $\omega(p)$. A line l of S_3 is then represented by a point l on this quadric. We shall use the same symbols in heavy type to denote the objects of S_5 which represent objects of S_3.

16·41. The condition that two lines p, p' should intersect is

$$p_{01}' \frac{\partial \omega}{\partial p_{01}} + \dots = 0,$$

and this is the condition that the points p, p' should be conjugate with respect to ω; all points on the line pp' then also lie on ω, so that pp' is a line of ω. The condition that two lines p, p' should intersect is therefore that the points p, p' should lie on the same line of ω.

16·42. A plane P in ω contains ∞^2 points, every two of which are conjugate. It therefore represents a doubly infinite system of lines of S_3 of which every two intersect, i.e. either a system of all lines lying in one plane (plane field of lines), or a system of all lines passing through one point (bundle of lines). These two systems correspond to the two systems of planes of ω; we shall call them *field-planes* and *bundle-planes*. Two planes of ω of the same system have always one point in common; this corresponds to the fact that two bundles of lines or two plane fields of lines have in each case one line in common. But a plane field of lines and a bundle of lines have no line in common unless the vertex of the bundle lies in the plane of the field, in which case they have a plane pencil in common. A line of ω, through which always two planes of different systems pass, therefore represents a plane pencil of lines.

16·43. If the lines p and p' intersect, $(p_{01} + \lambda p_{01}', \dots)$, or shortly $(p + \lambda p')$, represents for all values of λ a line which cuts both p and p'. These are represented by the points of the line pp'. Hence when p and p' are intersecting lines, $(p + \lambda p')$ represents the plane pencil of lines determined by p and p'. The vertex of the pencil is represented by the bundle-plane through pp' and the plane of the pencil by the field-plane through pp'.

Ex. Prove directly that when p and p' are intersecting lines, $(p + \lambda p')$ always represents a line and that it passes through the point of intersection of p and p' and lies in the plane determined by p and p'.

16·5. A linear complex $a_{01}p_{01} + \dots = 0$ is represented by the intersection of ω with a four-flat, i.e. by a V_3^2; a linear congruence by the intersection with a three-flat, i.e. by a quadric; and a linear series by the intersection with a plane, i.e. by a conic.

16·51. The linear complex $a_{01}p_{01} + \ldots = 0$ has an invariant, viz. the simultaneous invariant of this with ω, or the condition that the four-flat should touch ω. This is

$$\Omega \equiv a_{01}a_{23} + a_{02}a_{31} + a_{03}a_{12} = 0,$$

the tangential equation of ω.

16·52. When $\Omega = 0$ the complex is said to be *special* or *singular*. The condition $\Omega = 0$ shows that $[a_{23}, \ldots]$ are the coordinates of a line, and the equation of the complex expresses that the line p cuts this line. Hence a special linear complex consists of all the lines which cut a fixed line. It is represented in S_5 by the intersection of ω with a tangent four-flat, i.e. by a hypercone. The vertex of this hypercone represents the fixed line or *directrix* of the complex.

16·53. The polar of a three-flat is a line and this cuts ω in two points d_1, d_2, which are therefore conjugate to all points of the three-dimensional section. Hence a linear congruence is the assemblage of all lines which meet two fixed lines d_1, d_2, its directrices. The congruence is called *hyperbolic* or *elliptic* according as these two lines are real or imaginary.

16·531. If the directrices intersect, d_1 and d_2 are conjugate, and the line $d_1 d_2$ lies on ω. In this case the three-flat, whose intersection with ω represents the congruence, touches ω along this line. This line is a double-line on the quadric, which therefore becomes two planes. The congruence then consists of all the lines which lie in the plane $\alpha \equiv (d_1 d_2)$ together with all the lines which pass through the point of intersection P of d_1 and d_2. The directrices d_1 and d_2 really lose their identity and can be replaced by any two lines lying in α and passing through P. This congruence is said to be *singular*.

16·532. When the directrices coincide, the three-flat touches ω in a point, and its intersection with ω is a cone with vertex d. The congruence thus consists of all plane pencils of lines which have a line d in common. It is called a *parabolic* congruence.

16·54. A linear series is represented by the intersection of ω with a plane u. This plane is determined by three points, and therefore the linear series is determined by three of its lines. The polar of u is another plane u', which also is determined by three points. To this corresponds another linear series; and since every point of u is conjugate to every point of u', every line of the one series cuts every line of the other series. Hence a linear series is the assemblage of all lines which meet three fixed lines, i.e. a *regulus*. The two series, corresponding to the two mutually polar planes, are the two reguli which belong to the same quadric.

If the plane u touches ω at a point p, u' also touches ω at p. The two linear series have then a line p in common. u meets ω in a pair of lines l, m, and u' meets ω in a pair of lines l', m'. Each linear series consists of two pencils with a common line. Since l is conjugate to l' and to m', the planes ll' and lm' lie in ω; these are the bundle- and field-plane through l, and similarly ml' and mm' are the field- and bundle-plane respectively through m. Hence this linear series consists of two pencils (A, α) and (B, β), the line AB which joins the vertices coinciding with the line of intersection $\alpha\beta$ of the two planes; and the polar series consists of the two pencils (A, β) and (B, α).

If the plane u touches ω along a line l the series degenerates further to two coincident pencils. The polar plane u' touches ω along the same line and determines the same pencil. The vertex and plane of the pencil are represented by the bundle-plane and the field-plane through l.

Finally if u is a plane of ω it represents either a bundle of lines through a point, or a plane field of lines.

In general the lines common to two linear complexes form a linear congruence, and the lines common to three linear complexes form a regulus, but the congruence and the regulus may be specialised in one of the ways noted above.

16·55. Three points l, m, n of ω determine a conic on ω; three lines l, m, n, in S_3, determine a regulus. Four points on ω determine a three-flat cutting ω in a quadric; four lines in S_3 determine a congruence, the directrices of the congruence are the two transversals of the four lines. Five points on ω determine a four-flat; five lines in S_3 determine a linear complex.

16·6. Polar properties of a linear complex.

16·61. Let C denote the four-flat in S_5 which corresponds to a linear complex C. To the lines of C which pass through a point

P correspond the points common to the four-flat C and the bundle-plane P of ω. But P cuts C in a line l, and this represents a plane pencil. The plane of the pencil (*polar plane* of P) is represented by the field-plane p through l, the point P (pole) being represented by the bundle-plane P through l.

16·611. It will be more convenient to use the symbols $x_1, ..., x_6$ for the line-coordinates p_{ij}. The six equations $x_1 = 0$, ..., $x_6 = 0$ represent six four-flats of reference. These intersect in fifteen three-flats, twenty planes, fifteen lines, and six points, and form a figure called a *simplex*, the analogue of a triangle and a tetrahedron. A change of the simplex of reference is effected by a linear transformation of the coordinates, and the fundamental quadric ω may be represented by a general homogeneous equation $\omega(x) = 0$ of the second degree in $x_1, ..., x_6$.

16·621. A linear complex is represented by the two equations
$$\omega(x) = 0, \quad \Sigma ax = 0.$$
The condition that it should be singular is that the four-flat $\Sigma ax = 0$ should touch ω; this is represented by the tangential equation of the fundamental quadric with the coordinates a of the four-flat substituted, i.e. $\Omega(a) = 0$.

16·622. A linear congruence is represented by three equations
$$\omega(x) = 0, \quad \Sigma ax = 0, \quad \Sigma bx = 0.$$
The directrices are represented by the two points in which ω is cut by the line joining the poles of the four-flats $\Sigma ax = 0$ and $\Sigma bx = 0$. The coordinates of the pole of $\Sigma ax = 0$ are $A_i \equiv \dfrac{\partial \Omega}{\partial a_i}$, and freedom-equations of the line joining the two poles are
$$x_i = \lambda A_i + \mu B_i \quad (i = 1, ..., 6).$$
Substituting in $\omega(x) = 0$ we obtain a quadratic in λ/μ,
$$\lambda^2 \omega(A) + \lambda\mu \Sigma A \frac{\partial \omega}{\partial B} + \mu^2 \omega(B) = 0,$$
and this is equivalent to
$$\lambda^2 \Omega(a) + \lambda\mu \Sigma a \frac{\partial \Omega}{\partial b} + \mu^2 \Omega(b) = 0.$$
If $\Omega(a) = 0$, $\Omega(b) = 0$, and $\Sigma a \dfrac{\partial \Omega}{\partial b} = 0$, the directrices are indeterminate and the congruence is *singular*.

16·623. A linear series or regulus is represented by four equations $\qquad \omega(x)=0,\ \Sigma ax=0,\ \Sigma bx=0,\ \Sigma cx=0.$

The complementary regulus is determined by

$$\omega(x)=0,\quad x_i=\lambda\frac{\partial\Omega}{\partial a_i}+\mu\frac{\partial\Omega}{\partial b_i}+\nu\frac{\partial\Omega}{\partial c_i}\quad(i=1,\,\ldots,\,6).$$

Substituting these values for x_i in $\omega(x)=0$ we have a homogeneous quadratic in λ,μ,ν which may be represented by a conic in a (λ,μ,ν)-plane. When this conic breaks up into two lines the regulus breaks up into two pencils. The condition for this is therefore that the discriminant of the quadratic

$$\omega\left(\lambda\frac{\partial\Omega}{\partial a}+\mu\frac{\partial\Omega}{\partial b}+\nu\frac{\partial\Omega}{\partial c}\right)=0$$

should vanish.

16·63. Polar lines with respect to a linear complex.

Consider a linear complex C and any line l. The complex is represented by the intersection of ω with a four-flat C, and the line by a point l of ω. The points on l and the planes through l are represented respectively by the bundle-planes and the field-planes of ω which pass through the point l. These all lie in the tangent four-flat T to ω at l. Now T cuts the four-flat C in a three-flat A, and through A there is a second tangent four-flat T' to ω touching it at the point l'. A cuts ω in a quadric V_2^2, and the planes of ω which pass through either l or l' cut A in lines of this quadric. The bundle-planes through l and the field-planes through l' contain the lines of one regulus of V_2^2, the field-planes through l and the bundle-planes through l' contain the lines of the other regulus. Hence the polar-planes, with respect to the complex, of all points on the line l pass through l', and *vice versa*, so that l and l' are polar lines with respect to the linear complex.

The quadric V_2^2 in which A cuts ω represents the linear congruence consisting of lines of the complex which cut l; l is one directrix of this congruence and l' is the other.

The lines joining pairs of points l, l' which represent polar lines with respect to a linear complex C all pass through one point, the polar of the four-flat C with respect to ω. Any line of the complex is polar to itself.

Ex. Show that the two directrices of the linear congruence common to two given linear complexes are polar lines with respect to each of the complexes.

16·64. Conjugate linear complexes.

If the four-flats C and C' are conjugate with respect to ω the corresponding linear complexes C and C' are said to be *conjugate**.

If O and O' are the poles of the conjugate four-flats C and C', O is conjugate to O', C passes through O', and C' through O. *A self-conjugate complex is singular*, for if $O \equiv O'$, O lies on ω and C is the tangent four-flat at O. If one complex C is singular, so that O lies on ω, the four-flat C' corresponding to any conjugate complex passes through O, hence the directrix O of the singular complex is a line of the other. If both complexes are singular and conjugate, OO' lies in ω and represents a plane pencil belonging to both complexes; the two directrices O and O' intersect and the common pencil is that determined by the intersecting directrices.

16·65. If l is a line common to two linear complexes C and C', a homography is set up on the line l by the poles P, P', with respect to C and C', of a variable plane p through l. The homography is not in general symmetrical. If p' is the polar of P with respect to C', and P'' the pole of p' with respect to C, P'' will generally be different from P'. Project from the point l on to a four-flat π. The lines and planes of ω through l are projected into the points and lines of a quadric Q whose three-flat T is the projection of the tangent four-flat τ to ω at l. The four-flats C and C', which pass through l, are projected into three-flats which cut T in two planes C_1 and C_1'. If O is the pole of C it lies in τ as its projection O_1 lies in T. The polar of the line Ol is the three-flat $(C\tau)$; every point of Ol is conjugate to every point of $(C\tau)$ with respect to ω, and therefore, cutting these by π, O_1 is the pole of C_1 with respect to Q.

---

* Klein used the term "in involution," Hudson used "apolar," Baker uses "apolar" or "conjugate."

Through *l* take a field-plane *p* of *ω*. This cuts *C* and *C'* in lines *a* and *a'*, and through these we have the bundle-planes *P* and *P'*. The bundle-plane *P* cuts *C* and *C'* in *a* and *b'*, and the bundle-plane *P'* cuts *C* and *C'* in *b* and *a'*. In the projection (Fig. 47) *C* and *C'* are represented by plane sections of *Q*; the planes *p*, *P* and *P'* by generators of the quadric *Q*; and the lines *a*, *a'*, *b* and *b'* by points where these cut *C* and *C'*. Corresponding to the homography of points *P*, *P'* on a fixed line *l*, we have a homography of

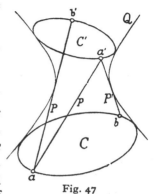

Fig. 47

generators *P*, *P'* of the quadric *Q* when the generator *p* is varied, and the condition that the homography be involutory is that *bb'* should be a generator of *Q*; then the lines *ab* and *a'b'* are polars with respect to *Q* and the planes *C* and *C'* are conjugate.

Hence *the homography on l is involutory only if the complexes are conjugate.*

Ex. 1. Two non-intersecting pairs of polar lines with respect to a linear complex belong to the same regulus.

For the corresponding points *l*, *l'* and *m*, *m'* of *ω* lie on two lines through *O* and therefore all lie in one plane.

Ex. 2. If two lines *l* and *m* intersect so also do their polars.

16·7. Just as the equation of a quadric can be expressed by the sum of four squares, by taking a self-polar tetrahedron as frame of reference, so a quadric in S_5 is expressed by the sum of six squares when referred to a self-polar simplex. By suitable choice of unit-point the equation can finally be reduced to the form
$$x_1^2 + x_2^2 + x_3^2 + x_4^2 + x_5^2 + x_6^2 = 0.$$
The expression for the identical relation connecting the six homogeneous coordinates of a line can, in fact, be reduced to this form by taking new coordinates which are linear functions of the old. One such transformation, which is due to Klein, is

$$p_{01} = x_1 + ix_4, \quad p_{02} = x_2 + ix_5, \quad p_{03} = x_3 + ix_6,$$
$$p_{23} = x_1 - ix_4, \quad p_{31} = x_2 - ix_5, \quad p_{12} = x_3 - ix_6.$$

Then the six linear complexes $x_1 = 0$, ..., $x_6 = 0$ are all conjugate in pairs. (It is not possible by a *real* transformation to reduce the equation to the *sum* of six squares. It can always be reduced to the form $a_1 x_1^2 + ... + a_6 x_6^2 = 0$; thus writing $p_{01} = x_1 + x_4$, $p_{23} = x_1 - x_4$, etc., we have

$$x_1^2 + x_2^2 + x_3^2 - x_4^2 - x_5^2 - x_6^2 = 0.$$

So long as the transformations are real, half the coefficients $a_1, ..., a_6$ must be positive and the other half negative. This is an instance of Sylvester's Law of Inertia. Geometrically, when the coordinates are real the quadric has real planes only when the signs are three + and three − ; when four are of one sign and two of the other, the quadric has real lines but no real planes; when five are of one sign and one of the other sign, the quadric has real points but no real lines or planes; and when the signs are all the same, the quadric has no real points.)

16·8. The Quadratic Complex.

A quadratic complex is represented by a homogeneous quadratic equation $U = 0$ in the six line-coordinates, and in S_5 is represented by the intersection of the quadric ω with another quadric U. The linear system of quadrics

$$U + \lambda \omega = 0$$

all correspond to the same quadratic complex, so that U by itself has no particular significance. We shall denote the quadratic complex by K, and the three-dimensional variety in S_5 which represents it by K. In general U and ω have a common self-polar simplex; taking this as frame of reference we can write

$$\omega \equiv \Sigma x^2, \quad U \equiv \Sigma k x^2.$$

16·81. The intersection K of U and ω is a three-dimensional variety which is cut by an arbitrary plane in four points. Hence *a quadratic complex has four lines in common with an arbitrary regulus*. A special quadratic complex consists of the tangent-lines to a quadric, hence, as a particular case, we have the result: *There are four generators of each system of one given quadric which touch another given quadric.*

All the lines of a quadratic complex which pass through a given point P form a quadric cone, and all the lines which lie in a given plane p envelop a conic. A plane P lying in ω cuts U in a conic, and this conic lies in all quadrics of the linear system $U+\lambda\omega$. If P is a bundle-plane the conic in P represents a quadric cone, the complex cone, and if it is a field-plane the conic represents a conic-envelope, the complex conic.

16·82. If the quadric cone through P has a double-line and therefore breaks up into two planes, the point P is called a *singular point*, and if the conic in the plane p has a double-tangent and therefore breaks up into two pencils, p is called a *singular plane*. The planes of ω which correspond to singular points and planes all touch K, and envelop in ω a certain variety S, which represents the locus of singular points and the envelope of singular planes. This surface S is called the *singular surface* of the quadratic complex. We shall show presently that it is of order 4 and class 4, i.e. that it is cut by an arbitrary line in four points, and has four tangent-planes through an arbitrary line.

16·83. **Polar of a line with respect to a quadratic complex.**

The polar four-flat Λ_U of a point $l \equiv (y)$ of ω with respect to the quadric U has the equation

$$\Sigma x \frac{\partial U}{\partial y} = 0,$$

while the polar four-flat of l with respect to ω is the tangent four-flat τ at l, and its equation is

$$\Sigma x \frac{\partial \omega}{\partial y} = 0.$$

The polar four-flats of l with respect to the quadrics of the linear system $U+\lambda\omega$ form a pencil of four-flats

$$\Sigma \left(x \frac{\partial U}{\partial y} + \lambda x \frac{\partial \omega}{\partial y} \right) = 0$$

all passing through the three-flat L of intersection of Λ_U and τ. This represents a linear congruence, the *polar congruence* of the line l with respect to the complex. The pole of τ with respect to ω is the point of contact l; let $l_U \equiv (u)$ be the pole of Λ_U with respect to ω. Then the directrices of the congruence are represented by the two points in which the line ll_U cuts ω, one of these being l itself. Taking the canonical equations $\omega \equiv \Sigma x^2$ and $U \equiv \Sigma k x^2$, $u_i = k_i y_i$. The points on the line $l l_U$ joining (y) and (u) are given by

$$x_i = y_i + \lambda k_i y_i,$$

and the points of intersection of this line with ω are determined by the quadratic equation

$$\Sigma (y + \lambda k y)^2 = 0.$$

But $\Sigma y^2 = 0$, hence one root is $\lambda = 0$, and the other is given by

$$\lambda \Sigma k^2 y^2 + 2 \Sigma k y^2 = 0.$$

16·831. If $(y) \equiv l$ is a line of the complex, $\Sigma k y^2 = 0$, and the second directrix of the polar congruence coincides with l. In this case the congruence (which is parabolic) is called the *tangent linear congruence* for the line l. l_U is conjugate to l with respect to ω and lies on τ, and the line $l l_U$ lies on τ.

16·84. Further, if l_U lies on ω, $\Sigma k^2 y^2 = 0$; the line $l l_U$ lies on ω and represents a plane pencil of S_3. The directrices of the tangent linear congruence become indeterminate; all the four-flats of the pencil

$$\Sigma \left(x \frac{\partial U}{\partial y} + \lambda x \frac{\partial \omega}{\partial y} \right) = 0,$$

i.e. $\Sigma k y x + \lambda \Sigma y x = 0$, are tangent to ω at points of the line $l l_U$. The tangent congruence is now singular, and consists of a plane pencil containing l, its vertex and plane being represented by the bundle-plane P and field-plane p through the line $l l_U$. In this case the line l is called a *singular line*. We shall see that P and p represent a singular point and a singular plane incident with this line.

Let (z) be a point on one of the two planes P, p, so that

$$\Sigma z^2 = 0, \ \Sigma y z = 0, \ \Sigma k y z = 0, \ \Sigma y^2 = 0, \ \Sigma k y^2 = 0, \ \Sigma k^2 y^2 = 0,$$

then all the points on one of these planes are represented by

$$x_i = \lambda z_i + \mu y_i + \nu k_i y_i.$$

This plane meets K or U at points (x) where $\Sigma k x^2 = 0$, i.e.

$$\lambda^2 \Sigma k z^2 + \nu^2 \Sigma k^3 y^2 + 2 \lambda \nu \Sigma k^2 y z = 0.$$

Since this quadratic in λ, μ, ν breaks up into factors the plane touches K and therefore corresponds to a singular point or a singular plane. A point (x) thus determined is conjugate to $l \equiv (y)$ both with respect to ω and with respect to U, for

$$\Sigma xy = \lambda \Sigma yz + \mu \Sigma y^2 + \nu \Sigma ky^2 = 0$$

and $$\Sigma kxy = \lambda \Sigma kyz + \mu \Sigma ky^2 + \nu \Sigma k^2 y^2 = 0.$$

Therefore the two planes P and p through ll_U both touch K

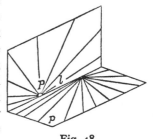

along lines passing through l. These lines of contact represent two pencils of lines of the complex, one with vertex P and the other with plane p, and the singular line l belongs to each pencil; the line ll_U represents another pencil of lines of the complex having vertex P and plane p. P is therefore a singular point, and p a singular plane. The complex cone at P breaks up into the

Fig. 48

plane p and another plane passing through the singular line l; the complex conic in p breaks up into the point P and another point lying on l.

The coordinates of singular lines (y) of the complex satisfy the three equations

$$\Sigma y^2 = 0, \ \Sigma ky^2 = 0, \ \Sigma k^2 y^2 = 0.$$

Hence the singular lines form a two-dimensional system or congruence represented by the points l common to these three quadrics, ω, U and U', say. (U' is the reciprocal of ω with respect to U.) They are tangents to the singular surface S.

16·85. We consider now the singular points on a given line l, and the singular planes through l. These are represented by bundle-planes and field-planes of ω passing through the point l, and touching K. The whole assemblage of lines and planes of ω which pass through l form a hypercone V lying in the tangent-four-flat τ to ω at l. Denote as before by L the three-flat common to all the polar four-flats of l with respect to the quadrics of the linear system $U + \lambda \omega$. L cuts V in an ordinary quadric V_1, and every quadric $U + \lambda \omega$ in a quadric U_1. The planes of ω which pass through l cut L in the generators of the quadric V_1, bundle-planes corresponding to one system of generators, and field-planes to the other. In order that one such plane should touch

K, the corresponding generator of V_1 must touch U_1. Now a regulus has four lines which touch a given quadric, hence there are four bundle-planes and four field-planes of ω through the point l which touch U_1 and therefore K, and for each the conic breaks up into two lines. Hence on any line l there are four singular points of the complex, and through l there are four singular planes. The singular surface S is therefore of order 4 and class 4.

16·86. Co-singular quadratic complexes.

The coordinates of a singular line of the quadratic complex

$$\Sigma x^2 = 0, \ \Sigma k x^2 = 0, \qquad \ldots \ldots (1)$$

are determined by these and the further equation $\Sigma k^2 x^2 = 0$. (kx) is then a line, and the two lines (x) and (kx) intersect and determine a point P and a plane p which are both singular. Now we can obtain a single infinity of quadratic complexes whose singular lines are lines of the pencil (P, p). If (y) is any line of the pencil determined by (x) and (kx)

$$y_i = k_i x_i - \lambda x_i \qquad \ldots \ldots (2)$$

and $\qquad x_i = (k_i - \lambda)^{-1} y_i.$

Now consider the quadratic complex

$$\Sigma y^2 = 0, \ \Sigma (k - \lambda)^{-1} y^2 = 0, \qquad \ldots \ldots (3)$$

in which λ is given a particular value. Its singular lines are determined by the two equations (3) and $\Sigma (k - \lambda)^{-2} y^2 = 0$.

Now $\quad \Sigma (k - \lambda)^{-2} y^2 = \Sigma x^2,$

$\Sigma (k - \lambda)^{-1} y^2 = \Sigma (k - \lambda) x^2 = \Sigma k x^2 - \lambda \Sigma x^2,$

$\Sigma y^2 = \Sigma k^2 x^2 - 2\lambda \Sigma k x^2 + \lambda^2 \Sigma x^2.$

Hence if (y) is a singular line of the complex (3), $\Sigma x^2 = 0$, $\Sigma k x^2 = 0$ and $\Sigma k^2 x^2 = 0$, so that (x) is a singular line of the complex (1); and *vice versa*. There is one line of the pencil

$$y_i = k_i x_i - \mu x_i,$$

viz. that for which $\mu = \lambda$, which is a singular line of the complex (3); P and p are a singular point and a singular plane for both complexes.

Hence all the quadratic complexes of the system

$$\Sigma x_i^2 = 0, \ \Sigma (k_i - \lambda)^{-1} x_i^2 = 0,$$

for all values of λ, have the same singular surface. These form a *co-singular system*. The original complexes must be included in this series. We have

$$\Sigma(k-\lambda)^{-1}x^2 = -\frac{1}{\lambda}\Sigma(1-k/\lambda)^{-1}x^2$$

$$= -\frac{1}{\lambda}\left(\Sigma x^2 + \frac{1}{\lambda}\Sigma kx^2 + \dots\right),$$

and another quadric of the linear system is

$$\Sigma(k-\lambda)^{-1}x^2 + \frac{1}{\lambda}\Sigma x^2 = -\frac{1}{\lambda^2}\left(\Sigma kx^2 + \frac{1}{\lambda}\Sigma k^2x^2 + \dots\right) = 0,$$

i.e. $$\Sigma kx^2 + \frac{1}{\lambda}\Sigma k^2x^2 + \dots = 0,$$

which becomes $$\Sigma kx^2 = 0$$

as $\lambda \to \infty$.

The co-singular complexes are represented in S_5 by the intersection of $\Sigma x^2 = 0$ with quadrics of the system

$$\Sigma(k-\lambda)^{-1}x^2 = 0.$$

These form a linear tangential system. Instead of this system we may take

$$\Sigma\{(k-\lambda)^{-1} + \lambda^{-1}\}x^2 = 0,$$

i.e. $$\Sigma k(k-\lambda)^{-1}x^2 = 0.$$

The tangential equation of this is

$$\Sigma(k-\lambda)k^{-1}\xi^2 = 0,$$

i.e. $$\Sigma\xi^2 - \lambda\Sigma\xi^2/k = 0.$$

$\Sigma\xi^2 = 0$ is the tangential equation of ω, and $\Sigma\xi^2/k = 0$ that of U or $\Sigma kx^2 = 0$. All the quadrics of this system touch all the four-flats which touch both ω and U. We may thus compare a co-singular system of quadratic complexes with a confocal system of quadrics.

16·9. We have considered the general quadratic complex in which the singular surface S is of order 4 and class 4, and possesses 16 nodes or points at which the tangent-lines form a quadric cone, and sixteen tropes or tangent-planes in which the tangent-lines envelop a conic. This type of surface is called a *Kummer surface*.

For quadratic complexes of special form the singular surface becomes specialised.

The complex being represented by two quadratic equations $\omega(x) = 0$ and $U(x) = 0$, consider the discriminant of the linear system

$$U - \lambda\omega = 0.$$

This is an equation of the sixth degree in λ, and for each root of the equation the quadric is specialised as a cone. The vertices of these six cones are the vertices of the common self-polar simplex. In the general case the six roots are all distinct.

16·91. Quadratic complex of tangent-lines to the quadric.

$$F \equiv ax^2 + by^2 + cz^2 + dw^2 = 0.$$

We have $U \equiv adx_1^2 + bdx_2^2 + cdx_3^2 + bcx_4^2 + cax_5^2 + abx_6^2 = 0$

with $\qquad \omega \equiv 2(x_1 x_4 + x_2 x_5 + x_3 x_6) = 0,$

where $\qquad x_1 \equiv p_{01}, \ldots, x_4 \equiv p_{23}, \ldots.$

The discriminant of $U - \lambda\omega$ is $(\lambda^2 - abcd)^3 = 0$, so that the roots are in two sets of three. Hence the six hypercones reduce to two. Taking $\lambda = + \sqrt{(abcd)}$, $U - \lambda\omega$ becomes

$$\{\sqrt{(ad)}\, x_1 - \sqrt{(bc)}\, x_4\}^2 + \{\sqrt{(bd)}\, x_2 - \sqrt{(ca)}\, x_5\}^2$$
$$+ \{\sqrt{(cd)}\, x_3 - \sqrt{(ab)}\, x_6\}^2 = 0$$

which represents a specialised hypercone having a double-plane C as edge determined by the three equations

$$\sqrt{(ad)}\, x_1 = \sqrt{(bc)}\, x_4, \quad \sqrt{(bd)}\, x_2 = \sqrt{(ca)}\, x_5, \quad \sqrt{(cd)}\, x_3 = \sqrt{(ab)}\, x_6.$$

The plane C cuts U and ω in the same conic; U and ω touch at all points of this conic. Similarly taking $\lambda = - \sqrt{(abcd)}$, we obtain another conic C' at all points of which U and ω have contact. The two planes C and C' are conjugate with respect to both ω and U, and the two conics represent the two reguli of the given quadric surface.

The singular surface is the quadric F itself, taken twice, and every tangent, i.e. every line of the complex, is singular.

16·92. The quadratic complex of tangent-lines to a cone, or of lines through a conic, is represented by the intersection of ω

with a hypercone having a double-plane as edge. In particular the complex of lines through the conic

$$x^2+y^2+z^2=0, \quad w=0$$

is (see 11·71)

$$U \equiv x_1{}^2+x_2{}^2+x_3{}^2=0, \quad \omega \equiv 2(x_1x_4+x_2x_5+x_3x_6)=0.$$

The discriminant of $U+\lambda\omega$ is $\lambda^6=0$, so that all six roots are equal. The double-plane $x_1=0$, $x_2=0$, $x_3=0$ is a plane of ω.

16·93. The tetrahedral complex.

There is one other quadratic complex of special interest, the complex of lines which cut the four planes of a given tetrahedron in a fixed cross-ratio. This is called the *tetrahedral complex*.

Taking the tetrahedron as frame of reference let (p) be a variable line determined by the two points (x', y', z', w') and (x'', y'', z'', w''). Freedom-equations of the line are

$$x=x'-\lambda x'', \text{ etc.}$$

This cuts the four planes $x=0$, $y=0$, $z=0$, $w=0$ where λ has the values

$$x'/x'', \quad y'/y'', \quad z'/z'', \quad w'/w'',$$

and the cross-ratio of these four numbers is

$$\frac{x'/x''-z'/z''}{x'/x''-w'/w''} \Big/ \frac{y'/y''-z'/z''}{y'/y''-w'/w''}=\frac{p_{13}}{p_{10}}\Big/\frac{p_{23}}{p_{20}}.$$

The tetrahedral complex is therefore represented by the equation

$$p_{02}p_{31}+kp_{01}p_{23}=0,$$

where k is constant. Taking with this

$$\omega \equiv p_{01}p_{23}+p_{02}p_{31}+p_{03}p_{12}=0$$

we can represent it symmetrically by

$$\omega=0, \quad U \equiv ap_{01}p_{23}+bp_{02}p_{31}+cp_{03}p_{12}=0.$$

If the cross-ratio of the range formed by the intersections of the line (p) with the planes $x=0$, $y=0$, $z=0$, $w=0$, in this order, is denoted by (XY, ZW) we have

$$(XY, ZW)=(a-c)/(b-c).$$

Through the line (p) there is a sheaf of four planes passing through the vertices of the tetrahedron. The equation of the plane through the given line and the vertex $[1, 0, 0, 0]$ is

$$u_1 \equiv p_{30}y+p_{02}z+p_{23}w=0.$$

Similarly
$$u_2 \equiv p_{30}x + p_{01}z + p_{13}w = 0,$$
$$u_3 \equiv p_{20}x + p_{01}y + p_{12}w = 0,$$
$$u_4 \equiv p_{23}x + p_{31}y + p_{12}z = 0.$$

u_3 and u_4 can be expressed in terms of u_1 and u_2, thus
$$p_{03}u_3 \equiv -p_{01}u_1 + p_{02}u_2,$$
$$p_{03}u_4 \equiv -p_{31}u_1 - p_{23}u_2.$$

Hence the cross-ratio
$$(u_1 u_2, \; u_3 u_4) = (0, \; \infty; \; p_{02}/p_{01}, \; -p_{23}/p_{31})$$
$$= -\frac{p_{02} p_{31}}{p_{01} p_{23}} = (XY, \; ZW).$$

The discriminant of $U - \lambda\omega$ is $(\lambda - a)^2(\lambda - b)^2(\lambda - c)^2 = 0$, so that the roots consist of three pairs. The three hypercones have each a double-line.

16·94. The harmonic complex of lines cut harmonically by the two quadrics
$$F \equiv ax^2 + by^2 + cz^2 + dw^2 = 0$$
and
$$F' \equiv a'x^2 + b'y^2 + c'z^2 + d'w^2 = 0$$
is
$$U \equiv (bc' + b'c)p_{23}^2 + \dots + (ad' + a'd)p_{01}^2 + \dots = 0,$$
$$\omega \equiv 2(p_{23}p_{01} + p_{31}p_{02} + p_{12}p_{03}) = 0.$$

The discriminant of $U + \lambda\omega$ is
$$\{\lambda^2 - (bc' + b'c)(ad' + a'd)\}\{\lambda^2 - (ca' + c'a)(bd' + b'd)\}$$
$$\times \{\lambda^2 - (ab' + a'b)(cd' + c'd)\} = 0.$$

The six roots are distinct, but they form three pairs equal but of opposite sign. This complex was studied by Battaglini as an example of the general quadratic complex; but it was shown by Klein that while the general quadratic complex involves nineteen independent constants, Battaglini's complex depends only on seventeen.

The singular surface of a harmonic complex is a particular form of Kummer's surface called the *Tetrahedroid*. A special case of the harmonic complex is afforded when one of the quadrics becomes the circle at infinity; it is then the locus of intersection of pairs of orthogonal tangent-planes to a quadric. This is called *Painvin's complex*, and its singular surface, when the quadric is an ellipsoid, is Fresnel's Wave Surface.

16·95. EXAMPLES.

1. Show that the assemblage of normals to a given quadric forms a congruence of order 6 and class 2.

Verify that on any plane section there are two points the normals at which to the curve of section are normals to the quadric: viz. the points where the given plane section is cut by the diametral plane conjugate to the normal to the given plane.

2. Show that the assemblage of all normals to quadrics of a confocal system form a tetrahedral complex, the planes of the tetrahedron being the three principal planes and the plane at infinity.

3. If (p), (q), (r) are the coordinates of three lines and the matrix

$$\begin{bmatrix} p_{01} \cdots p_{12} \\ q_{01} \cdots q_{12} \\ r_{01} \cdots r_{12} \end{bmatrix}$$

is of rank 2, show that the lines are coplanar.

4. If (p), (q), (r), (s) are the coordinates of four lines and the matrix

$$\begin{bmatrix} p_{01} \cdots p_{12} \\ q_{01} \cdots q_{12} \\ r_{01} \cdots r_{12} \\ s_{01} \cdots s_{12} \end{bmatrix}$$

is of rank 3, show that the lines belong to the same regulus.

5. If (p) are the Plücker coordinates of a line which is an axis of some plane section of the quadric $ax^2 + by^2 + cz^2 = 1$, prove that

$$\frac{p_{01} p_{23}}{a(b-c)} = \frac{p_{02} p_{31}}{b(c-a)} = \frac{p_{03} p_{12}}{c(a-b)}.$$

6. The condition of tangency of a line $[l, m, n, \lambda, \mu, \nu]$ and a quadric being given by $T = 0$, where T is quadratic in the coordinates of the line, show that the coordinates of the polar line are proportional to $\dfrac{\partial T}{\partial \lambda}, ..., \dfrac{\partial T}{\partial l}, ...$

Prove that if four lines are mutually conjugate (each meeting the polar of any other) then their two transversals are also conjugate. How may the double-six of lines be thus constructed?

(Math. Trip. II, 1913.)

7. Show that the lines common to three linear complexes given by

$$K_1 \equiv a_1\lambda + b_1\mu + c_1\nu + \alpha_1 l + \beta_1 m + \gamma_1 n = 0$$

and two similar equations, generate a quadric; and find linear complexes containing the other system of generators.

Show that the condition for a line to touch the quadric is

$$\begin{vmatrix} I_{11} & I_{12} & I_{13} & K_1 \\ I_{21} & I_{22} & I_{23} & K_2 \\ I_{31} & I_{32} & I_{33} & K_3 \\ K_1 & K_2 & K_3 & 0 \end{vmatrix} = 0,$$

where $\qquad 2I_{12} \equiv a_1\alpha_2 + b_1\beta_2 + c_1\gamma_2 + a_2\alpha_1 + b_2\beta_1 + c_2\gamma_1.$

(Math. Trip. II, 1913.)

CHAPTER XVII

ALGEBRAIC SURFACES

17·1. An algebraic surface is the locus of points (real or imaginary) whose homogeneous coordinates x, y, z, w satisfy an equation $F(x, y, z, w) = 0$, where F denotes a rational integral algebraic function. If F breaks up into rational factors the surface is *reducible*. We shall generally suppose that it is irreducible.

17·11. Considering first the equation written in non-homogeneous cartesian coordinates x, y, z, let it be arranged in groups of terms according to their degree

$$F_n + F_{n-1} + \dots + F_2 + F_1 + F_0 = 0,$$

each term being homogeneous in x, y, z, its degree being indicated by the subscript. Any line through the origin is represented by $x = l\rho$, $y = m\rho$, $z = n\rho$, where ρ is the radius-vector and l, m, n are direction-ratios. When these expressions are substituted for x, y, z the equation becomes

$$\rho^n \phi_n + \rho^{n-1} \phi_{n-1} + \dots + \rho^2 \phi_2 + \rho \phi_1 + \phi_0 = 0,$$

where ϕ_r denotes a homogeneous polynomial in l, m, n of degree r. Hence *an arbitrary line through O cuts the surface in n points.* This number n, the degree of the equation $F = 0$, is independent of the frame of reference, and is called the *order* of the surface.

Any plane section of the surface is an algebraic curve of order n, for any line in its plane meets the surface, and therefore the curve, in n points.

17·12. If $F_0 \equiv \phi_0 = 0$, one root of the equation in ρ is $\rho = 0$, and O is a point on the surface. A second root will be $\rho = 0$ if also $\phi_1 = 0$. This is a linear homogeneous equation in l, m, n, say $al + bm + cn = 0$, and represents an assemblage of directions through O belonging to one plane $ax + by + cz = 0$. Every line through O in this plane meets the surface in two coincident points at O, and the plane is called the *tangent-plane* at O. Thus when $F_0 = 0$ the surface passes through O and the equation of the tangent-plane at O is $F_1 = 0$.

17·13. *The tangent-plane at O cuts the surface in a curve which has a double-point at O,* for every line through O in this plane meets the surface, and therefore the curve, in two coincident points there.

17·14. There are in general two lines through O, in the tangent-plane, which meet the surface, and therefore the curve of section, in three coincident points at O. These are the tangents to the curve at its double-point. They are called the *inflexional* or *principal tangents* to the surface at O. Their directions are determined by the two equations $\phi_1 = 0$ and $\phi_2 = 0$. Now $F_2 = 0$ represents a quadric cone, therefore the tangents to the curve at O are the generators of this cone which lie in the tangent-plane. According as these lines are real and distinct, imaginary, or coincident, the point O is called a *hyperbolic, elliptic,* or *parabolic point.* A hyperboloid of one sheet and an ellipsoid are examples of surfaces whose points are *all* respectively hyperbolic or elliptic.

If we take the plane $z = 0$ as the tangent-plane at O, the equation of the surface is

$$F \equiv z + ax^2 + by^2 + cz^2 + 2fyz + 2gzx + 2hxy + \text{higher terms},$$

and the inflexional tangents at O are

$$z = 0, \quad ax^2 + by^2 + 2hxy = 0.$$

The point O is then hyperbolic, parabolic, or elliptic according as $h^2 - ab$ is positive, zero, or negative.

17·15. Equation of the tangent-plane at a given point.

Let $P' \equiv (x')$ be the given point and $P \equiv (x)$ any other point. Then a variable point on the line PP' is represented by $(\lambda x + x')$. We find where this line cuts the surface by substituting in the equation of the surface and then expanding by Taylor's theorem:

$$0 = F(\lambda x + x', \dots)$$
$$= F(x', y', z', w') + \lambda \left(x \frac{\partial F}{\partial x'} + \dots \right)$$
$$+ \frac{\lambda^2}{2!} \left(x^2 \frac{\partial^2 F}{\partial x'^2} + \dots + 2zw \frac{\partial^2 F}{\partial z' \partial w'} \right) + \dots + \lambda^n F(x, y, z, w).$$

This is the equation of the nth degree whose roots determine the n points in which the line cuts the surface. Since (x') is on the surface, $F(x', y', z', w') = 0$ and one root is $\lambda = 0$.

17·16. A second root is also $\lambda = o$ if, in addition,

$$x\frac{\partial F}{\partial x'} + y\frac{\partial F}{\partial y'} + z\frac{\partial F}{\partial z'} + w\frac{\partial F}{\partial w'} = o.$$

If $P \equiv (x)$ is a variable point, this equation represents the locus of points such that PP' is a tangent at P', i.e. the tangent-plane at (x').

17·17. Two tangents at O are said to be *conjugate* when they are harmonic conjugates with respect to the principal tangents at O. If we take the tangent-plane at O as $z = o$, and Ox, Oy as a pair of conjugate tangents, the equation of the surface is

$$F \equiv z + ax^2 + by^2 + cz^2 + 2fyz + 2gzx + \text{higher terms},$$

the principal tangents being $z = o$, $ax^2 + by^2 = o$. The tangent-plane at a point $P \equiv [\delta x, o, o]$ on Ox, very near to O, is (neglecting δx^2)

$$2a\,\delta x . x + (1 + 2g\delta x)\,z = o,$$

and this cuts $z = o$ in Oy.

Hence each of two conjugate tangents at O is the limiting position of the intersection of the tangent-plane at O with the tangent-plane at a point very near O on the other tangent.

In the case of a parabolic point O the two principal tangents at O coincide, and this tangent t is conjugate to any other tangent at O; in this case the tangent-plane at any point P very near to O ultimately passes through t, and if P lies on t, near O, the tangent-plane at P ultimately coincides with that at O. The tangent-plane at the parabolic point O is therefore said to be a *stationary plane*.

The inflexional tangents become indeterminate if the quadric cone F_2 breaks up into the tangent-plane and another plane, i.e. when F_2 contains F_1 as a factor. Then the three lines in which the cubic cone F_3 meets the tangent-plane meet the surface in four coincident points. The curve of section of the tangent-plane at O has then a triple point at O. The point O is called a *point of osculation*.

17·2. Curvature.

The distinction between hyperbolic and elliptic points can be explained with reference to curvature. Consider a section of the surface by a plane passing through a fixed line ON, and let OT, OT' be the two inflexional tangents through O. The plane NOT meets the surface in a curve which is met by OT in three coincident points at O, hence O is a point of inflexion on the curve, and the section has zero curvature at O. As the plane is rotated about ON, the curvature of the section becomes reversed, or changed in sign, when the plane passes through OT or OT'. In the case of an elliptic point all sections through O have curvature of the same sign.

Ex. If every point of a surface is a parabolic point, prove that the surface is a developable.

17·21. Meunier's Theorem.

We shall assume now rectangular coordinates with $z = 0$ as the tangent-plane at O, and therefore Oz the normal; and first we shall consider sections through a fixed tangent Ox.

Let the plane of section make an angle ϕ with the normal plane xOz. Take any point P on the curve of section, near O; let L be its projection on Ox and N on xOy. A definite circle is determined which passes through P and touches Ox at O, and in the limit when P approaches O this circle becomes the

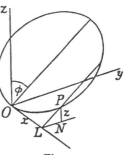

Fig. 49

osculating circle or circle of curvature of the curve of section. If its radius is ρ, then approximately $LP(2\rho - LP) = x^2$. But $LP = z \sec\phi$, hence

$$\rho = \operatorname{Lim}_{x \to 0} \tfrac{1}{2}(x^2/z) \cos\phi.$$

Substituting $y = z \tan\phi$ in the equation of the surface

$$z + ax^2 + by^2 + cz^2 + 2fyz + 2gzx + 2hxy + \ldots = 0,$$

dividing by z and letting x and $z \to 0$ we find

$$\rho = -\frac{1}{2a} \cos\phi.$$

Hence of all sections through a given tangent the normal section has the greatest radius of curvature, and if this is denoted by ρ_n

$$\rho = \rho_n \cos\phi.$$

This is known as *Meunier's Theorem**.

---

* Jean Baptiste Marie Charles Meusnier (1754–93): *Mémoire sur la courbure des surfaces*, 1776.

17·22. Measure of curvature.

We consider next the sections through the normal. Let the plane of section make an angle θ with the plane of xz; let K be the projection of P on the plane of xy and $OK = r$. Then we find by a similar method the radius of curvature

$$\rho = \operatorname*{Lim}_{r \to 0} \tfrac{1}{2} r^2 / z,$$

and the curvature of the section

$$\sigma = \rho^{-1} = -2 (a \cos^2\theta + 2h \cos\theta \sin\theta + b \sin^2\theta).$$

If O is an elliptic point, $h^2 - ab < 0$ and σ is always of the same sign; but if O is hyperbolic, σ vanishes and changes sign when the plane passes through one of the inflexional tangents OT, OT'. In either case σ acquires a maximum or minimum value when

$$\tan 2\theta = 2h/(a-b).$$

The two planes which correspond to these directions bisect the angles between ZOT and ZOT' and are at right angles. If σ_1 and σ_2 are the maximum and minimum values we find

$$\sigma_1 \sigma_2 = 4(ab - h^2) \quad \text{and} \quad \tfrac{1}{2}(\sigma_1 + \sigma_2) = -(a+b),$$

both of which are invariants of the quadratic expression

$$ax^2 + 2hxy + by^2$$

for orthogonal transformations. $\sigma_1 \sigma_2$ is called the *Gaussian measure of curvature** of the surface at O, and the positive value of $\tfrac{1}{2}(\sigma_1 + \sigma_2)$ is called the *mean curvature*. At an elliptic point the measure of curvature is positive, and at a hyperbolic point it is negative. At a parabolic point one of the curvatures is zero, corresponding to the section which contains the inflexional tangent, and the measure of curvature is zero.

A quadric surface has its measure of curvature everywhere of the same sign, but in general on a surface there are regions of positive curvature and regions of negative curvature, and these are separated by an *inflexional curve* or locus of parabolic points. A familiar example is the anchor-ring which is generated by a circle of radius b rotating about an axis in its plane at a distance

---

* Carl Friedrich Gauss (1777–1855): *Disquisitiones generales circa superficies curvas* (1827). English trans. by J. C. Morehead and A. M. Hiltebeitel, Princeton Univ. 1902.

a from the centre. A concentric sphere of radius $\sqrt{(a^2+b^2)}$ cuts the surface in two circles which separate the outer positively curved region from the inner region of negative curvature.

17·3. Polars.

The equation (17·16), whether the point (x') lies on the surface or not, represents a definite plane associated with the point (x'); this is called the *polar-plane* of (x'). The equation may be expressed in the following notation

$$(xD_{x'})F=0,$$

where $$(xD_{x'})\equiv x\,\frac{\partial}{\partial x'}+y\,\frac{\partial}{\partial y'}+z\,\frac{\partial}{\partial z'}+w\,\frac{\partial}{\partial w'}.$$

This is called the *polarising operator*. A repetition of this operator gives

$$(xD_{x'})^2\equiv x^2\,\frac{\partial^2}{\partial x'^2}+\ldots+2zw\,\frac{\partial^2}{\partial z'\partial w'},$$

since x, y, z, w are independent of x', y', z', w'. The equation (17·15) can then be written

$$F(x',y',z',w')+\lambda(xD_{x'})F+\tfrac{1}{2}\lambda^2(xD_{x'})^2F+\ldots=0$$

or $$e^{\lambda\,(xD_{x'})}F=0.$$

Equating to zero the coefficient of λ^2 we obtain a quadratic equation $(xD_{x'})^2F=0$, which represents a quadric called the *polar quadric* of (x'); and similarly a series of polar surfaces is obtained by equating to zero the coefficients of the various powers of λ. It is conventional to reckon the series of polars from the end, and the *first polar* is a surface of order $n-1$ represented by the equation

$$x'\,\frac{\partial F}{\partial x}+y'\,\frac{\partial F}{\partial y}+z'\,\frac{\partial F}{\partial z}+w'\,\frac{\partial F}{\partial w}=0$$

or $$(x'D_x)F=0.$$

The rth polar of (x') with respect to F is $(x'D_x)^r F=0$, and the sth polar of (x') with respect to this surface is $(x'D_x)^{r+s}F=0$, which is the $(r+s)$th polar of (x') with respect to F. In particular the polar plane of (x') with respect to F is also the polar plane of (x') with respect to any of the polar surfaces of (x'); and if (x') lies on F, so that its polar plane is the tangent-plane at (x'), this is also the tangent-plane to each of the polar surfaces. Hence all the polar surfaces of a point on F touch one another at this point.

17·31. *The first polar of* $P(x')$ *passes through the points of contact of all the tangent planes or lines through* P. Let (x_1) be the point of contact of a tangent-plane through P; the equation of the tangent-plane at this point is $(xD_{x_1})F = 0$. Since it passes through P, $(x'D_{x_1})F = 0$, hence (x_1) lies on the locus $(x'D_x)F = 0$ which is the first polar of P.

17·32. The Hessian.

Returning to the section of the surface by the tangent-plane at a point P, the tangents to the curve of section at the double-point P are lines which meet the curve, and therefore the surface, in three coincident points at P. These are determined therefore by the two equations

$$(xD_{x'})F = 0, \quad (xD_{x'})^2 F = 0,$$

which represent respectively the tangent-plane at P and the polar quadric of P. But since P lies on the surface the polar quadric touches the tangent-plane at P and its two generators through P are therefore the tangents to the curve of section at the double-point.

When P is a parabolic point the two generators through P coincide and the polar quadric becomes a cone. The condition for this is expressed by the vanishing of the discriminant of the quadric. Dropping the dashes, let F_{xx} denote $\dfrac{\partial^2 F}{\partial x^2}$, and so on, then the condition is

$$H \equiv \begin{vmatrix} F_{xx} & F_{xy} & F_{xz} & F_{xw} \\ F_{yx} & F_{yy} & F_{yz} & F_{yw} \\ F_{zx} & F_{zy} & F_{zz} & F_{zw} \\ F_{wx} & F_{wy} & F_{wz} & F_{ww} \end{vmatrix} = 0.$$

This equation represents a surface of order $4(n-2)$ which is called the *Hessian* of F. It is the locus of points whose polar quadrics are specialised as cones.

The parabolic or inflexional points on a surface therefore lie on a curve of order $4n(n-2)$ which is the intersection of the surface with its Hessian.

17·33. The Hessian of a given surface is one of a series of covariants, the *discriminant surfaces* whose equations are formed by equating to zero the discriminants of the various polars of a variable point. The Hessian is formed from the discriminant of the polar quadric and is the locus of points (y) whose polar quadrics are cones. If $(xD_v)^2 F = 0$ is a cone with vertex (x), the coordinates (x) satisfy the four equations (see 8·31)

$$x_0 \frac{\partial^2 F}{\partial y_0 \partial y_i} + x_1 \frac{\partial^2 F}{\partial y_1 \partial y_i} + x_2 \frac{\partial^2 F}{\partial y_2 \partial y_i} + x_3 \frac{\partial^2 F}{\partial y_3 \partial y_i} = 0 \quad (i=0, 1, 2, 3).$$

$$\ldots\ldots(1)$$

Eliminating (x) we get the equation of the Hessian in current coordinates (y). Eliminating (y) we get the locus of the vertices of those polar quadrics which are specialised as cones. But this eliminant is the discriminant of the first polar of (y), viz. $(xD_v)F = 0$. This locus, whose order is $4(n-2)^3$, is called the *Steinerian* of the given surface. Thus the Steinerian is either the locus of vertices of polar quadrics which are specialised as cones, or the locus of points whose first polars have a double-point. Similarly the Hessian is either the locus of points whose polar quadrics are cones, or the locus of double-points of those first polars which have a double-point.

Similarly we have a locus H' of points whose polar cubics have a double-point and this is also the locus of double-points of those second polars which have a double-point. Associated with this there is a locus S' of the double-points of those polar cubics which have a double-point, and this is also the locus of points whose second polars have a double-point. And so on.

Evidently for a cubic surface the Hessian and the Steinerian are one and the same, and it has no other covariants of this kind.

If $P(y')$ is a point on the Hessian there is a corresponding point $Q(x')$ on the Steinerian, the vertex of the quadric cone which is the polar quadric of P. The polar plane of P with respect to F is

$$x_0 \frac{\partial F}{\partial y_0'} + x_1 \frac{\partial F}{\partial y_1'} + x_2 \frac{\partial F}{\partial y_2'} + x_3 \frac{\partial F}{\partial y_3'} = 0.$$

For a variable point (y') on the Hessian the envelope of these polar planes is determined by eliminating (y') between the four consistent equations

$$\frac{\partial}{\partial y_i'} \left(x_0 \frac{\partial F}{\partial y_0'} + \ldots \right) = 0,$$

but the resultant is the equation of the Steinerian. Hence the polar plane with respect to F of any point on the Hessian is a tangent-plane to the Steinerian. The point of contact is in fact the corresponding point on the Steinerian, for multiplying the four equations (1) respectively by $y_0, \ldots, y_3$ and adding we obtain

$$x_0 \frac{\partial F}{\partial y_0} + x_1 \frac{\partial F}{\partial y_1} + x_2 \frac{\partial F}{\partial y_2} + x_3 \frac{\partial F}{\partial y_3} = 0,$$

i.e. the polar plane of (y) passes through the corresponding point (x) on the Steinerian.

We have seen that the Hessian of a surface F meets F in the locus of parabolic points. The polar planes of these points are the (stationary) tangent-planes to the surface. These therefore form a developable circumscribing the Steinerian. If F is a developable all its points are parabolic points and therefore, in this case, the Hessian contains F, breaking up into F and another surface which Cayley called the *Prohessian*.

17·4. The general homogeneous equation of the nth degree in four variables contains $\frac{1}{6}(n+1)(n+2)(n+3)$ terms; as only the ratios of the coefficients are significant an algebraic surface of order n is in general determined by

$$\tfrac{1}{6}(n+1)(n+2)(n+3) - 1 = \tfrac{1}{6}n(n^2 + 6n + 11)$$

constants. This is called the *constant-number* of the surface. Since the condition that a given point should lie on the surface is expressed by a linear homogeneous equation in the coefficients, an algebraic surface of order n is in general determined uniquely by $\tfrac{1}{6}n(n^2 + 6n + 11)$ points.

17·41. The *class* of a surface is equal to the degree of its tangential equation and is the number of tangent-planes which pass through an arbitrary line. Let the line be PQ. The point of contact of a tangent-plane through $P(x_1)$ lies on the first polar of P, say $u_1 = 0$, and the point of contact of a tangent-plane through $Q(x_2)$ lies on the first polar of Q, say $u_2 = 0$. The first polar of any point $(x_1 + \lambda x_2)$ on PQ is $u_1 + \lambda u_2 = 0$, hence the first polars of points on the given line all have a curve in common, of order $(n-1)^2$, and this cuts the given surface in $n(n-1)^2$ points. Hence *the class of a surface of order n cannot exceed $n(n-1)^2$*. The actual number may fall short of this for, as we shall see, some of the intersections may be double-points, etc., of the surface and not points of contact of tangent-planes.

The class of a surface is also equal to the class of the tangent-cone from an arbitrary point to the surface, and this is equal to the class of an arbitrary plane section of the cone.

17·42. Reciprocal surfaces.

If $F(x, y, z, w) = 0$ is the point-equation of a surface, $F(\xi, \eta, \zeta, \omega) = 0$ is the tangential equation of another surface called the *reciprocal* of the former. If $\Phi(\xi, \eta, \zeta, \omega) = 0$ is the tangential equation of the former, $\Phi(x, y, z, w) = 0$ is the point-equation of the latter. The relationship between the two surfaces is symmetrical; each is the reciprocal of the other. More generally, if $F = 0$ is a given surface and $f = 0$ a given quadric, the envelope of the polar planes, with respect to f, of points on F is a surface called the polar reciprocal or simply the reciprocal of F with respect to the quadric f. When $f \equiv x^2 + y^2 + z^2 + w^2$ the polar reciprocal of $F(x, y, z, w) = 0$ is $F(\xi, \eta, \zeta, \omega) = 0$. The order of the reciprocal surface is equal to the class of the original surface, and *vice versa*. Hence if m is the class, the order cannot exceed $m(m-1)^2$.

17·5. Double-points.

When the order is given the class is reduced, as in the analogous case of a plane curve, by the existence of certain *point-singularities*, and when the class is given the order is reduced by the presence of certain *tangential singularities*.

17·51. Consider again the equation in 17·11. If $F_0 = 0$ and also $F_1 = 0$ identically, so that the equation contains no terms of lower degree than the second in x, y, z, *every* line through the origin meets the surface in two coincident points there, and there is a quadric cone of lines $F_2 = 0$ which meet the surface in three coincident points. In this case the origin is a singular point (double-point) which is called a *node* or *conical point*.

17·52. From the equation (17·15) we see that (x) will be a node if the coordinates satisfy the five equations

$$F(x, y, z, w) = 0, \quad \frac{\partial F}{\partial x} = 0, \quad \frac{\partial F}{\partial y} = 0, \quad \frac{\partial F}{\partial z} = 0, \quad \frac{\partial F}{\partial w} = 0.$$

These are equivalent, however, to four only, since

$$xF_x + yF_y + zF_z + wF_w = nF.$$

These four equations cannot in general be satisfied simultaneously, hence a surface does not in general possess any double-points. Eliminating x, y, z, w between the four equations $F_x = 0$, $F_y = 0$, $F_z = 0$, $F_w = 0$ we obtain a relation between the coefficients which is called the *discriminant* of the surface.

17·53. When P is a node the cone of tangents at P may break up into two planes, distinct or coincident. P is then called a *biplanar* or a *uniplanar* node; these terms are sometimes contracted to *binode* and *unode*. Every plane through a conical point P cuts the surface in a curve having a double-point at P, but the proper tangent-planes at P are those which touch the cone, and these meet the surface in a curve having a cusp at P. When P is a biplanar node every plane containing the line of intersection of the two planes meets the surface in a curve having a cusp at P; either of the planes meets the surface in a curve having a triple-point at P.

17·531. When a surface has no double-points the tangent-cone from an arbitrary point is of class $n(n-1)^2$. We consider how this number is reduced when the surface has a double-point. Take the double-point as $D \equiv [0, 0, 0, 1]$, and the vertex of the cone as $A \equiv [1, 0, 0, 0]$. The equation of the surface is of the form

$$F \equiv (ax^2 + by^2 + cz^2 + 2fyz + 2gzx + 2hxy)w^{n-2}$$
$$+ \text{terms of lower degree in } w = 0.$$

The first polar of A is

$$F' \equiv (ax + hy + gz)w^{n-2} + \text{terms of lower degree in } w = 0.$$

The equation of the tangent-cone from A is obtained by eliminating x between these two equations and is of degree $n(n-1)$; it is generated by lines joining A to the points of intersection of F and F'. F' passes through D, and the tangent-plane there is

$$ax + hy + gz = 0.$$

This cuts $\quad ax^2 + by^2 + cz^2 + 2fyz + 2gzx + 2hxy = 0$

(the assemblage of tangents to F at the double-point) in general in two distinct lines, and A determines with them two planes α, β whose equation

$$-(hy+gz)^2 + a(by^2 + 2fyz + cz^2) = 0$$

or

$$(ab - h^2)y^2 - 2(gh - af)yz + (ca - g^2)z^2 = 0$$

is obtained by eliminating x. The left-hand side is the co-efficient of the highest power of w in the equation of the tangent-cone.

Any section of the cone by a plane γ has then a double-point, the tangents at which are the intersections of γ with α and β. If the double-point at D is a binode, $ax^2 + \ldots = 0$ breaks up into two planes and the discriminant Δ of this quadratic vanishes. But

$$(ab - h^2)(ca - g^2) - (gh - af)^2 = a\Delta = 0,$$

hence in this case the section of the cone has a cusp. If there is a unode at D, $ax^2 + \ldots = 0$ reduces to two coincident planes and $ab - h^2$, $gh - af$ and $ca - g^2$ all vanish. Hence the section of the cone has now a triple-point.

Since the class of a curve is diminished by 2 for a node, by 3 for a cusp, and by 6 for a triple-point, which is equivalent to three double-points, we deduce that *the class of a surface is diminished by 2 for every conical point, by 3 for every binode, and by 6 for every unode.*

Ex. 1. Show that the cubic surface

$$x^3 + y^3 + z^3 + (2p - 9)w^3 - 3x^2w - (2p - 3)xw^2 - 6y^2w$$
$$- (p - 12)yw^2 + qz^2w + pxyw = 0$$

has a double-point at $[1, 2, 0, 1]$ and determine its character.

Transform to a new tetrahedron of reference $XYZW'$ where $W' \equiv [1, 2, 0, 1]$. The equations of $W'YZ$, $W'ZX$, $W'XY$, XYZ are $x - w = 0$, $y - 2w = 0$, $z = 0$, $w = 0$, and the equations of transformation are

$$x' = x - w, \quad \text{therefore} \quad x = x' + w';$$
$$y' = y - 2w, \qquad\qquad y = y' + 2w';$$
$$z' = z, \qquad\qquad z = z';$$
$$w' = w, \qquad\qquad w = w'.$$

Substituting in the cubic equation we find the coefficients of w'^3 and w'^2 to be

$$1+8+(2p-9)-3-(2p-3)-24-2(p-12)+2p=0,$$
$$3x'+12y'-6x'-(2p-3)x'-24y'-(p-12)y'+p(2x'+y')=0,$$

and the coefficient of w' is

$$3x'^2+6y'^2-3x'^2-6y'^2+qz'^2+px'y'.$$

Hence the tangent-cone at the double-point is

$$px'y'+qz'^2=0.$$

In general it is a conical point; a binode if $q=0$ and a unode if $p=0$.

Ex. 2. Show that the cubic surface

$$ayzw+bxzw+cxyw+dxyz=0$$

has four conic nodes.

Ex. 3. Show that the tangential equation of the cubic surface in Ex. 2 is

$$(a\xi)^{\frac{1}{2}}+(b\eta)^{\frac{1}{2}}+(c\zeta)^{\frac{1}{2}}+(d\omega)^{\frac{1}{2}}=0,$$

and that it is of class 4.

Ex. 4. Show that the quartic

$$\lambda(x^2w^2+y^2z^2)+\mu(y^2w^2+z^2x^2)+\nu(z^2w^2+x^2y^2)=0,$$

where $\lambda+\mu+\nu=0$, has twelve conic nodes.

Ex. 5. In the assemblage of tangents at a node show that (if $n>3$) there are six which meet the surface in four coincident points.

Ex. 6. Show that the cubic surface $3xyz+kw^3=0$ has three binodes, and that it is of class 3.

17·54. When a surface F has a conical point, or singular point at which there are ∞ tangent-planes touching a cone, the reciprocal surface Φ has a singular plane which has ∞ points of contact lying on a conic. This singularity is called a *trope*. The tangential coordinates of a trope satisfy the equations

$$\frac{\partial\Phi}{\partial\xi}=0,\ \frac{\partial\Phi}{\partial\eta}=0,\ \frac{\partial\Phi}{\partial\zeta}=0,\ \frac{\partial\Phi}{\partial\omega}=0.$$

Corresponding to a binode we have a tangent-plane with two points of contact (*double tangent-plane*). A double tangent-plane in general meets the surface in a curve having two double-points.

Ex. Find the tangential equation of the anchor-ring

$$F \equiv \{x^2 + y^2 + z^2 + (a^2 - b^2) w^2\}^2 - 4a^2 (x^2 + y^2) w^2 = 0.$$

Show that it has two conical points and two tropes.

Ans. $\{(a^2 - b^2)(\xi^2 + \eta^2) - b^2\zeta^2 + \omega^2\}^2 = 4a^2\omega^2(\xi^2 + \eta^2).$

17·55. If a plane section of a surface is a curve having a double-point at O, either the surface has a double-point at O or the plane touches the surface at O. If the plane section has two double-points the plane either passes through two double-points of the surface, or touches the surface and passes through a double-point, or is a double tangent-plane, touching at two points. If the plane section has three double-points, and the plane does not pass through any double-points of the surface, it is a *triple tangent-plane* or *tritangent-plane*. One condition is required in order that a given plane may touch the surface, hence a surface has ∞^2 tangent-planes, and these envelop the surface. The condition, expressed in terms of the coordinates of the plane, is the tangential equation of the surface (11·3).

17·56. Hence also a surface without double-points has in general a single infinity of double tangent-planes; these form the planes of a developable doubly circumscribed to the surface, or *bitangent developable*. Further, a surface without double-points has in general a finite number of triple tangent-planes; and in general has no tangent-planes of higher multiplicity unless it is specialised. (A quadric surface has no double tangent-planes, for a conic, in order to have two double-points, must degenerate to two coincident straight lines. A cubic surface may have double tangent-planes, but the line joining two points of contact must lie entirely in the surface.)

17·57. The reciprocal relations may be deduced from the tangential equation. The assemblage of planes through the point $P \equiv [x', y', z', w']$ tangent to the surface $\Phi(\xi, \eta, \zeta, \omega) = 0$ is represented by the simultaneous equations

$$\Phi(\xi, \eta, \zeta, \omega) = 0, \quad x'\xi + y'\eta + z'\zeta + w'\omega = 0,$$

and forms the tangent-cone from P to the surface, a cone of order m equal to the class of the surface. If the point P lies on the surface, the tangent-cone has the tangent-plane at P as a double-plane; and conversely if the tangent-cone from P has a double-plane, either this plane is a double tangent-plane for the surface or P lies on the surface. If P does not lie on any double tangent plane and the tangent-cone from P has two double-planes, P is a double-point on the surface; and if the tangent-cone has three double-planes, P is a triple-point. A surface without double tangent-planes has in general a single infinity of double-points (binodes), forming a double-curve on the surface; and a finite number of triple-points.

17·6. The curves which lie on a surface are of interest; we shall consider in particular straight lines and conics.

We have seen that a quadric has two singly infinite systems of straight lines, a cubic surface (not ruled) has a finite number, while a surface of higher order does not in general possess any straight lines.

A conic in space is determined by eight conditions: three to determine its plane and five to determine it in the plane. A conic cuts a surface of order n in $2n$ points, and if it contains $2n+1$ points of the surface it will lie entirely in the surface. For $n=2$ therefore only five conditions are given in order that a given conic may lie on a given quadric surface, hence there are ∞^3 conics on a quadric, one for each plane in space. For $n=3$, seven conditions are required, hence a single infinity of conics lie on a cubic surface. For $n>3$, more than eight conditions are required, therefore in general no conics lie on a quartic or any surface of higher order. The plane which contains a conic lying on a cubic surface cuts the surface also in a straight line; the conic and the line form a cubic curve having two double-points, hence the plane is a double tangent-plane. Thus, as we have seen already, a cubic surface in general has a single infinity of double tangent-planes. There is, however, only a finite number of lines since through each line pass a single infinity of planes, thus accounting for the single infinity of conics which lie on the surface. A ruled cubic has a single infinity of lines and ∞^2 conics.

A quartic surface which has a double-line contains ∞ conics, since every plane through the double-line meets the surface again in a conic. There is a remarkable quartic surface, the

Steiner Surface, which contains ∞^2 conics. It is known that the only surfaces which contain ∞^2 conics are the quadric, the ruled cubic and the Steiner surface.

We shall consider now some general properties of ruled surfaces, and in particular the ruled surfaces of the third and fourth orders.

17·7. Ruled surfaces.

17·71. A ruled surface is generated by a straight line having one degree of freedom. It is therefore the complete or partial intersection of three complexes. Let the degrees of the three complexes be n_1, n_2, n_3, then, since in general there is one generating line through every point of the surface*, in order to determine the order of the ruled surface we have to find the number of generators which meet an arbitrary line. Let (a) be the line-coordinates of the line, and (p) those of a generator which meets it. Then $\Sigma ap = 0$. But (p) also satisfy the equations of the three complexes

$$\phi_1(p) = 0, \quad \phi_2(p) = 0, \quad \phi_3(p) = 0$$

which are of degrees n_1, n_2, n_3 respectively. They also satisfy the fundamental quadratic equation $\omega(p) = 0$. These five equations determine the ratios of the p's and give $2n_1 n_2 n_3$ solutions. The order of the ruled surface is therefore $2n_1 n_2 n_3$.

Ex. 1. The lines common to three linear complexes form one regulus of a quadric.

Ex. 2. Show that for all values of u and v the linear complex $up_{02} + vp_{31} + p_{03} + p_{12} = 0$ contains one regulus of the quadric $xy - zw = 0$ and $up_{01} + vp_{23} + p_{03} - p_{12} = 0$ contains the other.

Ex. 3. If the two pairs of generators which the regulus $x = \lambda z$, $w = \lambda y$ has in common with the linear complexes $\Sigma a_{ij} p_{ij} = 0$, $\Sigma b_{ij} p_{ij} = 0$ are harmonic, show that

$$2(a_{01} b_{23} + a_{23} b_{01}) = (a_{03} - a_{12})(b_{03} - b_{12}).$$

Ex. 4. The lines which meet each of three fixed curves, plane or skew, of orders n_1, n_2, n_3 respectively, is in general a ruled surface of order $2n_1 n_2 n_3$.

---

* In the case of a quadric there appear to be two generating lines through every point, but only one set (a regulus) belongs to three given linear complexes; with regard to these the other set must be considered as directrices (see 17·77).

17·72. Let $F(x, y, z, w) = 0$ be the equation of the surface and (x') an arbitrary point on it. Through this point there is a generating line; let (x'') be any point on it, then $(x' + \lambda x'')$ must be a point on the surface for all values of λ. Expanding by Taylor's theorem

$$F(x' + \lambda x'', \ldots) = F(x', \ldots) + \lambda \left(x'' \frac{\partial F}{\partial x'} + \ldots \right) + \ldots = 0.$$

$(x'' D_x) F = 0$ is the condition that (x'') should lie on the tangent-plane at (x'). Hence *the generating line through any point P lies in the tangent-plane at P*, and conversely the tangent-plane at any point P contains the generating line through P. There is therefore a $(1, 1)$ correspondence between the range of points on a generating line and the pencil of planes consisting of the tangent-planes at these points.

17·73. An arbitrary line l cuts the surface in n points; through each of these points there passes a generating line, and each of these with l determines a tangent-plane passing through l. Also every tangent-plane which contains l contains one of these generating lines. Hence there are n tangent-planes through the arbitrary line l, and therefore *the class of the ruled surface is equal to its order*. Either of these may therefore be called the *degree* of the surface. The reciprocal of a ruled surface is a ruled surface of the same degree. The degree of the equation in point-coordinates is equal to the degree of the equation in plane-coordinates.

17·74. The order of any plane section of a surface is equal to the order of the surface, and the class of any tangent-cone is equal to its class. Consider the class of a plane section. This is equal to the number of tangent-lines through an arbitrary point and lying in the plane. But this is equal to the degree of the complex of tangent-lines to the surface, or the degree of the equation in line-coordinates. Again, the order of a tangent-cone is equal to the number of generating lines of the cone which lie in a plane through the vertex; this is also the number of tangent-lines of the surface lying in a given plane and passing through a given point of the plane. This number is called the *rank* of the surface. (This applies whether the surface is ruled or not.)

17·75. Consider any plane section of a ruled surface, not containing a line of the surface. There is a (1, 1) correspondence between the points of the section and the generating lines of the surface, since each line cuts the plane in one point. Hence there is a (1, 1) correspondence between the points of any two plane sections. All plane sections are therefore of the same genus, and this is called the *genus* of the ruled surface. Thus, if any section of the surface is a rational curve, every plane section is rational and the surface is rational; a point on the surface can be rationally represented by two parameters, one for the generator through the point and one for its position on the generator.

17·761. A plane (tangent-plane) containing a generator g of a ruled surface of order n meets the surface again in a plane curve of order $n-1$, and this curve cuts g in $n-1$ points; these are all double-points on the complete curve of intersection of the plane with the surface, and are either points of contact of the plane with the surface or else double-points on the surface. Now at an ordinary point on the surface there is a unique tangent-plane, and reciprocally an ordinary tangent-plane has a unique point of contact. Of the $n-1$ double-points on the curve of intersection with the tangent-plane, one is then the point of contact and the remaining $n-2$ points are double-points on the surface. By varying the tangent-plane we obtain a locus of double-points which form a *double-curve* on the surface.

17·762. Similarly the tangent-cone from a point P of the surface consists of the pencil of tangent-planes having as axis the generator g through P together with a cone of class $n-1$. Through g there are $n-1$ tangent-planes of this cone. One of these is the tangent-plane to the surface at P, the remaining $n-2$ are double tangent-planes of the surface. By varying P we obtain an assemblage of double tangent-planes which form a *bitangent developable* of the surface. Every generating line cuts the double-curve in $n-2$ points, and through every generating line there are $n-2$ planes of the bitangent developable.

17·763. If the surface is rational an arbitrary plane section is a rational algebraic curve of order n, and this has $\frac{1}{2}(n-1)(n-2)$ double-points. This must therefore be the order of the double-curve on the surface. More generally, if the surface is of genus p, an arbitrary plane section has $\frac{1}{2}(n-1)(n-2)-p$ double-points, i.e. *the order of the double-curve on a ruled surface of degree n and genus p is $\frac{1}{2}(n-1)(n-2)-p$.* Reciprocally, this is also equal to the class of the bitangent developable.

17·77. A curve on the surface which is met by every generator is called a *directrix curve*; reciprocally, a developable on the surface, i.e. all of whose planes are tangent-planes of the surface and such that through every generator there is at least one plane of the developable, is called a *directrix developable.* The double-curve is a directrix curve, and the bitangent developable is a directrix developable. In general any plane section, not containing any generator, is a directrix curve.

A ruled surface may be determined by three directrix curves, and if these are of orders m, n, p, the degree of the ruled surface is in general $2mnp$. If, however, the curves intersect or have multiple points the degree of the surface is lowered. In particular if the surface has two line-directrices a, b, and a plane curve K of order m which cuts each of the lines in a single point, A, B respectively, part of the complete assemblage of lines which meet a, b and K consists of the planes aB and bA, and the order of the surface is reduced by 2. More generally, if A, B are multiple points on K of multiplicities α and β, the order is reduced by $\alpha + \beta$. For a ruled surface of order n, which has two line-directrices a, b, it is always possible to choose a plane section (curve of order n) as a third directrix. The condition is that the curve of section should have multiple points where it is cut by a and b, of multiplicities α and β such that $\alpha + \beta = n$; the two lines must then themselves have these multiplicities on the surface. Thus for a cubic surface with double-line l as directrix (nodal directrix) and a single directrix-line l', an arbitrary plane section is a cubic curve cutting l' and having a double-point where it cuts l.

17·78. Ruled cubics.

For a ruled cubic the double-curve is a straight line l, and reciprocally the bitangent developable is of class one, i.e. it consists of a pencil of planes through a line l'. We have to consider whether l and l' may be (a) intersecting, (b) skew, or (c) coincident.

l and l' are both directrix-lines, hence (a) if they intersect, every generator lies either in their plane or passes through their point of intersection. The surface would thus degenerate to either a plane curve or a cone.

We shall consider as the general case (b) that in which l and l' are skew. A plane ϖ through l meets the surface again in a generating line g; every such plane is a tangent-plane, its point of contact being the point P in which g cuts l. Every general point P on l is a binode, and there is a second tangent-plane at P. The pairs of tangent-planes which pass through l form an involution, and the double-elements of this involution are tangent-planes at two unodes, C and D, *cuspidal points* or *pinch-points* on l. These may be real or imaginary.

The tangent-cone from a point P' on l' breaks up into the pencil of planes with axis l' (counted twice) and another pencil of planes through a generator g'. P' lies on the surface and the tangent-plane ϖ' at P' is the plane ($g'l'$). Every plane through l' is a bitangent, and there is a second point of contact of the plane ϖ' on l'. These pairs of points on l' form an involution, and the double-points of this involution are two points, A and B, on l'.

Let P be any point on l, and let the two tangent-planes at P cut l' in P' and Q'; PP' and PQ' are generating lines, and the plane $PP'Q'$ meets the surface in a cubic curve consisting of three straight lines and having three double-points. P is the intersection of this plane with the double-line l, therefore P' and Q' are the two points of contact of the bitangent plane $PP'Q'$. Again if P' is any point on l' there is a unique tangent-plane at P' which cuts l in a unique point P. Hence there is a $(1, 2)$ correspondence between the points of l and the points of l'. Through every point of l there pass two generators, and through every point of l' one. The pairs of points on l' which correspond to points of l form an involution which may be either hyperbolic or elliptic. The simplest algebraic expression of a $(1, 2)$ correspondence in which the involution is hyperbolic is $t = u^2$. Taking l as $x = 0 = y$, l' as $z = 0 = w$, $P \equiv [0, 0, 1, t]$ and $P' \equiv [u, 1, 0, 0]$, so that $t = w/z$ and $u = x/y$, the equation of the ruled surface is $x^2 z - y^2 w = 0$.

This (1, 2) correspondence may be produced geometrically as follows. Let C be a conic which cuts l in one point O, but does not cut l'. The ruled cubic is generated by lines which meet l, l' and the conic. Let P' be any point on l'; the lines through P' which cut l generate a plane which cuts the plane of C in a straight line through O, and this line cuts the conic again in a unique point Q, and $P'Q$ cuts l in a unique point P. Starting with P, the lines through P which meet l' generate a plane which cuts C in two points Q, R distinct from O; QP and RP cut l' in two distinct points P', P''.

17·781. In the case (c) where l' coincides with l, as before every point of l is a binode and every plane through l is a bitangent. But the involutions of pairs of tangent-planes at points on l, and pairs of points of contact on l of bitangent-planes through l, are degenerate. One plane of each pair is fixed and one point of contact of each bitangent is fixed, i.e. there is one plane, say $x=0$, which is a tangent at all points of l, and one point, say $[0, 0, 0, 1]$, which is a point of contact for every plane through l. The general equation of a cubic surface with $x=0=y$ as a double-line is

$$(a_1x + b_1y + c_1z + d_1w)x^2 + (a_2x + b_2y + c_2z + d_2w)y^2$$
$$+ 2(c_3z + d_3w)xy = 0.$$

Every line in the plane $x=0$ meets the surface in three coincident points on l. Hence $c_2 = 0 = d_2$. Also $[0, 0, 0, 1]$ is a unode, therefore every plane section through this point has a cusp there. Hence $d_3 = 0$. $a_2x + b_2y = 0$ is then any plane through l, except $x=0$, and we may take this for the plane $y=0$, thus choosing $a_2 = 0$. Also $a_1x + b_1y + c_1z + d_1w = 0$ is any plane not passing through $[0, 0, 0, 1]$, and we may take this for the plane $w=0$. Then by suitable choice of unit point the equation of the surface reduces to the form

$$wx^2 - y^3 + xyz = 0.$$

This is known as *Cayley's ruled cubic*.

Ex. Show that for the surface $z(x^2 - y^2) - 2xyw = 0$ the tangent-planes at all points on the double-line $x=y=0$ are real, while for the surface $x^2z - y^2w = 0$ the tangent-planes may be real or imaginary.

17·79. Ruled quartics.

17·791. Among ruled quartics has to be included the developable whose curve is the general space cubic curve. There is no developable, other than a cone, of lower order than the fourth. The class of the quartic developable is three. If the freedom-equations of the cubic curve are

$$x : y : z : w = t^3 : t^2 : t : \mathrm{1},$$

the equation of the developable is

$$(xw - yz)^2 - 4(y^2 - xz)(z^2 - yw) = \mathrm{o}.$$

17·792. The non-developable ruled surfaces of the fourth order may be classified according to the nature of the double-curve and the bitangent developable.

When the surface is rational the double-curve is of order 3. It may be

(*a*) *A space cubic.* (It could not be a plane cubic, for then a line in this plane would meet the surface in six points.)

(*b*) *A conic and a straight line.* The plane of the conic cannot contain any other points of the surface, for any line in this plane already meets the surface in four points. Hence the straight line must meet the conic in one point.

(*c*) *Three distinct straight lines.* For the same reason as in (*a*) the three lines cannot be coplanar. Nor can they be all mutually skew. In fact if a quartic surface has two mutually skew double-lines *a* and *b*, and *c* is a third line on the surface, skew to both, all the transversals of *a*, *b*, *c* meet the surface in more than four points and are therefore generators. But these form a regulus; hence in this case the surface resolves into two quadric surfaces. If all three lines are concurrent we shall see (17·93) that the surface is the Steiner surface and is not ruled. If two of the double-lines intersect these form the complete intersection of their plane with the surface and the third line must meet one of them. We have then two skew lines *l*, *l'* and a third line *g* meeting both. No generating line cuts *g*, for a plane through *g* meets the surface again in a conic and this cannot break up into two lines since the surface has no other double-points. But every generating line meets either l or g and either l' or g, therefore it meets both l and l'; these are therefore directrix lines, g itself is one of the generating lines.

(d) *Three straight lines of which two are coincident* (directrices), and one double generator meeting it.

(e) *Three coincident lines*, i.e. *a triple line.*

The bitangent developable, in general (a) a quartic developable on a space cubic curve, may be specialised similarly as (b) a quadric cone and a straight line (i.e. pencil of planes), or three straight lines, (c) distinct, (d) two coincident, or (e) all coincident.

In the case of the elliptic ruled surfaces of the fourth order the double-curve is of order 2 and can only be either two skew directrices, or a directrix counted twice. If a quartic surface has a double-conic it is necessarily rational (see 17·98).

This classification is carried out by W. L. Edge, *The theory of ruled surfaces* (Cambridge, 1931), who enumerates ten rational ruled quartics and two elliptic. Cayley at first enumerated only eight, but later added two which he had overlooked. His memoirs on ''skew surfaces, otherwise scrolls'' are contained in vols. v and vi of his *Collected Papers*; they were written between 1863 and 1868.

Ex. 1. Show that the locus of bisecants of the cubic curve

$$x : y : z : w = t^3 : t^2 : t : 1$$

which belong to the linear complex $\Sigma\Sigma c_{ij} p_{ij} = 0$ is

$$c_{12}\phi_1{}^2 - c_{01}\phi_2{}^2 - c_{03}\phi_3{}^2 - c_{02}\phi_2\phi_3 + (c_{01} + c_{23})\,\phi_3\phi_1 - c_{31}\phi_1\phi_2 = 0,$$

where $\phi_1 \equiv xz - y^2$, $\phi_2 \equiv xw - yz$, $\phi_3 \equiv yw - z^2$,

and that the linear complex to which the ruled quartic $\Sigma\Sigma a_{ij}\phi_i\phi_j = 0$ belongs is

$$a_{22}p_{01} + 2a_{23}p_{02} + a_{33}p_{03} - (a_{22} + 2a_{31})\,p_{23} + 2a_{12}p_{31} - a_{11}p_{12} = 0.$$

Ex. 2. In Ex. 1 if the complex is special, so that

$$a_{22}{}^2 + 2a_{22}a_{31} - 4a_{23}a_{12} + a_{11}a_{33} = 0,$$

show that the surface has also a directrix-line l; through every point of l there passes one generating line and through every point of the cubic curve two. Show also that the bitangent developable is the pencil of planes through l counted three times.

Ex. 3. Show that

$$x^2y^2 = x^2z \, (ax + by) + y^2w \, (cx + dy)$$

is a ruled quartic with a triple line and that the bitangent developable is a proper quartic.

Show that the tangential equation is

$$\{(ad - bc) \, \xi^2 - \omega \, (b\xi + a\eta)\} \, \{(ad - bc) \, \eta^2 - \zeta \, (d\xi + c\eta)\}$$
$$= \{(ad - bc) \, \xi\eta - c\xi\zeta - b\eta\omega - \zeta\omega\}^2.$$

Ex. 4. Show that

$$x^2y^2 = (ax + by) \, (x^2z + y^2w)$$

has a triple line, and that the bitangent developable consists of a pencil of planes through this line, and a quadric cone.

Ex. 5. Show that

$$x^2z \, (ax + by) + y^2w \, (cx + dy) = 0$$

has a triple line l and that the bitangent developable is a pencil of planes through another line l' counted three times. (Tangential equation

$$\xi^2\zeta \, (d\xi - c\eta) - \eta^2\omega \, (b\xi - a\eta) = 0.)$$

Ex. 6. Show that

$$x^2y^2 - (Ax^2 + 2Bxy + Cy^2) \, (xz + yw) = 0$$

has a triple line l and that the bitangent developable is a pencil of planes through l counted three times. (Tangential equation

$$\zeta^2\omega^2 - (A\zeta^2 + 2B\zeta\omega + C\omega^2) \, (\xi\zeta + \eta\omega) = 0.)$$

Ex. 7. Show that the lines joining corresponding points of two conics which are in $(1, 1)$ correspondence generate a ruled quartic.

Ex. 8. If C is a conic and l a line meeting it in one point P, and their points are connected by a $(2, 2)$ correspondence in which P corresponds to itself doubly on both loci, show that the lines joining corresponding points generate a ruled quartic.

Ex. 9. Show that

$$x^2z^2 + axyzw + (bx + cy) \, yw^2 = 0$$

has three double-lines.

Ex. 10. Show that a ruled quartic with three double-lines (one a generator and two directrices) is generated by the lines which meet two skew lines and a conic which has no point in common with either of the two lines.

Ex. 11. Show that
$$(xz - y^2)^2 + a (xz - y^2) yw + (bx + cy) yw^2 = 0$$
has a double-line and a double-conic.

Ex. 12. If a cubic surface has a double-conic it degenerates to a quadric and a plane.

Ex. 13. Show that $z (x^2 - y^2) - 2xyw = 0$ is a ruled cubic in which the involution of pairs of points of contact of bitangent planes is elliptic, and show that if $P \equiv [0, 0, 1, t]$ and $P' \equiv [u, 1, 0, 0]$ are corresponding points on the double and the single directrix respectively, $u^2 - 2tu - 1 = 0$.

Ex. 14. Show that the equation of a ruled cubic can be expressed in the form
$$\alpha x^2 + \beta y^2 + 2\gamma xy = 0,$$
where α, β, γ are expressions of the first degree in x, y, z, w.

Ex. 15. Find the tangential equations of the cubics

(i) $zx^2 - wy^2 = 0$, (ii) $z (x^2 - y^2) - 2xyw = 0$, (iii) $wx^2 + xyz - y^3 = 0$.

Ans.   (i) $\zeta\xi^2 + \omega\eta^2 = 0$, (ii) $\omega(\xi^2 - \eta^2) + 2\xi\eta\zeta = 0$,
                (iii) $\xi\omega^2 + \eta\zeta\omega + \zeta^3 = 0$.

Ex. 16. Show that a quadric surface is generated by a straight line which meets a fixed conic and two straight lines each of which meets the conic in one point.

Ex. 17. Show that the lines joining corresponding points on two skew lines which are in $(2, 2)$ correspondence generate a ruled quartic (in general irrational).

Ex. 18. Show that
$$(xw + yz + azw)^2 = zw(x + y)^2$$
has a double-line and a double-conic, and that the bitangent developable is a pencil of planes counted three times.

Ex. 19. Show that
$$(yz - xy + axw)^2 = xz(x - z)^2$$
has two coincident double directrix-lines and a double generator.

17·8. Cubic surfaces.

Analytically, the general cubic surface is the locus of the general homogeneous equation of the third degree in x, y, z, w. Geometrically, there are several ways in which the surface may be generated. We know that a conic can be generated by the intersection of corresponding lines of two related pencils in a plane; a quadric surface is generated by the line of intersection of corresponding planes of two related pencils with axes mutually skew; a space cubic curve is generated by the point common to a set of corresponding planes of three related pencils whose axes have no common point. If in the last case the planes have each just one fixed point so that they have two degrees of freedom, but are still connected mutually in $(1, 1)$ correspondence, the locus of their common point is a surface, and, as we shall show, it is a cubic surface.

17·81. If O is a fixed point and $\alpha = 0$, $\beta = 0$, $\gamma = 0$ represent three planes through O, any plane through O is represented by the equation $l\alpha + m\beta + n\gamma = 0$. If O' is a second point the planes through O', $l'\alpha' + m'\beta' + n'\gamma' = 0$ say, are correlated to those through O when l', m', n' are connected with l, m, n by linear homogeneous equations of the form $l' = a_1 l + b_1 m + c_1 n$, where a_1, b_1, c_1 are constants. To each plane through O corresponds uniquely a plane through O' and *vice versa*. If α', β', γ' are the planes which correspond respectively to α, β, γ the equations of correlation are simply $l' = al$, $m' = bm$, $n' = cn$. Changing the notation, let O_1 and O_2 be the two fixed points, α_1, β_1, γ_1 three given planes through O_1 and α', β', γ' the corresponding planes through O_2 so that to the plane $l\alpha_1 + m\beta_1 + n\gamma_1 = 0$ through O_1 corresponds $l a\alpha' + m b\beta' + n c\gamma' = 0$ through O_2. Then if we write α_2, β_2, γ_2 for $a\alpha'$, $b\beta'$, $c\gamma'$ the corresponding plane through O_2 is $l\alpha_2 + m\beta_2 + n\gamma_2 = 0$. Similarly we have a corresponding plane $l\alpha_3 + m\beta_3 + n\gamma_3 = 0$ through a third fixed point O_3. Eliminating l, m, n between these three equations we obtain the cubic equation

$$\begin{vmatrix} \alpha_1 & \beta_1 & \gamma_1 \\ \alpha_2 & \beta_2 & \gamma_2 \\ \alpha_3 & \beta_3 & \gamma_3 \end{vmatrix} = 0,$$

which represents the locus of points common to three corresponding planes. The surface passes through each of the points O_1, O_2, O_3, since the determinant vanishes when $\alpha_1 = 0 = \beta_1 = \gamma_1$, etc.

17·811. We have still to discover a geometrical determination of the correlations between the bundles of planes. Consider the quadrics $\beta_1\gamma_2 - \beta_2\gamma_1 = 0$, $\gamma_1\alpha_2 - \gamma_2\alpha_1 = 0$ and $\alpha_1\beta_2 - \alpha_2\beta_1 = 0$. Each pair has a line in common, viz. the first pair has the line $\gamma_1 = 0 = \gamma_2$ in common, and the residual intersection is a space cubic curve which is common to the three quadrics and lies on the cubic surface; it also passes through the points $\alpha_1 = 0 = \beta_1 = \gamma_1$ and $\alpha_2 = 0 = \beta_2 = \gamma_2$, i.e. O_1 and O_2. An arbitrary plane

$$l\alpha_1 + m\beta_1 + n\gamma_1 = 0$$

through O_1 cuts the curve of intersection of the two quadrics

$$\beta_1\gamma_2 - \beta_2\gamma_1 = 0, \quad \gamma_1\alpha_2 - \gamma_2\alpha_1 = 0$$

in four points; one of these is its intersection with the line

$$\gamma_1 = 0 = \gamma_2,$$

one is the point $\alpha_1 = 0 = \beta_1 = \gamma_1$, i.e. O_1; eliminating α_1 and β_1 (γ_1 and γ_2 being $\neq 0$) we find to determine the other two

$$l\alpha_2 + m\beta_2 + n\gamma_2 = 0.$$

Hence the corresponding planes

$$l\alpha_1 + m\beta_1 + n\gamma_1 = 0$$

and $$l\alpha_2 + m\beta_2 + n\gamma_2 = 0$$

cut the cubic curve in the same two points.

Finally, to generate the cubic surface we take two space cubic curves C and C' both passing through a point O. An arbitrary plane through O cuts the first cubic in two points A, B and the second in two points A', B'. The cubic surface is the locus of the intersection of AB and $A'B'$.

We may verify as follows that there is a unique cubic surface which contains two space cubic curves having one point O in common. If nine other points are taken on each cubic curve we obtain nineteen points and these determine a unique cubic surface; but this surface must contain each of the cubics, for a cubic surface can intersect a cubic curve in only nine points while this has ten points in common with each.

17·82. We can now prove that *every cubic surface is rational*, i.e. can be rationally represented on a plane. To every plane

$$l\alpha_1 + m\beta_1 + n\gamma_1 = 0$$

through O_1 there corresponds a unique point P on the surface and also a unique point P' with coordinates $[l, m, n]$ on a fixed plane. Conversely, if P is any point of the cubic surface there is through P a single bisecant of each of the cubic curves and each of these determines with O_1 the same unique plane.

Algebraically, the three equations

$$l\alpha_1 + m\beta_1 + n\gamma_1 = 0, \quad l\alpha_2 + m\beta_2 + n\gamma_2 = 0, \quad l\alpha_3 + m\beta_3 + n\gamma_3 = 0$$

which are linear and homogeneous in x, y, z, w can be solved for the ratios of x, y, z, w, and thus the coordinates of any point on the surface are expressed by homogeneous polynomials of the third degree in l, m, n.

17·83. We have seen that a cubic surface has a finite number of straight lines. It has also, like any algebraic surface, a finite number of tritangent-planes. A tritangent-plane cuts the surface in a cubic curve having three double-points and therefore reducing to three straight lines. If l is a line lying on the surface a plane through l cuts the surface in this line and a conic; these form a plane cubic curve having two double-points. Hence if l does not pass through a double-point of the surface every plane through l is a double tangent-plane. Conversely, a double tangent-plane meets the surface in a straight line and a conic. The bitangent developable therefore consists of a finite number of pencils of planes whose axes are lines of the surface.

17·84. Through each line of the surface there pass a finite number of tritangent-planes which are determined by forming the condition that the residual conic should break up into a pair of straight lines.

If we take $x = 0 = w$ as one line of the surface, the equation of the surface is of the form

$$x\phi + w\psi = 0,$$

where ϕ and ψ are quadratic expressions. A plane $w = \mu x$ through this line cuts the surface again in a conic which is the intersection of the plane $w = \mu x$ with the quadric cone

$$\phi' + \mu\psi' = 0,$$

ϕ' and ψ' being the expressions obtained by substituting $w = \mu x$ in ϕ and ψ. ϕ' and ψ' are homogeneous quadratics in x, y, z, and in each the coefficients of x^2, y^2, z^2, yz, zx, xy contain μ to the powers 2, 0, 0, 0, 1, 1 respectively. The elements of the determinant of $\phi' + \mu\psi'$, whose vanishing is the condition for factorisation, are functions of μ of degrees according to the scheme

$$\begin{matrix} 3 & 2 & 2 \\ 2 & 1 & 1 \\ 2 & 1 & 1 \end{matrix}.$$

Equating this determinant to zero we obtain an equation of the fifth degree in μ. Hence *through any line of the surface there are five tritangent-planes.*

17·85. Consider now the double-points on a cubic surface. Writing the equation in the form

$$F_3 + wF_2 + w^2 F_1 + w^3 F_0 = 0,$$

the origin $O \equiv [0, 0, 0, 1]$ lies on the surface if $F_0 = 0$, and is a double-point if also $F_1 = 0$; it cannot be a triple-point unless the surface is a cone. When O is a double-point, $F_2 = 0$ represents the quadric cone of tangents at O; this is cut by the cubic cone $F_3 = 0$ in six lines which meet the surface in four points at O and therefore lie entirely in the surface. Hence *through a double-point there pass six lines of the surface.*

Conversely, *if three lines of the surface, not in a plane, pass through a point, this point is a double-point.* Let OX, OY, OZ be lines of the surface. Then since the equation is satisfied identically by $y = 0 = z$ and by $z = 0 = x$ and by $x = 0 = y$, it is of the form

$$F_3 + wF_2 = 0,$$

where $F_2 \equiv \Sigma ayz$ and $F_3 \equiv \Sigma cy^2 z + dxyz$. Hence O is a double-point.

17·861. We can now find the number of lines on a general cubic surface without double-points. Starting with a tritangent-plane we obtain as its intersection with the surface three lines a, b, c forming a triangle ABC. Through each of the lines a, b, c there pass four other tritangent-planes, each meeting the surface in three lines, and since no other lines besides a, b, c can pass through A, B or C we have $3 \times 4 \times 2 = 24$ lines in addition to the first three. And besides these 27 lines there are no more, for if l is any line of the surface other than a, b or c, it cuts the plane of

abc in a point lying on one of these lines and has therefore been enumerated among the 24. Hence *there are precisely 27 lines on the surface.* These lines may not all be real.

The following equation* represents a cubic surface with 27 real and distinct lines:

$$\left(\frac{x}{x_2}+\frac{y}{y_2}+\frac{z}{z_2}+\frac{w}{w_2}\right)\left(\frac{xz}{x_1z_1}-\frac{yw}{y_1w_1}\right)$$
$$=\left(\frac{x}{x_1}+\frac{y}{y_1}+\frac{z}{z_1}+\frac{w}{w_1}\right)\left(\frac{xz}{x_2z_2}-\frac{yw}{y_2w_2}\right).$$

17·862. The number of tritangent-planes is now easily determined. Through each line there are five tritangent-planes, and each tritangent-plane contains three lines, hence the number of tritangent-planes $=\frac{1}{3}\times 5\times 27=45$.

The 27 lines and 45 planes and their points of intersection form a configuration which is represented by the scheme

$$\begin{array}{|ccc|}\hline 135 & 10 & 27 \\ 2 & 27 & 3 \\ 9 & 5 & 45 \\ \hline \end{array}$$

Each line is met by five pairs of other lines in the five tritangent-planes through the line, therefore $N_{01}=10$. Through each point there are two lines a, b and through each of these there are four other planes besides the plane (ab), hence $N_{20}=9$. Also on each line there are ten points, each of which belongs to two lines, hence the total number of points is 135.

17·863. Schläfli's notation for the lines on a cubic surface.

Let a_1 and b_1 be two non-intersecting lines on the surface. Through each of these there are five tritangent-planes, each containing two lines which cut the given line. Denote the pairs of lines which cut a_1 by b_2, c_{12}; b_3, c_{13}; b_4, c_{14}; b_5, c_{15}; b_6, c_{16}. Any other line must cut each of the five planes, and must therefore cut one of each of these five pairs of lines; hence b_1 cuts, say, c_{12}, c_{13}, c_{14}, c_{15} and c_{16}. Hence *any two non-intersecting lines of the surface have five common transversals on the surface.* Denote the

* See A. Henderson, *The twenty-seven lines upon the cubic surface*, Cambridge Tracts, No. 13 (1911). Also Cayley, *Collected Math. Papers*, VII, 316–30 and VI, 359–455.

remaining five lines which intersect b_1 by a_2, a_3, a_4, a_5, a_6, pairing these respectively with c_{12}, ..., c_{16}.

We have now 17 lines such that a_1 cuts b_2, ..., b_6, c_{12}, ..., c_{16}, and b_1 cuts a_2, ..., a_6, c_{12}, ..., c_{16}, and they form 10 tritangent-planes. Each line of the surface cuts one and only one line of each triad. Hence since a_2 cuts b_1 and c_{12}, it does not cut a_1, a_3, ..., a_6, c_{13}, ..., c_{16}, b_2; and since a_2 does not cut a_1 it must cut b_3, ..., b_6. Similarly b_2 cuts a_1, a_3, ..., a_6 and c_{12}. Thus of these 17 lines we have 12, a_i and b_i $(i = 1, ..., 6)$, such that a_i meets b_j $(i \neq j)$ but does not meet b_i. Of the other five lines $c_{1i}(i = 2, ..., 6)$, each meets both a_1 and b_1, a_i and b_i.

Now each line of the surface is met by five pairs of inter-secting lines; but so far a_2 is met by just one pair, b_1 and c_{12}, and four single lines b_3, b_4, b_5, b_6, no two of which intersect. Hence there are four other lines, say c_{23}, c_{24}, c_{25}, c_{26}, pairing respectively with these. Now b_2 meets one of each of the triads: a_2, b_3, c_{23}; a_2, b_4, c_{24}; a_2, b_5, c_{25}; a_2, b_6, c_{26}; and since it does not meet a_2, b_3, b_4, b_5 or b_6 it must meet c_{23}, c_{24}, c_{25} and c_{26}. Further, a_3 does not meet a_2 or b_3, therefore it meets c_{23}; similarly a_4 meets c_{24}, a_5 meets c_{25}, and a_6 meets c_{26}.

We have now a_3 cut by the two pairs b_1, c_{13} and b_2, c_{23}, and also by b_4, b_5, b_6, no two of which intersect. Hence we have three more lines, say c_{34}, c_{35}, c_{36}, paired with these. Then by the same reasoning b_3 meets these three lines, and as a_4 does not meet a_3 or b_4 it must meet c_{34}; similarly a_5 meets c_{35} and a_6 meets c_{36}.

a_4 is now cut by the three pairs b_1, c_{14}; b_2, c_{24}; b_3, c_{34} and by b_5 and b_6 which do not intersect. Hence we have two more lines, say c_{45} and c_{46}, paired with these.

Lastly we have the line c_{56} meeting a_5, a_6, b_5 and b_6.

We have now obtained the 27 lines a_i, b_i $(i = 1, ..., 6)$, c_{ij} $(i \neq j = 1, ..., 6)$, and the whole scheme of intersections is given by the statements that

$$a_i \text{ meets } b_j \ (i \neq j),$$
$$a_i \text{ and } b_i \text{ meet } c_{ij},$$
$$c_{ij} \text{ meets } c_{kl} \ (i \neq j \neq k \neq l).$$

The 45 triads which determine the tritangent-planes are

30 of $a_i b_j c_{ij}$, and 15 of $c_{ij} c_{kl} c_{mn}$.

17·87. The set of 12 lines

$$\left.\begin{array}{c} a_1 \ a_2 \ldots a_6 \\ b_1 \ b_2 \ldots b_6 \end{array}\right\}$$

which are such that each one intersects only the five which do not lie in the same row or column is called a *double-six*. There are 36 of these, the others being of the types

$$\left.\begin{array}{cccccc} a_1 & b_1 & c_{23} & c_{24} & c_{25} & c_{26} \\ a_2 & b_2 & c_{13} & c_{14} & c_{15} & c_{16} \end{array}\right\} \quad (15)$$

and

$$\left.\begin{array}{cccccc} a_1 & a_2 & a_3 & c_{56} & c_{64} & c_{45} \\ c_{23} & c_{31} & c_{12} & b_4 & b_5 & b_6 \end{array}\right\} \quad (20)$$

To determine a double-six take two non-intersecting lines, say a_1 and a_2; write down the pairs which intersect them:

$$\begin{array}{cccccc} a_1 \text{ meets} & b_2 & (b_3) & (b_4) & (b_5) & (b_6) \\ & (c_{12}) & c_{13} & c_{14} & c_{15} & c_{16} \\ a_2 \text{ meets} & b_1 & (b_3) & (b_4) & (b_5) & (b_6) \\ & (c_{21}) & c_{23} & c_{24} & c_{25} & c_{26} \end{array}$$

Then delete the symbols of the lines which are common to the two sets. a_1 is then taken with the remainder of the lines which meet a_2, and *vice versa*.

17·88. Classification of cubic surfaces according to the reality of the 27 lines.

The surface being assumed to be general (without double-points), and the 27 lines all distinct, the reader may verify that there are the following five cases:

(1) All the 27 lines real and all the 45 planes real.

(2) [Every imaginary line of the first species, i.e. meeting its conjugate.]

Three real lines (forming a triangle), 13 real planes, 15 real points (12 elliptic and three hyperbolic).

(3) [Some imaginary lines of the second species, i.e. not meeting their conjugates.]

(3*a*) [The five transversals of two non-intersecting conjugate imaginary lines all real.]

15 real lines, 15 real planes, 45 real points (all hyperbolic).

(3*b*) [Three real transversals.]

Seven real lines, five real planes, 11 real points (two elliptic and nine hyperbolic).

(3*c*) [One real transversal.]

Three real lines, seven real planes, nine real points (six elliptic and three hyperbolic).

17·89. The projective classification of cubic surfaces, without regard to the reality of the lines, is based on their singularities. A cubic surface cannot have more than four conical points, for the class is diminished by two for every conical point, and the class of the general cubic surface is $n(n-1)^2 = 12$. If there were five nodes the class would be reduced to two, but a surface of class 2 is a quadric. There are 21 species when we distinguish biplanar and uniplanar nodes, and whether the planes of a binode contain lines of the surface, and so on. A conical point is represented by C_2, an ordinary binode by B_3, an ordinary unode by U_6, other varieties by B_4, B_5, B_6, U_7, U_8. The suffix in each case denotes the number by which the class (12) is reduced. (All combinations are possible except $2B_4$, $C_2 + U_6$, $C_2 + U_7$ and $B_3 + U_6$.)

Ex. 1. The surface
$$w(x+y+z)(lx+my+nz) - kxyz = 0$$
has a single binode B_3 at [0, 0, 0, 1]. Find the equations of the planes at the binode and the six lines through it.

Ex. 2. The surface
$$xzw - (x+z)(x^2 - y^2 + z^2) = 0$$
has a binode B_4 at [0, 0, 0, 1]. Show that the plane $x+z=0$ touches the surface at all points of the edge of the binode (i.e. the edge is *torsal*; it is a line of the surface).

Ex. 3. The surface
$$xzw + y^2z + x^2y - z^3 = 0$$
has a binode B_5 at [0, 0, 0, 1]. Show that the edge of the binode is torsal and the tangent-plane at any point of it coincides with one of the planes of the binode.

Ex. 4. The surface
$$xzw + y^2z + x^3 - z^3 = 0$$
has a binode B_6 at [0, 0, 0, 1]. Show that the edge of the binode is *oscular*, i.e. one of the planes of the binode meets the surface in three coincident lines.

Ex. 5. The surface
$$w(x+y+z)^2 + xyz = 0$$
has a unode U_6 at $[0, 0, 0, 1]$ whose plane meets the surface in three distinct lines.

Ex. 6. The surface
$$wx^2 + xz^2 + y^2z = 0$$
has a unode U_7 at $[0, 0, 0, 1]$ whose plane meets the surface in three lines, two of which are coincident.

Ex. 7. The surface
$$wx^2 + xz^2 + y^3 = 0$$
has a unode U_8 at $[0, 0, 0, 1]$ whose plane meets the surface in three coincident lines.

17·9. Quartic surfaces.

17·91. There is a very great variety of surfaces of the fourth order and we shall consider only a few types. A quartic surface does not in general possess any lines; at the other extreme we have ruled quartics having an infinity of lines. It is known that a quartic surface, not ruled, cannot have more than 80 lines, but whether a quartic surface can possess so many lines without being ruled is not known. The Weddle surface (locus of vertices of quadric cones through six given points) contains 25 lines, and Richmond* has given an example
$$x^4 - 6x^2y^2 + y^4 = z^4 - 6z^2w^2 + w^4$$
which contains 48, only 24, however, being real.

17·911. *Ex. The Weddle Surface.* Show that the locus of vertices of quadric cones which pass through the six points $[1, 0, 0, 0]$, $[0, 1, 0, 0]$, $[0, 0, 1, 0]$, $[0, 0, 0, 1]$, $[1, 1, 1, 1]$, $[a, b, c, d]$ is
$$w^2 \Sigma a(b-c)yz - w\Sigma(a-d)x(cy^2-bz^2) - dxyz\Sigma(b-c)x = 0.$$
Prove that each of the six points is a conical node and that the surface contains the 15 lines joining these points, and the 10 lines of intersection of pairs of planes each containing three of the points.

17·92. A quartic surface is not in general rational, and there is no very simple criterion for its rationality. There are three main types of rational quartics, viz. quartics with

    (1) a triple-point, (2) a double-line, (3) a double-conic.

---

* *Edinburgh Math. Notes*, October, 1930.

But in addition to these there are some other isolated forms. We shall confine our discussion of quartics to examples of these three types of rational surfaces.

17·921. A surface of order n which has a multiple point O of order n is necessarily a cone, for any line through O and one other point of the surface meets the surface in more than n points and therefore lies entirely in the surface.

A surface of order n which has a multiple point O of order $n-1$ was called by Cayley a *Monoid*. A monoid of any order is a rational surface, for any line through O meets the surface in just one other point; there is therefore a $(1, 1)$ correspondence between the points of the surface and the lines through O and therefore the points of a plane. A quartic surface with a triple point is a particular case of a monoid.

Ex. Show that if P and Q are multiple points of orders r and s on a surface of order n, and $r+s>n$, the line PQ is a multiple line on the surface, of order $r+s-n$.

17·93. The Steiner surface.

A quartic monoid of special interest is one which has three double-lines passing through the triple-point. Taking as the triple-point $[0, 0, 0, 1]$ and as the double-lines $y=0=z$, $z=0=x$, $x=0=y$, the equation of the surface is of the form

$$ay^2z^2 + bz^2x^2 + cx^2y^2 + xyz(fx+gy+hz+kw)=0.$$

Changing the plane of reference $w=0$, and choosing the unit-point suitably, the equation can be reduced to the simpler form

$$y^2z^2 + z^2x^2 + x^2y^2 - 2xyzw = 0.$$

The surface is named after Steiner who studied it during a visit to Rome, and it is sometimes called the Roman surface.

17·931. To obtain parametric equations write $\rho x = 2\mu\nu$, $\rho y = 2\nu\lambda$, $\rho z = 2\lambda\mu$, then we find $\rho w = \lambda^2 + \mu^2 + \nu^2$. From these a symmetrical form of the equation can be obtained, referred to another tetrahedron. We have

$$\rho(w+x+y+z) = (\lambda+\mu+\nu)^2,$$
$$\rho(w+x-y-z) = (-\lambda+\mu+\nu)^2, \text{ etc.}$$

Hence writing

$$w + x - y - z = X, \quad w - x + y - z = Y,$$

$$w - x - y + z = Z, \quad w + x + y + z = W,$$

we have
$$X^{\frac{1}{2}} + Y^{\frac{1}{2}} + Z^{\frac{1}{2}} + W^{\frac{1}{2}} = 0.$$

17·932. This shows that each of the four planes of reference $X = 0$, etc., meets the surface in a conic twice. These are not double-conics on the surface, but each of the four planes is a singular tangent-plane or trope touching the surface at all points of the conic. The triple-point is [1, 1, 1, 1] and the double-lines are $Y = Z$, $X = W$; $Z = X$, $Y = W$; $X = Y$, $Z = W$.

Ex. 1. Show that the tangential equation of the Steiner surface $\Sigma x^{\frac{1}{2}} = 0$ is $\Sigma \xi^{-1} = 0$, and that it is therefore of class 3.

Ex. 2. Show that the Steiner surface is the reciprocal of a cubic surface with four conic nodes.

Ex. 3. Show that the four conics at which the surface $\Sigma x^{\frac{1}{2}} = 0$ is touched by the planes $x = 0$, etc., touch the edges of the tetrahedron of reference, and that the six points of contact, one on each edge, are unodes.

17·933. Every plane section of the Steiner surface is a quartic curve with three nodes, where the plane cuts the three double-lines. The section by a tangent-plane has an additional node and therefore breaks up into two conics. As there are ∞^2 tangent-planes the surface contains ∞^2 conics.

17·934. In the (λ, μ, ν)-plane when the parametric equations are
$$x : y : z : w = 2\mu\nu : 2\nu\lambda : 2\lambda\mu : \lambda^2 + \mu^2 + \nu^2$$

a plane section $\xi x + \eta y + \zeta z + \omega w = 0$ is represented by a conic

$$\omega(\lambda^2 + \mu^2 + \nu^2) + 2\xi\mu\nu + 2\eta\nu\lambda + 2\zeta\lambda\mu = 0.$$

For the tangent-planes this breaks up into two straight lines. The condition for this gives a homogeneous equation of the third degree in ξ, η, ζ, ω which is the tangential equation of the surface, viz.

$$\omega^3 - \omega(\xi^2 + \eta^2 + \zeta^2) + 2\xi\eta\zeta = 0.$$

17·935. The parametric equations of the Steiner surface are all of the second degree in the parameters, and conversely freedom-equations of the second degree in general represent a Steiner surface. Let $\rho x = U_1$, $\rho y = U_2$, $\rho z = U_3$, $\rho w = U_4$, where U_i are homogeneous quadratic expressions in λ, μ, ν, and therefore $U_i = 0$ represent four conics in the (λ, μ, ν)-plane. We shall assume that these conics have no point common to all four. There is then a pencil of conic-envelopes all in-polar to each of these. When we choose the triangle of reference so that the four common tangents of the pencil are $\pm \lambda \pm \mu \pm \nu = 0$ the tangential equation of the pencil is $A\xi^2 + B\eta^2 + C\zeta^2 = 0$ with $A + B + C = 0$. The conic

$$U \equiv a\lambda^2 + b\mu^2 + c\nu^2 + 2f\mu\nu + 2g\nu\lambda + 2h\lambda\mu = 0$$

is out-polar to every conic-envelope of the pencil if $a = b = c$. Hence U_i are linear homogeneous functions of $\lambda^2 + \mu^2 + \nu^2$, $\mu\nu$, $\nu\lambda$, $\lambda\mu$, and by changing the tetrahedron of reference we can express the freedom-equations in the form

$$\rho x = 2\mu\nu, \quad \rho y = 2\nu\lambda, \quad \rho z = 2\lambda\mu, \quad \rho w = \lambda^2 + \mu^2 + \nu^2.$$

We have seen (9·731) that if the four conics have one point in common the parametric equations represent a ruled cubic; if they have two points in common they represent a quadric; with three points in common they represent a plane; and with four points in common a straight line.

17·94. The surface of Veronese.

The plane sections of a Steiner surface represent the conics of a three-parameter system. The system of *all* conics in a plane depends upon five parameters and would require space of five dimensions for its representation. If $x_i (i = 1, 2, ..., 6)$ are homogeneous coordinates in S_5, and U_i are homogeneous quadratic expressions in λ, μ, ν, the equations

$$\rho x_i = U_i \quad (i = 1, ..., 6)$$

are parametric equations of a two-dimensional surface in S_5. This is called the *Surface of Veronese*. Simpler parametric equations can be obtained by solving these equations for λ^2, μ^2, ν^2, $\mu\nu$, $\nu\lambda$, $\lambda\mu$, considering them as six equations linear in these six quantities; each is then expressed as a linear homogeneous function of $x_1, \ldots, x_6$, and then by a change of the frame of reference we may express the parametric equations in the form

$$\rho x_1 = \lambda^2, \;\; \rho x_2 = \mu^2, \;\; \rho x_3 = \nu^2, \;\; \rho x_4 = 2\mu\nu, \;\; \rho x_5 = 2\nu\lambda, \;\; \rho x_6 = 2\lambda\mu.$$

To every point in the $(\lambda\mu\nu)$-plane corresponds a unique point in S_5, and to every point on the surface of Veronese corresponds a unique point in the $(\lambda\mu\nu)$-plane.

Any conic in the $(\lambda\mu\nu)$-plane is represented by a homogeneous linear equation in x_i, and is therefore represented in S_5 by the curve of section of the surface with a four-flat. A three-flat cuts the surface in points which correspond to the four points of intersection of two conics in the $(\lambda\mu\nu)$-plane. Hence the surface of Veronese is cut by an arbitrary three-flat in four points, i.e. it is of order 4.

An arbitrary plane does not in general meet the surface in any point, but since any three points determine a plane there are planes which meet the surface in three points. A plane cannot in general meet the surface in more than three points, for if the plane α cuts the surface in four points, then through these four points and one other point on the surface there is determined a three-flat meeting the surface in five points.

An arbitrary line does not in general meet the surface, but there are lines which meet the surface in two points. No line can meet the surface in more than two points.

Ex. Show that there are no straight lines lying on the surface.

17·941. There are special planes, however, which meet the surface in a curve, and since a line cannot cut the surface in more than two points these curves are conics. These conics are represented by straight lines in the $(\lambda\mu\nu)$-plane; for a straight line in this plane is represented by a linear homogeneous equation in λ, μ, ν. Substituting for ν in terms of λ and μ in the parametric equations we express the coordinates as quadratic functions of the single parameter λ/μ; the locus is therefore a conic. The surface therefore possesses ∞^2 conics, corresponding to the lines of the $(\lambda\mu\nu)$-plane, and it contains no other plane curves.

Ex. Show that the equations of the plane of the conic corresponding to $a\lambda + b\mu + c\nu = 0$ are expressed by equating to zero the determinants of the fourth order in the matrix

$$\begin{bmatrix} a^2x_1 & b^2x_2 & c^2x_3 & bcx_4 & cax_5 & abx_6 \\ 1 & 1 & 0 & 0 & 0 & -2 \\ 0 & 1 & 0 & -1 & 1 & -1 \\ 0 & 1 & 1 & 2 & 0 & 0 \end{bmatrix}.$$

17·942. An arbitrary four-flat cuts the surface in a quartic curve, but if the four-flat contains a given conic of the surface the rest of the intersection is another conic. The four-flat

$$\Sigma \xi_r x_r = 0$$

cuts the surface in points whose parameters λ, μ, ν are connected by the equation

$$\xi_1\lambda^2 + \xi_2\mu^2 + \xi_3\nu^2 + 2\xi_4\mu\nu + 2\xi_5\nu\lambda + 2\xi_6\lambda\mu = 0.$$

The curve of intersection breaks up into two conics if the left-hand side of this equation factorises, and the condition for this is

$$\begin{vmatrix} \xi_1 & \xi_6 & \xi_5 \\ \xi_6 & \xi_2 & \xi_4 \\ \xi_5 & \xi_4 & \xi_3 \end{vmatrix} = 0.$$

Hence the four-flats which cut the surface in pairs of conics envelop a variety of class 3. The point-equation of this variety is easily obtained, for denoting the determinant by Δ the co-ordinates x_i are proportional to $\dfrac{\partial\Delta}{\partial\xi_i}$. For x_1, x_2, x_3 these are the cofactors of ξ_1, ξ_2, ξ_3, and for x_4, x_5, x_6 they are double the cofactors of ξ_4, ξ_5, ξ_6. But the determinant formed from the cofactors $= \Delta^2 = 0$, hence

$$\begin{vmatrix} 2x_1 & x_6 & x_5 \\ x_6 & 2x_2 & x_4 \\ x_5 & x_4 & 2x_3 \end{vmatrix} = 0.$$

The variety is therefore of order 3. As a locus it may be denoted by M_4^3 and as an envelope by Φ_4^3; the surface of Veronese itself is denoted by V_2^4.

17·943. Two conics $a\lambda + b\mu + cv = 0$, $a'\lambda + b'\mu + c'v = 0$ intersect in one point whose parameters are

$$(bc' - b'c, \; ca' - c'a, \; ab' - a'b);$$

and this is the only point common to their planes. Through any point of V_2^4 there pass ∞ conics. A four-flat which contains two conics touches V_2^4 at their common point. The tangents to the two conics at this point then determine a plane, the tangent-plane to V_2^4 at this point.

Ex. The equations of the tangent-plane at $[\lambda, \mu, v]$ are expressed by equating to zero the determinants of the fourth order in the matrix

$$\begin{bmatrix} x_1 & x_2 & x_3 & x_4 & x_5 & x_6 \\ \lambda & 0 & 0 & 0 & v & \mu \\ 0 & \mu & 0 & v & 0 & \lambda \\ 0 & 0 & v & \mu & \lambda & 0 \end{bmatrix}.$$

Three equations determining the tangent-plane are therefore

$$v^2 x_2 + \mu^2 x_3 - \mu v x_4 = 0,$$

$$\lambda^2 x_3 + v^2 x_1 - v\lambda x_5 = 0,$$

$$\mu^2 x_1 + \lambda^2 x_2 - \lambda\mu x_6 = 0.$$

17·944. If the four-flat (ξ) meets V_2^4 in two coincident conics

$$u\lambda + v\mu + wv = 0,$$

$$\xi_1 : \xi_2 : \xi_3 : \xi_4 : \xi_5 : \xi_6 = u^2 : v^2 : w^2 : vw : wu : uv.$$

These four-flats form a two-dimensional assemblage Φ_2^4 of class 4, the exact reciprocal of V_2^4. The four-flats of this assemblage which pass through an arbitrary point form a one-dimensional quartic assemblage (reciprocal of a quartic curve), which reduces to two quadric cones when the point lies on M_4^3, and to two coincident quadric cones when the point lies on V_2^4. A four-flat which meets V_2^4 in two coincident conics touches V_2^4 at all points of this conic. M_4^3 contains not only all the points of V_2^4 but also all its tangent-planes. The tangent four-flat to M_4^3 at the point (y) is

$$(4y_2 y_3 - y_4^2) x_1 + \ldots + (y_5 y_6 - 2y_1 y_4) x_4 + \ldots = 0,$$

and if (y) is a point on V_2^4 this becomes indeterminate. Hence V_2^4 is a double-surface on M_4^3.

17·95. Normal varieties.

A curve of order r always lies in a space of r dimensions or fewer, for the r-flat determined by $r+1$ points on the curve would meet the curve in more than r points and must therefore contain the curve. A curve of order r which cannot be contained in a space of fewer than r dimensions is called a *normal curve*. Examples are: straight line, conic, space cubic, etc.

Similarly a surface of order r, V_2^r, always lies in a space of $r+1$ dimensions or fewer, for if it be supposed to lie in an $S_n(n>r+1)$, the r-flat determined by $r+1$ arbitrary points on the surface would meet the surface in more than r points.

More generally, a variety of p dimensions and of order r, V_p^r, is always contained in a flat space of $r+p-1$ dimensions or fewer; for if it lies in S_n but not in S_{n-1} an arbitrary S_{n-p} cuts it in r points. But the S_{n-p} may be determined by $n-p+1$ points on the variety, hence $n-p+1 \leqslant r$, i.e. $n \leqslant r+p-1$.

A V_p^r which cannot be contained in a space of fewer than $r+p-1$ dimensions is called a *normal variety*.

17·951. *A normal variety V_p^r in S_{r+p-1} is rational.* To prove this we observe that a V_p^r in S_{r+p-1} is cut by an arbitrary $(r-1)$-flat in r points. Also the $(r-1)$-flat is determined by r points. If $r-1$ of these points are fixed points on V_p^r and the remaining point is on a fixed p-flat we have a $(1, 1)$ correspondence between this variable point and the rth point in which the $(r-1)$-flat cuts the variety. That is, the points of the variety are in $(1, 1)$ correspondence with the points of a given p-flat.

17·952. *A normal variety has no double-points,* for an $(r-1)$-flat passing through a double-point and $r-1$ other points of the variety would meet the variety in more than r points.

The surface of Veronese is a normal surface in S_5.

17·96. Projections of the surface of Veronese on space of three dimensions.

A figure in space of five dimensions may be projected from a point on to a four-flat by lines passing through the point. It may be projected on to a three-flat *from a line* by planes passing through the line. There are different cases according as the line does not meet the surface or meets it in one or in two points.

(1) *Projection on to a three-flat α from a line a not meeting the surface.* If P is any point on the surface the plane Pa does not in general meet the surface again and cuts α in a point P'. An arbitrary three-flat through a cuts the surface in four points and α in a line. Hence the projection is a surface of order 4. A four-flat through a and a conic-plane of the surface gives a conic in the projection. Hence the projection is a quartic surface having ∞^2 conics. Let the line a be the join of the points

$$A[0,\ 1,\ -1,\ 0,\ 0,\ 0] \quad \text{and} \quad B[1,\ 0,\ -1,\ 0,\ 0,\ 0],$$

and let the three-flat α be $x_1 = 0 = x_2$. Taking any point $P[\lambda,\ \mu,\ \nu]$ on the surface, freedom-equations of the plane ABP with parameters u and v are

$$\rho x_1 = v + \lambda^2, \qquad\qquad \rho x_4 = 2\mu\nu,$$
$$\rho x_2 = u + \mu^2, \qquad\qquad \rho x_5 = 2\nu\lambda,$$
$$\rho x_3 = -u - v + \nu^2, \qquad \rho x_6 = 2\lambda\mu.$$

Hence freedom-equations of the projection are

$$\rho x_3 = \lambda^2 + \mu^2 + \nu^2, \quad \rho x_4 = 2\mu\nu, \quad \rho x_5 = 2\nu\lambda, \quad \rho x_6 = 2\lambda\mu,$$

and these represent a Steiner surface.

(2) *If the line a meets the surface in one point A* an arbitrary three-flat through a meets the surface in just three other points, hence the projection is a cubic surface. The tangent-plane at A cuts α in a point A'. Through A there are ∞ conics on the surface. The three-flat determined by a and the plane of one of these conics cuts α in a straight line. Hence the projection is a ruled cubic surface.

(3) *If the line a meets the surface in two points A, B,* the projection is a quadric surface, and its two sets of generating lines are the projections of the conics which pass through A and those which pass through B.

The quadric, the ruled cubic, and the Steiner surface (all projections of the surface of Veronese) are the only surfaces in three dimensions which possess ∞^2 conics.

17·97. Quartic surfaces having a double-line.

If a quartic surface possesses a double-line l every plane through this line cuts the surface again in a conic, and so there is a single infinity of conics lying on the surface. If the double-line l is $x=o=y$ the equation of the quartic surface is of the form

$$Px^2 + 2Qxy + Ry^2 = o,$$

where P, Q, R are expressions of the second degree. A plane through the double-line, $y = tx$, cuts the surface where $x^2 = o$ and again in a conic, the intersection of l with

$$P' + 2Q't + R't^2 = o,$$

where P', Q', R' are quadratics in x, z, w obtained from P, Q, R by substituting $y = tx$. In this equation the coefficients of x^2, z^2, w^2, zw, wx, xz contain t to the powers 4, 2, 2, 2, 3, 3 respectively, hence its determinant is of degree 8 in t. There are therefore eight planes through l which cut the surface in pairs of lines, and hence in addition to l there are 16 lines on the surface, all of which cut l. There are in general no other lines on the surface.

Now an algebraic surface has a certain number of tritangent-planes. Any one of these meets the surface in a curve having the three points of contact as double-points, and there is in the present case a fourth double-point at its intersection with l. This quartic curve therefore reduces to either two conics or a nodal cubic with a straight line; in the former case one of the points of intersection of the two conics lies on l, and in the latter case one of the points of intersection of the line with the cubic lies on l.

Now let ϖ be any plane and P any point on it, and let C be a conic in a tritangent-plane. The plane Pl cuts C in two points, one of which is its intersection with l. There is therefore just one variable point Q and the line PQ cuts the surface in one other point P' which is associated with P. Conversely P' determines P uniquely. Hence there is (with certain exceptions) a $(1, 1)$ correspondence between the points P' of the quartic surface and the points P of the plane. The surface is therefore rational.

17·98. Quartic surfaces having a double-conic.

If a quartic surface possesses a double-conic C, a tritangent-plane cuts the surface in a quartic curve having five double-points and therefore breaking up into a conic and two straight lines. Two of the five double-points, A and B, are the intersections of the plane with C, and since neither of the lines can lie in the plane of C the two lines must pass one each through A and B. Let l be the line through A. Take any plane ϖ and any point P on it. The plane Pl cuts C in two points, one of which is A; let Q be the other point. The line PQ cuts the surface in two points at Q, one point on l, and one remaining point P'. Conversely, P' determines a plane with l; this plane cuts C in A and another point Q, and $P'Q$ cuts ϖ in P. Hence there is a $(\mathrm{1}, \mathrm{1})$ correspondence between the points P' of the surface and the points P of the plane ϖ. The surface is therefore rational.

Ex. 1. Show that the anchor-ring

$$\{x^2 + y^2 + z^2 + (a^2 - b^2)\, w^2\}^2 = 4a^2 w^2\, (x^2 + y^2)$$

has the circle at infinity as a double-conic.

Ex. 2. Show that the anchor-ring contains the four lines

$$x \pm iy = \mathrm{o} = z \pm cw,$$

where $c^2 = b^2 - a^2$.

Ex. 3. Show that the equation of the anchor-ring can be reduced to the form

$$(XY + ZW)^2 - kXY\, (Z - W)^2 = \mathrm{o}$$

by the transformation

$$x + iy = X, \quad z + cw = Z,$$
$$x - iy = Y, \quad z - cw = W,$$

where $c^2 = b^2 - a^2$ and $k = a^2/c^2$

Ex. 4. Show that the surface

$$(XY + ZW)^2 - kXY\, (Z - W)^2 = \mathrm{o}$$

has freedom-equations

$$\rho X = k\lambda^2 \mu^2,$$
$$\rho Y = \nu^2\, (\mu + \nu)^2,$$
$$\rho Z = k\lambda\mu^2\, (\mu + \nu),$$
$$\rho W = \lambda\mu\nu\, (k\mu - \nu).$$

Ex. 5. Show that freedom-equations of the anchor-ring are

$$\rho x = 2c^2 t\, (u^2 - \mathrm{1}),$$
$$\rho y = c^2\, (t^2 - \mathrm{1})\, (u^2 - \mathrm{1}),$$
$$\rho z = 2bcu\, (t^2 + \mathrm{1}),$$
$$\rho w = (t^2 + \mathrm{1})\, \{a\, (u^2 - \mathrm{1}) + b\, (u^2 + \mathrm{1})\}.$$

[If $a > b$ write $c = ic'$ and $u = iu'$.]

17·981. The general equation of a quartic surface having a double-conic

$$w = 0, \quad F_2 = ax^2 + by^2 + cz^2 + 2fyz + 2gzx + 2hxy = 0$$

is
$$F_2^2 + 2wF_2G_1 + w^2G_2 = 0,$$

where G_1 and G_2 are homogeneous expressions in x, y, z, w of degree one and two respectively. [G_1 indeed need not contain w.] An arbitrary plane section is a quartic curve with two double-points.

In the special metrical case where the conic is the circle at infinity, so that the equation in rectangular cartesian co-ordinates is

$$(x^2 + y^2 + z^2)^2 + 2wG_1(x^2 + y^2 + z^2) + w^2G_2 = 0,$$

$w = 0$ being the plane at infinity, the surface is called a *Cyclide*, and an arbitrary plane section is a quartic having double-points at the circular points in its plane, i.e. a bicircular quartic.

17·982. A binodal plane quartic curve is the projection of the quartic curve of intersection of two quadrics. A quartic surface which possesses a double-conic has an analogous property of being the projection of the surface of intersection of two quadric loci in space of four dimensions.

Let $Q = 0$ and $R = 0$ represent two quadric loci in S_4, Q and R being homogeneous quadratic expressions in x_0, x_1, x_2, x_3, x_4. Then $Q + \lambda R = 0$ represents a linear system of quadric loci all containing the quartic surface V_2^4 common to Q and R. If this is projected from any point O on to a three-flat S_3 we obtain a quartic surface F in S_3. Through O there passes one quadric, say Q, of the system, and the tangent three-flat at O meets Q in a cone with vertex O which is cut by S_3 in a conic C and this conic lies on F. But every generating line of the cone meets the other quadrics of the system, and therefore V_2^4, in two points, and each such pair of points is projected into one point. Hence C is the locus of double-points or a double-conic on F.

17·99. EXAMPLES.

1. Show that the constant-number of a rational algebraic curve of order n in a plane is $3n - 1$.

[The parametric equations contain $3(n + 1)$ coefficients; but by the fundamental theorem of projective geometry to any three points may be assigned arbitrary values of the parameter, and, further, only the ratios of the coefficients are significant, so that the number of essential constants is reduced by four. Otherwise: the constant-number of the general plane curve of order n is $\frac{1}{2}n(n+3)$, and the rational curve has $\frac{1}{2}(n-1)(n-2)$ double-points; subtracting these we get $3n-1$.]

2. Show that the constant-number of a rational curve of order n in space is $4n$.

3. Show that on an algebraic surface of order n there are in general $\infty^{4r-(rn+1)}$ rational algebraic curves of order r; in particular ∞^{2r-1} rational r-ics on a quadric surface and ∞^{r-1} on a cubic surface.

Deduce also that on a general surface of order 4 or more there are no rational curves of any order.

4. Find the number of conditions in order that a surface should possess (i) a conical point, (ii) a triple point.

Ans. (i) 1, (ii) 7.

5. Show that nine conditions are required in order that a quartic surface should possess a double-line, and find the number of conditions in the case of a surface of order n.

Ans. $3n-3$.

6. Show that the constant-number of the general ruled cubic surface is 13.

7. Show that the constant-number of the Steiner surface is 15.

8. Show that a cubic surface may have as many as four nodes but cannot have more; also that if it has four nodes the tangent-cone from any point of the surface consists of two quadric cones.
                                        (Math. Trip. II, 1914.)

9. Show that upon a cubic surface there are two families of skew cubic curves associated with any double-six of lines of the surface, and that the quadric surfaces which pass through these curves are all linear functions of nine of them. Prove also that

the first polar of any point in regard to the cubic surface is a linear function of these nine quadrics.

Find the general form of these cubic curves, and of these quadric surfaces, so far as they exist, for the surface ·

$$x^2 + y^2 + z^2 + w^2 = 0.$$

(Math. Trip. II, 1914.)

10. Of five non-intersecting lines in space the five pairs of transversals of each set of four of these lines are constructed. Prove that the five transversals of these pairs which can be drawn from an arbitrary point of space are coplanar.

(Math. Trip. II, 1915.)

11. Prove that a ruled surface of order n has in general a double-curve cutting each generator in $n-2$ points.

Show that the normals of an ellipsoid at the points of a given plane section are chords of a twisted cubic curve, and generators of a ruled surface of order 4. Prove that if the plane of the section touch a certain surface of the fourth class, the cubic curve is replaced by a straight line; and investigate the character of this line upon the ruled surface.   (Math. Trip. II, 1914.)

INDEX

Printed in the United States
By Bookmasters